The Historiography of Contemporary Science, Technology, and Medicine

Writing recent science

The Historiography of Contemporary Science, Technology, and Medicine: Writing recent science investigates the new challenges facing those who explore the history of recent science, technology, and medicine, looking at what makes their tasks distinct from those faced by historians of earlier times. This book offers a rare and engaging look at writing contemporary history, when key individuals are still alive and crucial historical records more likely linger on fragile computer hard-drives than reside in secure filing cabinets.

Writing recent science features original contributions by sixteen distinguished writers and historians who face these challenges. It includes a work of historical fiction and the fascinating story of how a review criticizing a contract history of English medicine was nearly barred from publication, in a case that reverberated all the way to the British Cabinet Office. Among the important questions this book raises are: how does secrecy affect modern science? Should biography now confront the ethical dimensions of scientists' lives? Are science journalists, seduced by Carl Sagan, ignoring the somber lessons of Thomas Kuhn? Have historians failed to consider what oral and photographic evidence can reveal? How have politicians inappropriately used history to justify contemporary policy for victims of eugenics practice? Are historians ignoring the profound explosion of "South–North" relations? This book will prove a fascinating read for those interested by the role of science, technology, and medicine in contemporary society.

Ronald E. Doel is Associate Professor in the Department of History, with a joint appointment in the Department of Geosciences, at Oregon State University. He writes on the modern environmental sciences, on the integration of scientists into foreign policy, and on historical methodology.

Thomas Söderqvist is Professor in the History of Medicine and Director of the Medical Museion at the University of Copenhagen, Denmark. His research focuses on the role of individuals and material objects in the historiography of recent biomedicine.

Routledge Studies in the History of Science, Technology and Medicine

Edited by John Krige
Georgia Institute of Technology, Atlanta, USA

Routledge Studies in the History of Science, Technology and Medicine aims to stimulate research in the field, concentrating on the twentieth century. It seeks to contribute to our understanding of science, technology, and medicine as they are embedded in society, exploring the links between the subjects on the one hand and the cultural, economic, political and institutional contexts of their genesis and development on the other. Within this framework, and while not favoring any particular methodological approach, the series welcomes studies which examine relations between science, technology, medicine, and society in new ways, for example, the social construction of technologies, large technical systems, etc.

The Historiography of Contemporary Science, Technology, and Medicine

Writing recent science

Edited by
Ronald E. Doel and
Thomas Söderqvist

Routledge
Taylor & Francis Group

LONDON AND NEW YORK

First published 2006
by Routledge

2 Park Square, Milton Park, Abingdon, Oxon OX14 4RN

Simultaneously published in the USA and Canada
by Routledge
270 Madison Ave, New York, NY 10016

Routledge is an imprint of the Taylor & Francis Group, an informa business

Typeset in Garamond by
Newgen Imaging Systems (P) Ltd, Chennai, India

British Library Cataloguing in Publication Data
A catalogue record for this book is available
from the British Library

Library of Congress Cataloging in Publication Data
    The historiography of contemporary science, technology, and medicine:
  writing recent science / edited by Ronald E. Doel and Thomas Söderqvist.
      p. cm.
    Includes bibliographical references and index
    1. Science–Historiography. 2. Technology–Historiography. 3. Medicine–
  Historiography. I. Doel, Ronald Edmund. II. Söderqvist, Thomas.

  Q175.H496 2006
  506–dc22   2005031144

ISBN10: 0–415–27294–7 (hbk)
ISBN10: 0–415–39142–3 (pbk)
ISBN10: 0–203–32388–2 (ebk)

ISBN13: 978–0–415–27294–0 (hbk)
ISBN13: 978–0–415–39142–9 (pbk)
ISBN13: 978–0–203–32388–5 (ebk)

www.writingrecentscience.com

Transferred to Digital Printing 2010

Contents

Illustrations

Figures

Tables

Contributors

David Cantor works as a historian at the National Cancer Institute and the History of Medicine Division of the National Library of Medicine, Bethesda, Maryland. After completing his Lancaster University PhD (1987) on the history of interwar British radiobiology, he held various research and teaching appointments at a number of British universities before moving to the United States. He has published on the histories of cancer, the rheumatic diseases, drug innovation, and a holistic movement in early twentieth-century Britain, called neo-Hippocratism. He is the editor of *Reinventing Hippocrates* (2001). His current research is on the history of US cancer control.

Keay Davidson has been a science writer for American newspapers since 1979. A native of Columbus, GA, he covered science and the space program at the Orlando, FL. *Sentinel Star*, then worked at the San Diego bureau of the *Los Angeles Times*, where he reported on topics including brain research and the early days of the biotech revolution and the AIDS epidemic. From 1986 to 2000 he was science writer for the *San Francisco Examiner*. His reporting won him the top two awards in American science writing: the American Association for the Advancement of Science-Westinghouse award and the National Association of Science Writers' Science-in-Society award. He has also received the Responsibility in Journalism award of the Committee for Scientific Investigation of Claims of the Paranormal. Since 2000 he has been a science writer at the *San Francisco Chronicle*. He is author of three books, including the first biography of Carl Sagan; his forthcoming publications include a biography of Thomas S. Kuhn tentatively titled *Reluctant Revolutionary: Thomas S. Kuhn and the Discovery of Paradigm Shifts*, for Oxford University Press.

Alexis DeGreiff Acevedo, a native of Bogotá, Colombia, studied Physics and received his MSc in Theoretical Physics before becoming a historian of science. He holds an MSc and a PhD in History of Science from London University. He worked as consultant to the Departamento Nacional de Planeación in Colombia, served as director of the Office of International Relations of Universidad Nacional, and is currently Associate Professor and Vice-Rector of the same institution. He has published on discourses and practices of South–North scientific exchange in the twentieth century, particularly in the framework of the United Nations system. He is director of the interdisciplinary research group on Science Studies of Science at the Universidad Nacional. He has been visiting professor at the Università degli Studi di Milano.

Michael Aaron Dennis, who has written on the history of science from the seventeenth to the twentieth centuries, is best known for his research on twentieth-century science. His honors include an appointment to the Harvard Society of Fellows, the American Historical Association's Aerospace Fellowship, the Schuman Prize from the History of Science Society, and the Young Scholar's Publication Prize from the Forum for History of Science in America. In addition to a forthcoming history of scientific practice in twentieth century America, titled *A Change of State: Political Culture, Technical Practice and the Origins of the Cold War* (Johns Hopkins University Press, forthcoming), his recent publications include "Reconstructing Sociotechnical Order: Vannevar Bush and US Science Policy," in Sheila Jasanoff, ed., *States of Knowledge: The Co-Production of Science and Social Order* (2004). He has taught most recently at Cornell University.

Ronald E. Doel is Associate Professor of History of Science at Oregon State University in Corvallis, Oregon, and serves the Department of History and the Department of Geosciences. He is writing new books on the integration of scientists into foreign policy after the Second World War and on the rise of the environmental sciences in America during the twentieth century; with Henson, he is writing a book-length essay on photographs as evidence in the history of recent science. His recent articles include "Constituting the Postwar Earth Sciences: The Military's Influence on the Environmental Sciences in the USA After 1945," *Social Studies of Science*, 2003 and with Kristine C. Harper, "Prometheus Unleashad: Science as a Diplomatic Weapon in the Lyndon B. Johnson Administration," *Osiris 21*, (2006).

Anne Fitzpatrick until recently was a technical staff member in Los Alamos National Laboratory's Computer and Computational Sciences Division, where she worked on computing policy and social informatics issues. She is also an instructor at the University of New Mexico and is Associate Editor-in-Chief of the *IEEE Annals of the History of Computing*. Fitzpatrick gained a doctorate from Virginia Tech's Science and Technology Studies (STS) program in 1998. She is writing a book comparing the origins of scientific supercomputing in the United States and Commonwealth of Independent States and is editor of a forthcoming book on the history of computing technology in the Soviet Union. She has worked and lived in several of the former Soviet republics.

Pamela M. Henson is director of the Institutional History Division of the Smithsonian Institution Archives. As historian for the history of the Smithsonian, she is responsible for research, writing, a documentary editing project, exhibits, and public programs. She uses oral history interviews with Smithsonian staff to document their careers and also curates the Smithsonian Videohistory Collection. Her research interests focus on the history of evolutionary biology, natural history, museums, and the role of women in science. In her work, she emphasizes the integration of traditional archival documents with oral histories and visual materials to more fully analyze the history of science.

Arne Hessenbruch worked on the project called History of Recent Science and Technology, described in this book, at the Dibner Institute, MIT, between 2000 and 2003. He has descended further down the ethical slippery slope and now lobbies Congress for more money to do materials research. He currently teaches a science studies course in collaboration with the Department of Materials Science and Engineering at MIT. He is the editor of the *Reader's Guide to the History of Science* (2000) and has published on many topics including radiation standards and science in peripheral Europe. He holds a PhD in History and Philosophy of Science from the University of Cambridge and an MSc in physics from the University of Freiburg.

Lillian Hoddeson, Professor of History at the University of Illinois, regularly teaches a graduate seminar on oral history as a research tool. She has been employing oral history in all her research over the past three decades, including her studies, with collaborators, of the atomic bomb: *Critical Assembly: A History of Los Alamos During the Oppenheimer Years, 1943–1945* (1993) and studies of solid-state physics: *Out of the Crystal Maze: A History of Solid State Physics, 1900–1960* (1992); *Crystal Fire: the Birth of the Information Age* (1997); and *True Genius: The Life and Science of John Bardeen* (2002).

Lene Koch is Professor at the Department of Health Services Research at the Institute of Public Health, the University of Copenhagen. She has written extensively on the social uses of genetic and reproductive technology; her work has appeared in several edited collections as well as in *Social Science and Medicine* and the *Scandinavian Journal of History*. Her current work is on the uses of modern genetics in the health service system, particularly the development of prenatal diagnosis. Koch served on the Danish Ethical Council from 1994 to 2000, the year she received the Danish National Radio Rosenkjær Prize. Since 2002 she has been visiting professor at Rokkansenterett, Bergen University.

John Krige is the Kranzberg Professor in the School of History, Technology and Society at the Georgia Institute of Technology in Atlanta. In academic year 2004–2005 he was also the Charles Lindbergh Fellow in Aerospace History at the National Air and Space Museum of the Smithsonian Institution in Washington DC. Before moving to the United States in 2000, he was a research fellow at the European University Institute in Florence, Italy, and the Director of the Centre de Recherche en Histoire des Sciences et des Techniques in Paris. Krige has researched and published extensively on the place of science and technology in the postwar reconstruction of Europe, notably in the domains of physics and space. He is currently working on a book exploring the role of the United States in this process, tentatively entitled *American Hegemony and the Postwar Reconstruction of Science in Europe*. His next project will focus on technology, and the efforts by the United States to help and to hinder the development of a European civilian satellite launcher.

Bruce V. Lewenstein is Associate Professor of Science Communication in the Departments of Communication and of Science and Technology Studies at Cornell University in Ithaca, New York. He received his PhD in History and Sociology of Science in 1987 from the University of Pennsylvania. He teaches and does research on various aspects of public understanding of science. He is a co-author of *The Establishment of American Science: 150 Years of the AAAS* (1999, with Sally Gregory Kohlstedt and Michael M. Sokal), and co-editor of *Creating Connections: Museums and the Public Understanding of Research* (2004, with Dave Chittenden and Graham Farmelo). From 1998 to 2003, he was editor of the journal *Public Understanding of Science.*

Mauricio Nieto Olarte received his master's and doctorate in History of Science at London University. He was also visiting scholar at Cambridge University in the Department of History and Philosophy of Science. He has been Professor of History and Sociology of Science at Universidad de los Andes and Universidad Nacional in Bogotá. From 1995 to 1998, he served as Director of the Department of History at Universidad de los Andes, as well as editor of the journal *Historia Crítica*. He has also worked at the Departamento Nacional de Planeación and at Colciencias (Colombian Institute for Science and Technology Development), in the area of Science and Technology Policy. Currently he is Professor in the Department of History at Universidad de los Andes. Among his notable publications is the book, *Remedios para el imperio: historia natural y la apropriacion del nuevo mundo* (2000), winner of the 2001 award, Premio Silvio Zavala de Historia Colonial de América.

Thomas Söderqvist is Professor of History of Medicine and Director of the Medical Museion at the University of Copenhagen. His major research interests are the history of twentieth-century life sciences and the history and poetics of the genre of scientific biography. His books include *The Ecologists: From Merry Naturalists to Saviors of the Nation* (1986), *The Historiography of Contemporary Science and Technology* (ed., 1997), *Science as Autobiography: The Troubled Life of Niels Jerne* (2003) and *The History and Poetics of Biography in Science, Technology, and Medicine* (Ashgate, forthcoming, 2007). He is now coordinating a research project aimed at integrating medical museology and the historiography of recent biomedicine, and he is also trying to find some spare time to work on a book on the history of the genre of scientific biography.

E. M. "Tilli" Tansey is Reader in the History of Modern Medical Science at the Wellcome Trust Centre for the History of Medicine at University College, London and Convenor of the History of Twentieth Century Medical History Group. After completing a PhD on the neurochemistry of the octopus brain, she worked for many years as a research neuroscientist before becoming a historian of medicine, finishing a PhD on the career of Sir Henry Dale in 1990. In addition to running the Twentieth Century Group and initiating, organizing, and editing the Witness Seminars, she has worked on a number of research projects in recent medical history, including the history of the pharmaceutical industry.

Alfred I. Tauber is Professor of Medicine and Professor of Philosophy and directs the Center for Philosophy and History of Science at Boston University. Aside from his research publications in immunology, Tauber has published extensively on nineteenth and twentieth century biomedicine, contemporary science studies, and ethics. He is the author of *The Immune Self, Theory or Metaphor?* (1994), *Confessions of a Medicine Man, An Essay in Popular Philosophy* (1999), and *Henry David Thoreau and the Moral Agency of Knowing* (2001), as well as co-author of *Metchnikoff and the Origins of Immunology* (1991) and *The Generation of Diversity: Clonal Selection Theory and the Rise of Evolutionary Biology: Papers of Elie Metchnikoff* (2000).

1 Introduction

What we know, what we do not— and why it matters

Ronald E. Doel and Thomas Söderqvist

Every generation thinks it discovered sex, according to an old witticism. Likewise, a historically minded reader of today's news media might conclude that every generation thinks it invented scientific scandals, biomedical hucksterism, and scientific-technological disasters. Postmodern "pop" culture, in particular, seems oblivious to historical context. Perhaps this is because the postmoderns are too dazzled by the glittery spectacles of the present and the promissory notes of the future to heed the lessons of the past. Consider the almost total lack of historical references in most recent media reports on the Korean biotech scandal of 2006; on the utopian-worlds-to-be advertised by stem-cell and nanotech researchers; and on the space shuttle Columbia disaster in 2003 and the levee failures in New Orleans during Hurricane Katrina in 2005.[1]

With few exceptions (especially in the United States, where history is one of the least popular subjects among high school students and the label "outdated" is culturally akin to a death sentence), media commentary treated these developments as if they had fallen from the sky, as if they had no past or prehistory to better illuminate their meanings. One need not invoke Santayana's by-now clichéd dictum about those who forget the past fail to recognize the dangers that await civilizations, which voyage at light speed into the cosmos with no memory of where they've been—and, thus, of who they are and of what they value.

At the very least, awareness of the history of recent science, technology, and medicine can benefit the wise cynic or jaded undergraduate, who greet news developments like these with a shrug and a smirk: "What did you expect? So what else is new?" Better, an awareness of the history of science can enlighten both taxpayers and policymakers, who, in pondering (and perhaps bankrolling) the sales pitches of today's scientists, must ask themselves the same question that so many Americans asked themselves during the "Moon race" of the 1960s: "Is this trip necessary?" Better still: we *need* history to decide the fortunes of scientific research and biomedical pursuits—to help present and future generations to understand the larger temporal and social contexts of these contemporary transformations.

Moreover, historians of recent science, technology, and medicine now need to tell new stories, and face new challenges in doing so. The Cold War ended almost two decades ago. Even before the terrorist attack of September 11, 2001, physics had ceased to be the queen of the natural sciences in most scientifically advanced nations. In 1993, the US Congress had voted against continued funding of the

Superconducting Super Collider, the largest particle accelerator then in planning, while backing the Human Genome project.[2] Already by then the biosciences had become the new favorites not only of governments and industrial interests but also of universities, overturning dominant patterns in place for nearly a century.[3] In 1996 the life sciences (biology, medicine, and agriculture) dominated university research and development in the United States, with 55% of $21 billion spent.[4] This trend intensified after the September 11th attack. After anthrax-filled letters contaminated offices of the US Congress, biological warfare research joined stem cell research and human-induced global warming as key issues in international science policy.[5]

This volume is about what we know—and don't yet know—about science, technology, and medicine in the recent past. This volume also addresses new methods that historians can use to explore these developments. It is the second volume in this series to address these critical themes.[6]

What questions should we ask? Certainly many challenges that historians of science, technology, and medicine face in studying this period are reassuringly familiar. The Limited Nuclear Test Ban Treaty of 1963 (involving seismologists) was preceded by the international Migratory Bird Treaty of 1919, where biologists helped guide negotiations.[7] Big science undertakings at CERN and the US national laboratories had antecedents in the major "factory observatories" in Britain and the United States in the late nineteenth century, whose hard-driving directors and highly specialized staffs recalled their industrial counterparts.[8] Technology transfer was already a concern for Britain and the United States soon after the American Revolution, even if US officials now are more concerned about advanced technology flowing to less developed nations than with circumventing British restrictions on exporting powerloom technologies.[9] Certain methodological issues are familiar as well: the same social history approaches that have broadened our understanding of the roles of women, minorities, skilled artisans and technicians in recent history of medicine and science also have challenged traditional accounts of the Scientific Revolution.[10]

Yet other themes and circumstances have a new feel to them, and occupy unfamiliar ground. They require historians of recent science, technology, and medicine to journey beyond familiar conceptual coastlines into largely uncharted historiographic waters. Physics no longer stands as the exemplar for all fields of recent science. Moreover, because science, technology, and medicine increasingly operate on global scales, comprehensive accounts of these activities now need to cover a much larger-than-traditional geographic canvas, including the People's Republic of China, India, South America, and South-East Asia.[11] Relationships between the largely industrialized northern hemisphere nations and less-developed southern hemisphere nations ought be as carefully examined as the better-studied tensions and exchanges between the East Bloc and West Bloc scientific communities.[12] Scientific intelligence-gathering became a high-priority concern for Washington, Moscow, and Beijing after the Second World War, another novel departure.[13] When in the early 1960s a conservative American diplomat sniffed that the US embassy in Brazil needed a science attaché "like a cigar store Indian needs a brassiere," he called attention to another fundamental shift: the growth of science, technology, and medicine as aspects of diplomacy and foreign policy.[14] Historical judgments on many major issues remain elusive: did the Cold War hinder international scientific interactions, or did increased avenues of communication "denationalize science," as Elisabeth Crawford, Terry Shinn, and Sverker Sörlin have argued?[15]

Even apparently familiar landscapes in the history of recent science, technology, and medicine now require perceptive re-examination. If at the start of the twenty-first century biological warfare has largely replaced thermonuclear war as a dominant Western anxiety, so have the security restrictions and intimate ties to the state once characteristic of nuclear physics come to characterize several domains of modern biology. In the US, restrictions on access to academic biological laboratories—already restricted by controversial privacy agreements with pharmaceutical firms and other major donors—have grown tighter since the anthrax attacks in October 2001.[16] Federal contracts with universities increasingly require pre-publication reviews of scientific papers. Expanded restrictions on sharing data (including a new "Sensitive but Unclassified Information" clause) have limited the numbers of participants in research.[17] How these issues will shape biological research remains unclear. A senior official for research and development in the newly formed US Office of Homeland Security lamented in 2002 that "We've never owned the biology community in the way we own the physicists."[18] It is too early to know if growing state interest in biology in the early twenty-first century will come to be seen as a historical milestone on a par with the influence of the Second World War and the Cold War. But it does suggest large new questions that historians need to consider.[19]

Historians are also increasingly concerned with the ethics and morality of science. After the Berlin Wall fell in 1989, many individuals around the world finally felt free to address issues of moral conduct during the four decades of the Cold War. Popular interest in ethical transgressions by the state soared. This interest meshed with the *Vergangenheitspolitik* movement that sought to explore silences in historical accounts throughout the twentieth century, resulting from the state's heightened power to shape historical narratives and to repress memory (from South African apartheid to the appalling Tuskegee syphilis experiments involving black Americans).[20] The Advisory Committee on Human Radiation Experiments, established by US Energy Secretary Hazel O'Leary in 1994, sought to discover the extent to which American citizens had been subject to covert efforts by the US government to assess individual responses to radiation as the Cold War intensified.[21]

This pioneering undertaking was not the first to probe morals in scientific research. Historian Jan Sapp's 1990 analysis of research ethics (centered on the mid-twentieth century biologist Franz Moewus) helped to pave the way for Daniel J. Kevles's study of David Baltimore, the Nobel Laureate and immunology researcher accused of fabricating data in a published paper who became the poster-boy of academic fraud within the US Congress in the mid-1990s.[22] Baltimore was later exonerated. But a darker judgment has fallen on the Stanford physicist Victor Ninov, whose boastful 1998 claim that he had discovered two new trans-uranium elements now seems "a result of fabricated research data and scientific misconduct."[23] Contemporary fascination with morality in science runs deeper still. Michael Frayn's

1998 play *Copenhagen* (about the conflict between the Danish physicist Niels Bohr and his former student Werner Heisenberg, leader of the German atomic bomb project, during their fall 1941 meeting in Nazi-occupied Copenhagen) played to sold-out audiences in London, Berlin, Paris, Toronto, New York and elsewhere—including Copenhagen.[24] When in 2002 Bohr's previously unknown letters to Heisenberg surfaced (denouncing Heisenberg's subsequent benign recollections about his famous 1941 visit, asserting instead that Heisenberg was then looking forward to a German-dominated Western Europe), the story made the front page of the *New York Times*.[25]

These are only some of the issues historians have recently addressed. Consider the problem of sources. The once-stable world of typewritten and handwritten letters preserved in university archives, together with bound periodicals lining library shelves, is yielding to the realm of email, e-journals, weblogs, and other web-based reports. How stable are these new sources of historical information? Studies in 2003 found that up to half of web-based citations (URLs, or *uniform resource locators*) became inaccessible within four years.[26] Archivists and historians worry about the long-term implications of these trends. They are also concerned about the fate of classified documents—an issue that extends beyond the formal end of the Cold War—since entire archives of classified documents, many stored without traditional archival safe-guards, have gone missing in the United States and several Western European nations. This complicates the task of writing traditional historical accounts.[27]

A further challenge in writing this history is that scientists increasingly work in large multi-disciplinary teams. Experimental papers in biomedical journals like *Cell* frequently involve ten or more authors with different disciplinary backgrounds, each contributing a particular methodological skill to the common outcome. The epochal papers in *Nature* on February 16, 2001 that reported on the draft human genome sequence were crafted by more than 2500 authors from 20 laboratories around the world (the Craig Venter *et al.* paper in *Science* the day before "only" had about 250 authors). If the relevant archives for writing the history of recent science, technology, and medicine are no longer the papers of individual scientists but of collaborative groups, how confident can we be that the records of these collaborations will survive?[28] Since contemporary scientists have deployed new writing and authorship strategies, do historians of recent science, technology, and medicine need to adopt new methodologies as well? One approach—extending the collaborative, multidisci-plinary, social history-oriented research projects favored by the *Annales* School—involves employing teams of historians to explore the largest undertakings of modern science, each researcher tackling a manageable part: Big History for Big Science. Several notable historical accounts, including those of nuclear physics laboratories, have utilized this approach.[29] In the late 1980s and 1990s, the Sloan Foundation pursued pioneering efforts to develop historical information on significant yet largely unexplored fields of recent science. They did so by inviting grassroots participation, allowing scientist-participants in activities such as GATE (the Atlantic Tropical Experiment of the larger Global Atmospheric Research Program, or GARP, conducted during the summer of 1976) to write online narratives about their expe-riences. But these projects foundered. Collaborations of more than two historians remain difficult, and few scientists responded to the Sloan Foundation's invitation to contribute online.[30] The challenge of writing histories of rapidly burgeoning science and medical fields remains immense, although the last contributor to this volume (Hessenbruch) offers bold new suggestions.

Finally, historians of recent science, technology, and medicine also confront the challenge of relevancy: who reads their works? Who profits by them, beyond colleagues and undergraduates? Should new historical accounts of these fields be more accessible to broader public realms? Heated controversies involving recent museum exhibits—such as the ill-fated Smithsonian Institution plan in 1995 to develop a comprehensive exhibit around the *Enola Gay*, the plane that had dropped the first atomic bomb on Hiroshima a half century before—remind us that our work can stimulate (or infuriate) large audiences.[31] Should historical research also be relevant to policy studies, since (to make an obvious point) bad history inspires misguided policy? A quarter century ago the eminent historian A. Hunter Dupree lamented that no credible means existed to bring historical insights into science policy discussions.[32] More recently historian Jane Maienschein—who has served as science advisor for the US House of Representatives—has argued that such links need to be forged.[33] If ethical issues (such as past medical experiments involving unwitting subjects) and political-scientific controversies (such as global warming) continue to dominate front page headlines, then historians of recent science, technology, and medicine can further contribute to large debates at the core of contemporary life.[34]

The universe of possible topics for a book such as this is vast. Some are addressed in the current volume. Many more still need to find their voice.

* * *

We divide these essays into six parts. They address large themes: "Where Are We Now? The Challenges of Writing Recent Science," "Whose History? Ethics, Lawsuits, Natural Security, and the Writing of Contemporary History," "Witnesses to History: Grasping the Big Picture," "Secrecy and Science: Probing the Meaning of the Cold War," "History Detectives: New Ways of Approaching Modern Science, Technology, and Medicine," and "New Voices: Neglected and Novel Perspectives."

Why are science historians not more widely read? What lessons can science journalists learn from the history of science; what can historians do to bring their knowledge into mainstream reporting and policy discussions? KEAY DAVIDSON, a veteran science journalist at the *San Francisco Chronicle*, argues that far too few science journalists understand the history of science, leading them to become science propagandists rather than insightful critics of its emerging concepts and social impacts. Author of a well-received biography of Carl Sagan and a forthcoming biography of Thomas Kuhn, Davidson decries that too many science correspondents would agree that the history of science should be rated "X," drawing on Stephen G. Brush's classic 1974 essay.[35] Rather than be seduced by Sagan's optimistic vision of science, Davidson wants science journalists to wrestle with the pessimistic pragmatism of Kuhn, arguing that public understanding of science (and informed policy decisions) depend on well-informed citizens who understand the process of science.

In his own provocative essay, BRUCE V. LEWENSTEIN raises a different theme: how do existing archival collections shape what historians of recent science, technology, and medicine write about? What important stories and issues are *not* being addressed? Lewenstein, an Associate Professor of Communications Studies in the Department of Science and Technology Studies at Cornell University, has written extensively on media coverage of science. He explores the challenge of (and controversy over) creating archives for pivotal developments in contemporary science and technology, including the 1989 Cold Fusion controversy and the Y2K computer scare of the late 1990s, where real-time information and opinions can be preserved before memories are edited and refined in later years.

How has the modern state—not just through national security restrictions but also the courts and the news media—shaped, restricted, and warped what we know about recent science, technology, and medicine? Are historical accounts blunted by the power of the modern state to shape what we know of vital issues, including the history of medical institutions and the history of recent nuclear weapons developments? Have historians failed to find an appropriate moral and ethical language to depict twentieth century experimentation with eugenics and larger efforts to shape the human race?

In Part 2, "Whose History? Ethics, Lawsuits, National Security, and the Writing of Contemporary History," three historians—each working in very different environments—explore aspects of how the state shapes historical narratives. British historian DAVID CANTOR offers a gripping, eye-opening account of his late 1980s attempt to review a book on the Imperial Cancer Research Fund, a contracted history he considered compromised by its sponsorship. "Is the history of medicine in Britain for sale?" began Cantor's hard-hitting review. Because British defamation law is broader than that in the US and in many European nations, he and his intended publisher—the journal *Social History of Medicine*—were threatened with lawsuits if they actually published his review. For more than a year the "Cantor affair" echoed from Oxford to the Cabinet Office of the Prime Minister in London. Though his review ultimately appeared in print (it is reproduced here), the issue has hardly gone away.

Addressing this large theme from a different perspective, ANNE FITZPATRICK writes about doing history from "behind the fence": in this case, preparing a history of nuclear weapons development as an employee of the Los Alamos National Laboratory. Emphasizing the insights gained by having a Q clearance—which allows access to highly classified nuclear information—Fitzpatrick explores post-Cold War efforts to comprehend the global development of nuclear devices. She also narrates her research in the archives in the once-secret city of Sarov, the Soviet Union's mirror equivalent of Los Alamos. A 1998 PhD from Virginia Tech (some 2000 words from her dissertation on the Los Alamos Thermonuclear Weapons Project ultimately were classified by government censors), Fitzpatrick followed standard security procedures in writing for this volume. Her chapter, reviewed before publication, is officially the publication LA-UR 038579 of the Los Alamos National Laboratory.

In the last chapter of this section, LENE KOCH addresses the profound challenges of writing about state-sanctioned eugenics policies in Scandinavia. These efforts persisted through the 1960s, so close to the present that many scientists and physicians involved with the effort remain alive. A Professor in the Department of Health Services Research at the University of Copenhagen who has published extensively on twentieth century sterilization practices in Denmark, Koch explores the moral dilemma of imposing our own contemporary morality inappropriately on the past, and the ways that such histories have been used by the state to justify new remedies for past policies.

Part 3, "Witnesses to History: Issues in Biography and Ethics," addresses two quite different yet interwoven themes. What can we learn from a historian's thoughtful self-examination of his own work? What can we say about the ethical dimension in the historiography of recent science, technology, and medicine? The two essays in this section have much in common. Both authors review their own heterodox writing on the history of immunology over the past fifteen years. They both intersected with the life of Niels Jerne (1911–1994), the British-Danish-Dutch theoretical immunologist who set a bold agenda for the emerging discipline of immunology in the 1960s and 1970s (and shared the Nobel Prize for Physiology or Medicine in 1984). Both believe that historians (and even scientists) inscribe their own lives in their work, intellectually and existentially. And while both sought to use the history of immunology to address problems in epistemology, in the course of their work they discovered ethical questions that could not be ignored.

Drawing on his writing about Jerne, THOMAS SÖDERQVIST analyzes seven reasons for writing biographies of recent scientists. Classical uses of the genre—providing the public with means for understanding science—remain valued. More recently, historians have regarded biography as a way to constitute fine-grained epistemological case studies or to write contextual history of science. Söderqvist argues, however, that biographers of scientists must ask a broader range of questions so that the whole *life*, rather than just the science, takes central place in the narrative. This, in turn, raises an ethical question: are biographies of recent scientists fundamentally an ethical genre? "Care of self"—an ancient notion cast in modern form by Pierre Hadot and Michel Foucault—is becoming a central category for both biographers and historians of recent science. (Readers interested in these issues should also read John Krige's essay in the following section, where he boldly takes up Söderqvist's challenge to explore the human dimensions of science and science policy.)

In the second essay of this section, ALFRED TAUBER, a physician-turned-historian and philosopher of science, traces his own intellectual journey in the 1980s as he wrestled with the fundamental question: What defines an organism? Eventually recognizing that his own ecologically-oriented perspective reflected an orientation towards immunological research in vogue in the late nineteenth century (but largely discarded in favor of more reductionist approaches in his own time), Tauber reminds historians that a postmodernist perspective is required. Failing to consider larger issues at stake, such as moral agency and the moral dimensions of "natural" categories, will likely lead us to historical accounts that fail to ask the poignant and significant questions. In retrospect he discovered that philosophical historiography was driven by twin concerns: a self-evident epistemological exercise, and in a more latent form, an exploration in moral philosophy. Tauber thus identifies a close connection between epistemology and moral agency that has served as a dominant theme of his writings on science in culture.

What do we ought to remember about the Cold War? For individuals born after 1980—today a significant fraction of the world's population—the Cold War at best is a dim and hazy memory, ending with the collapse of the Berlin Wall in November 1989. Yet virtually all of the scientific agencies created after 1947, including the Science and Technology division of the CIA, remain in place. Space exploration, spurred by the space race following the launch of Sputnik in 1957, continues unabated. Thus the Cold War casts a significant shadow on science and society today, and much contemporary scholarship continues to examine its influence. In Part 4, "Secrecy and Science: Probing the Meaning of the Cold War," two distinguished historians of recent science and technology focus on distinct aspects of the Cold War. In an innovative essay that blends diplomatic history, history of science, and historical fiction, JOHN KRIGE explores how the politics of the Marshall Plan played out in the case of a dying worker in occupied Trieste in the late 1940s. Krige, the Kranzberg Professor at the Georgia Institute of Technology, argues that historians—by focusing too much on issues trumpeted in Washington and in Moscow—have missed how official policies influenced the attitudes of residents of Western Europe in the early Cold War. Yet this was precisely what Cold War foreign policies sought to do, and hence Krige's essay points to important studies that remain unwritten.

In the second essay in this section, MICHAEL A. DENNIS tackles the fundamental question of what can be known about secret science: To what extent can anyone confidently know how knowledge production worked during the Cold War? Was secrecy antithetical to science or fundamental to the practice of research, essential to the maintenance of democratic traditions within the Soviet Union? Dennis says: No. Contrary to common perceptions, secrecy may be essential to science, and the entire matter needs to be rethought. A thoughtful scholar of scientific practice, Dennis argues that we need to reevaluate the reflexive categories of openness and secrecy: only then will we be able to better assess the impact and significance of the Cold War.

How might we better understand what took place in recent science, technology, and medicine? Historians of science have had a much easier time of it than social historians: scientists leave behind bulging boxes and file cabinets, brimming with letters, reports, memoirs, and photographs. These seemed reliable and enduring sources of history until the start of the electronic age. Since the 1980s, email has largely replaced typed or handwritten letters, but it is far more ephemeral and easily lost. Photographs—whose manipulation once took skill, time, and persistence—also seem less certain witnesses in the new digital age than in the past. The one great advantage that historians of recent science continue to enjoy is the chance to talk with their subjects. Part 5, "History Detectives," addresses new options available to historians. In her essay, LILLIAN HODDESON tackles a central issue: the conflict between memory and documents, better understood as the battle between memory and history.[36] A physicist-turned-historian and one of the best-known historians of twentieth-century physics, Hoddeson revisits her extensive oral interviews with Edward Teller, Richard Feynman and other key figures in twentieth century physics, and maps out the construction of memory.

Also in this section, one of us (RONALD E. DOEL), together with PAMELA M. HENSON of the Smithsonian Institution, jointly explores what can be learned from considering photographs themselves as evidence for writing the history of science. Scenes of laboratories, science exhibits in county fairs, and science classrooms, as well as photographs of scientists on field expeditions and negotiating bilateral treaties, sometimes contradict accounts based entirely on written records. Doel and Henson argue that, despite the value of photographs as sources of evidence, few historians of science have paused to consider what images might teach them.

Where are the most promising new frontiers? What new challenges do historians face in writing about recent science, when the Internet and email have largely replaced typed letters stored in desk-drawer folders? What questions have we not yet asked? Part 6, "New Voices: Neglected and Novel Perspectives," provides three perspectives on current and future challenges. In their thought-provoking essay, ALEXIS DE GREIFF AND MAURICIO NIETO argue that historians of recent science, technology, and medicine, by reflexively focusing on East–West relations, have neglected South–North technoscientific exchanges. De Greiff and Nieto, university-based historians in Bogotá, Colombia, provide a wide-ranging roadmap to future studies, including a comprehensive examination of current insights about the Green Revolution.

TILLI TANSEY's chapter assesses the Witness Seminar series of the Wellcome Trust Centre. Established more than a dozen years ago, it has brought together Nobel laureates, senior scientists, technicians and medical practitioners in lively, sometimes heated sessions. In her account, Tansey, a biologist-turned-historian, addresses a different issue: what can we learn about key episodes in the history of recent medicine by doing *group* interviews with individuals intimately connected with the development of oral contraceptives, genetic testing, and post-penicillin antibodies, to name a few topics?

Finally, in one of the more provocative pieces within this volume, ARNE HESSENBRUCH recounts efforts by the Sloan Foundation and the Dibner Institute beginning in the late 1980s to encourage the history of recent science, technology, and medicine online, using techniques pioneered in social history. While this project revealed limits to the approach, Hessenbruch argues that it nevertheless represents the future of historical work. As the infrastructure of electronic media strengthens, making online materials as secure as printed books, Hessenbruch envisions a time when scientists and historians will both collaborate in determining what historical data ought be gathered and how it is interpreted. Indeed, he argues, historians of recent science will necessarily become mongrels, or "mutts." Not all readers will agree. But all the chapters in the volume seek to raise significant questions about the past, present, and future of this field.

* * *

All of these essays appear in print for the first time. Several were first conceived for a conference entitled "Problems in the Historiography of Recent Science, Technology and Medicine," organized by Thomas Söderqvist and Jeff Hughes, with an international Advisory Program Committee (Finn Aaserud, Horace F. Judson, Lene Koch,

Helge Kragh, Timothy Lenoir, and Hans-Jörg Rheinberger). This mid-August 1998 gathering took place at the Magleas Conference Center, a former country estate outside Copenhagen. In this congenial and informal setting, ideas about writing recent science were raised, critiqued, and further refined. Essays in this book were selected to address topics distinct from those raised in the related volume in this series edited by Thomas Söderqvist, *The Historiography of Contemporary Science and Technology* (1997), based on an earlier conference in Gothenburg in 1994.[37] We are grateful to all participants in the meeting for their comments and ideas, and to Jeff Hughes for his contribution to the early stage of the editorial process. We are also grateful for editorial assistance provided by Roger Turner, currently a doctoral student in history of science at the University of Pennsylvania, and especially by Kristine C. Harper (MIT / New Mexico Institute of Mining and Technology).

We acknowledge support from the Danish Humanities Research Council for organizing the Magleas conference in 1998; Doel acknowledges NSF grants DIR-9112304 and NSF SBR-9511867. Söderqvist is grateful for support from a special program for science studies financed by the combined Danish Research Councils in 1997–2002. We also appreciate additional editorial comments and criticisms from Keay Davidson, Lawrence J. Friedman, Kris Harper, and Pamela M. Henson, as well as the many individuals and institutions who supplied us photographs and illustrations, including Neville Young. Finally, we would like to thank the encouraging staff at Routledge who saw this book through to completion, particularly Amrit Bangard, Catherine Carpenter, Terry Clague, and Emma Hart; we are also grateful to John Clement and his staff at NewGen for diligent copy-editing.

Notes

1 For missing historical context on Hurricane Katrina, see Ted Steinberg, "Opinion: Natural Disaster, A Man-made Catastrophe, and a Human Tragedy," *Chronicle of Higher Education*, September 2, 2005, http://chronicle.com/free/2005/09/2005090906n.htm (September 9, 2005).

2 Daniel J. Kevles, "Big Science and Big Politics in the United States: Reflections on the Death of the SSC and the Life of the Human Genome Project," *Historical Studies in the Physical and Biological Sciences* 27, no. 2 (1997): 269–298; and Kenneth H. Keller, "Science and Technology," *Foreign Affairs* 69, no. 4 (1990): 123–138; see also Daniel J. Kevles, *The Physicists: The History of a Scientific Community in Modern America*, 3rd ed. (Cambridge, MA: Harvard University Press, 1995): ix–xii.

3 Philip R. Sloan, ed., *Controlling our Destinies: Historical, Philosophical, Ethical, and Theological Perspectives on the Human Genome* (Notre Dame, IN: University of Notre Dame Press, 2000).

4 Nathan Rosenberg, "America's Entrepreneurial Universities," *The Emergence of Entrepreneurship Policy*, ed. David M. Hart (Cambridge, MA: Cambridge University Press, 2003), 113–137.

5 Context for the fall 2001 anthrax attacks in the United States appears in Matthew Meselson, "Bioterror: What Can Be Done?," *New York Review of Books* 48, no. 20 (December 20, 2001): 38–41.

6 See Thomas Söderqvist, ed., *The Historiography of Contemporary Science and* Technology (Amsterdam: Harwood Academic, 1997); see also Arnold Thackray, ed., *Science After '40, Osiris* 7 (1992).

7 Kai-Henrik Barth, "The Politics of Seismology: Nuclear Testing, Arms Control, and the Transformation of a Discipline," *Social Studies of Science* 33, no. 5 (2003): 743–781;

Kurkpatrick Dorsey, *The Dawn of Conservation Diplomacy: U.S.-Canadian Wildlife Protection Treaties in the Progressive Era* (Seattle: University of Washington Press, 1998).

8 John Lankford, *American Astronomy: Community, Careers, and Power, 1859–1940* (Chicago, IL: University of Chicago Press, 1997).

9 Stephen H. Cutcliffe and Terry S. Reynolds, eds, *Technology and American History* (Chicago, IL: University of Chicago Press, 1997), 6.

10 Andrew Cunningham, "De-Centering the 'Big Picture': The Origins of Modern Science and the Modern Origins of Science," *British Journal for the History of Science* 26 (1993): 407–432; Margaret C. Jacob, *Scientific Culture and the Making of the Industrial West* (New York: Oxford University Press, 1997); and Margaret J. Osler, *Rethinking the Scientific Revolution* (New York: Cambridge University Press, 2000).

11 New work on these issues remains scarce, but see Zuoyue Wang and Peter Neushul, "Between the Devil and the Deep Blue Sea: C.K. Tseng, Mariculture, and the Politics of Science in Modern China," *Isis* 91, no.1 (2000): 59–88, and Danian Hu, "Organized Criticism of Einstein and Relativity in China, 1968–1976," *Historical Studies in the Physical and Biological Sciences* 34, no. 2 (2004): 311–338.

12 Alexis De Greiff, "A History of the International Centre for Theoretical Physics, 1960–1980: Ideology and Practice in a United Nations Institution for Scientific Co-operation for the Third World Development," (PhD dissertation, Imperial College, London, 2001); Alexis De Greiff, "The Tale of Two Peripheries: The Creation of the International Centre for Theoretical Physics in Trieste," *Historical Studies in the Physical and Biological Sciences* 33, no. 1 (2002): 33–60; see also De Greiff and Nieto, this volume.

13 Jeffrey Richelson, *The Wizards of Langley: Inside the CIA's Directorate of Science and Technology* (Boulder, CO: Westview Press, 2001); and Ronald E. Doel and Allan A. Needell, "Science, Scientists, and the CIA: Balancing International Ideals, National Needs, and Professional Opportunities," in *Eternal Vigilance: Fifty Years of the CIA*, eds Rhodri Jeffreys-Jones and Christopher Andrew (London: Frank Cass, Publishers, 1997), 59–81.

14 Quoted in Council on Foreign Relations, Discussion Group on Science and Foreign Policy, February 7, 1964, Box 2, Ludwig Audrieth papers, University of Illinois Archives.

15 Elisabeth Crawford, Terry Shinn, and Sverker Soerlin, eds., *Denationalizing Science: The Contexts of International Scientific Practice* (Dordrecht: Kluwer, 1992).

16 Judith Reppy, "Regulating Biotechnology in the Age of Homeland Security," *Science Studies* 16, no. 2 (2003): 38–51; and National Research Council, *Biotechnology Research* in an Age of Terrorism (Washington, DC: National Academy of Sciences, 2004).

17 Arnold Thackray, *Private Science: Biotechnology and the Rise of the Molecular Sciences* (Philadelphia: University of Pennsylvania Press, 1998); Susan Wright, ed., *Biological Warfare and Disarmament: New Problems/New Perspectives* (Lanham, MD: Rowman & Littlefield, 2002); and Meselson, "Bioterror."

18 Comment by Department of Homeland Security official Penrose Albright at the April 1, 2002 open meeting of the National Academy of Sciences' Committee on Research Standards and Practices to Prevent the Destructive Applications of Biotechnology, Washington, DC. (we thank participant Judith Reppy for this information); see also the National Research Council's Committee report, *Biotechnology Research in an Age of Terrorism* (Washington, DC.: National Academies Press, 2004), 85, and John Dudley Miller, "National Academy Proposes Scientists Self-Police," *The Scientist*, October 9, 2003, http://www.biomedcentral.com/news/20031009/04/ (March 23, 2004). Creation of the Office of Homeland Security in 2001 was the largest restructuring of US federal government agencies since the consolidated Department of Defense and the CIA were created in 1947; on the former, see the "Special Issue on 9/11," *History and Technology: An International Journal* 19, no. 1 (2003); on the latter, Rhodri Jeffreys-Jones, *The CIA and American Democracy* (New Haven, CT: Yale University Press, 1989).

19 For instance, recent opinion surveys indicate that less than 30% of Americans accept the theory of evolution, while more than half regard the Book of Genesis and young-Earth creationism as literally true; see Jennifer Harper, "Most Americans Take Bible Stories

Literally," *Washington Times*, February 17, 2004, http://www.washtimes.com/national/20040216-113955-2061r.htm (March 23, 2004) and Nicholas D. Kristof, "Believe It, or Not," *New York Times*, August 15, 2003, sec. A, p. 29. The "poisonous" rift that Kristof finds "between intellectual and religious America" has obvious and profound importance for US science policy.

20 Norbert Frei, *Vergangenheitspolitik: Die Anfänge der Bundesrepublik und die NS-Vergangenheit* (Munich: Beck, 1996); see also Michael Lynch and David Bogen, *The Spectacle of History: Speech. Text and Memory at the Iran-Contra Hearings* (Durham, NC: Duke University Press, 1996).

21 US Advisory Committee on Human Radiation Experiments, *Final Report of the Advisory Committee on Human Radiation Experiments* (New York: Oxford, 1996); more detailed information appears online http://www.eh.doe.gov/ohre/roadmap/achre/ (March 23, 2004).

22 Jan Sapp, *Where the Truth Lies: Franz Moewus and the Origins of Molecular Biology* (New York: Cambridge University Press, 1990); Daniel J. Kevles, *The Baltimore Case: A Trial of Science, Politics, and Character* (New York: Norton, 1998), and Horace F. Judson, *The Great Betrayal: Fraud in Science* (New York: Harcourt Books, 2004).

23 Keay Davidson, "Berkeley Lab found Research Fabricated; Scientist Accused of Misconduct Fired," *San Francisco Chronicle*, July 13, 2002, sec. A, p. 1.

24 For the response of historians to Frayn's play, see the *Proceedings* of the Copenhagen Symposium, Washington, DC., March 2, 2002, http://web.gc.cuny.edu/ashp/nml/artsci/symposium.html (March 23, 2004).

25 James Glanz, "New Twist on Physicist's Role in Nazi Bomb," *New York Times*, February 7, 2001, sec. A, p. 1. These documents are available online at the Niels Bohr Archive website http://www.nbi.dk/NBA/release.html (March 23, 2004).

26 Rick Weiss, "On the Web, Research Work Proves Ephemeral," *Washington Post*, November 24, 2003, p. A8; see also Joab Jackson, "Digital Dark Ages: Many Federal E-Records Lost," *Washington Technology*, June 21, 2002, www.washingtontechnology.com/news/1.1/daily_news/10461-1.html (March 23, 2004) and Kevin Coughlin, "Groups Seeking Ways to Preserve History on Electronic Documents," *Sunday Oregonian*, June 22, 2003, sec. A, p. 5.

27 See Ronald E. Doel, "Oral History of American Science: A Forty Year Review," *History of Science* 41 (2003), 349–378; for extended discussions see Fitzpatrick and Hessenbruch (this volume).

28 American Institute of Physics, *AIP Study of Multi-Institutional Collaborations. Final Report: Highlights and Project Recommendations* (College Park, MD: AIP, 2001).

29 Lillian Hoddeson, "The Unfulfilled Promise of Collaboration in the History of Science" (we thank Hoddeson for kindly supplying this draft manuscript to us).

30 Background on the Sloan Foundation's Recent History of Science and Technology on the Web project is online at http://www.sloan.org/programs/scitech_historysci.shtml (March 23, 2004); see also Hessenbruch, this volume.

31 Edward T. Linenthal, ed., *History Wars: The Enola Gay and other Battles for the American Past* (New York: Metropolitan Books, 1996).

32 A. Hunter Dupree, "A Historian's View of Advice to the President on Science: Retrospect and Prescription," in *Science Advice to the President*, ed. William T. Golden (New York: Pergamon, 1980), 175–190, on 180–181.

33 Jane Maienschein, "Why Study History for Science?," *Biology and Philosophy* 15 (2000): 339–348.

34 John Krige and Kai-Henrik Barth, "Introduction" in *Global Power Knowledge: Science, Technology, and International Affairs, Osiris* 21 (2006), 1–21.

35 Keay Davidson, *Carl Sagan: A Life* (New York: Wiley, 1999) and Davidson, *The Death of Truth: Thomas S. Kuhn and the Evolution of Ideas* (New York: Oxford University Press, forthcoming); Stephen G. Brush, "Should the History of Science be Rated X?," *Science* 183 (1974): 1164–1172.

36 This important issue transcends the history of science, technology and, medicine; see for instance Dominick LaCapra, *History and Memory after Auschwitz* (Ithaca, NY: Cornell University Press, 1998), 20.

37 Söderqvist, *Contemporary Science*.

Part I

Where are we now? The challenges of writing recent science

2 Why science writers should forget Carl Sagan and read Thomas Kuhn

On the troubled conscience of a journalist

Keay Davidson

Mad Scientist's Plot Thwarted by Budget Cuts.

(Headline in *The Onion*)

At age 52, I belong to the last generation of American newspaper reporters who wrote at a time when newspapers still mattered, really mattered, as a cultural force. I entered this business the summer that Nixon was driven from office. At that time, Woodward, Bernstein, and Sy Hersh were heroes to an angry young generation. Now legends of their Watergate and My Lai exposés sound like radio signals from a dying star system. Newspapers still make money, sort of, yet they are increasingly timid, colorless, corporate-controlled contributors to post-Millennium discourse. Once, the sharpest newspaper reporters were high-school educated, hard-drinking, working-class souls who read Mencken, James T. Farrell, and *The Masses*, and prided themselves on their hostility to "the comfortable." Now leftists rightly deride us as "stenographers to power," as mouthpieces of the white, college-educated, sober, non-smoking, privileged American status quo who prefer politicians who are "centrists" and "pragmatists" and who buy *The New Yorker* for the cartoons.

In other words, we are boring. Newspaper circulation has been steadily declining for decades, and will continue to do so. Young people do not read us at all; eventually, no one will. And why should they? What do we offer them that they cannot find on TV or online, and in a more vivid, entertaining form? True, TV and the Net are shallow; but so are we. When was the last time a major American newspaper challenged you mentally? Defied your expectations? Exposed you to radically dissenting voices? Shook your paradigms? (The sole exception, naturally, is my present employer, which is flawless and beyond reproach.)

The problems of American newspaper journalism are mirrored, in microcosm, by those of its rarely noticed little subsidiary, science reporting. Most of my career I have worked in science journalism; and in this essay, I must say some unkind things about its present state. But before I bite the hand that feeds me, let me explain why I love this career despite its shortcomings. Before I bash it, let me clarify why I have remained a science journalist for a quarter of a century, well past the age at which most newspaper journalists, exhausted by the long hours and hounded by their children's orthodontist bills, have left for jobs that actually pay living wages.

Mine has been a happy career, a rare shrewd choice in a bumpy life. I cannot imagine a career that is better suited to a person of my interests, tastes, and (to be frank) my rather bookish, eccentric personality. In fact, I highly recommend science reporting to any young writer who has strong intellectual inclinations but prefers to avoid the bitchiness of academia and the penury of bohemia, and who wishes to experience the newspaper business while it is still afloat. Besides cerebral stimulation, science reporting offers moments of—I cannot think of an alternative term—metaphysical bliss. We are the intellectual adventurers of the news business who, day after day, pound out reports from the interior of the cell and the edge of the cosmos. Most other reporters grow cynical and depressed covering the human saga: corporate greed, legislative cretinism, police cruelties, wars and genocides, the vacuity of public debate. By comparison, we science writers are the pampered Little Lord Fauntleroys of the news business, the staffers who seem to be having an unconscionably merry time while everyone else is condemned to cover stomach-churning crimes, profile cliché-spouting CEOs, interview nitwit candidates for the school board, and report how the cops and firefighters are planning for the next terrorist assault. I jokingly tell student journalists that I hope physicists never discover the Theory of Everything, because if they do I will be out of work "and I'll have to get a real job."

This accounts for a demographic fact: Science reporters tend to stay on the job until they are, well, dead. I mean, they are the oldest active reporters you will ever meet. If you are intellectually inclined, why on Earth would you leave for a better-paying but dumber career? So we science writers stick around the newsroom, covering quarks and quasars, until they have to haul out our carcasses—sometimes literally. One of my predecessors at another San Francisco newspaper, an India-born gentleman named Gobind Behari Lal, kept working until his death at 92; at each day's end, a clerk escorted him across the street to his car. The great Walter Sullivan of the *New York Times* pounded out copy well into his early 80s, until mortality had its say. My colleague Dave Perlman of the *San Francisco Chronicle* is even older than Sullivan was when he died, and is still going strong (some claim that Dave covered the discovery of fire). Such men are the Homers of our time—men who covered the early atomic age, interviewed the first astronauts, peppered Salk and Einstein and Gallo and Guth and Prusiner with questions, and scribbled notes as paradigms collapsed in physics, geology, and astronomy.

I am 52, yet most non-science journalists of my age have long since left newspapers for better-paying work, for example, in public relations. "Sellouts," we call them. They cannot really be blamed, though. Creditors cannot be eluded indefinitely; and there is the matter of feeding one's children. Besides, newspaper hours are unpredictable, a few editors are psychotics, and the pressures can be spiritually depleting. (As Stewart Alsop noted in announcing his retirement, the legs are the first to go.)

And so they leave—the cop reporters, the foreign correspondents, the business writers, the movie and opera critics, the obituary prosesmiths, good people all, a finer and smarter and wittier bunch than you will find at most any gathering of the college-educated American middle-class. Left behind are we aging science writers, the newsroom oddballs and kooks, who will cling to our desks to the bitter end, legs or no legs, credit-worthy or not, until the last giant media merger has shuttered the last newspaper. The great science writer Isaac Asimov, the only human who could make egomania seem adorable, was a freelancer most of his life; someone asked him if he planned to retire. His approximate reply: "I expect they'll find me with my nose in the keyboard." I hope to go the same way.

But love of career has its dark side. In an important sense, we science writers cannot be trusted: we love science too much to cover it with total, brass-tacks honesty. In the 1970s, when Thomas S. Kuhn's iconoclastic view of science history was the rage, Stephen G. Brush wrote a tongue-in-cheek essay titled, "Should the History of Science Be Rated X?"[1] We newspaper science writers usually write as if we assume the answer is "Yes." You can count on us to explain the technical details of scientific research accurately (assuming that we ex-history majors managed to figure it out in the first place). We almost always fail, though, in explaining how "science" as a process really works. I am as guilty of this as any of my peers. As anyone who has read Kuhn knows, that process rarely jibes with textbook recipes for how it "should" work. We science writers also usually fail to explain how scientific ideas bumpily evolve over time, how they are influenced by the larger sociocultural context, and how science—this is the dark side—often serves the interests of those other than the citizens and taxpayers who bankroll it.

Of course, it is not uncommon for scientists to criticize science journalists, and often for good reasons. One of my favorite examples is a letter to *Nature* written in 1901 by the great Jacques Loeb. (I stumbled across Loeb's letter while researching my forthcoming biography of Kuhn, to be published by Oxford University Press.) The era of yellow journalism was in full swing, and Loeb—who later inspired Sinclair Lewis' *Arrowsmith*—was one of its victims. He wrote *Nature*:

> In the interest of the dignity of scientific research I venture to hope you will print the following statement. Some American papers have recently published sensational and absurd reports of physiological theories and experiments whose authorship they attributed to me. These reports, which in America nobody takes seriously, were reprinted and discussed in European papers. I hardly need to state that I am in no way responsible for the journalistic idiosyncrasies of newspaper reporters and that for the publication of my experiments or views I choose scientific journals and not the daily Press.[2]

In the century since Loeb vs. yellow journalism, science writing has improved. For one thing, we are more accurate now. Science journalism has not had a really juicy scandal since the 1970s, when David Rorvik fabricated a tale of human cloning.[3] Even so, many scientists remain wary of journalists, and understandably so. Scientists' core missions do not quite jibe with ours. They seek timeless truth; we seek a gripping story. The results can be some, well, pretty hairy science writing. In recent years, as part of my one-man campaign to "prick balloons" in science, I occasionally asked scientists to criticize science coverage for the record. A European high-energy physicist told me that if he could address newspaper readers directly, he would say: "If you want a recipe for minimizing 'illusions', here it is: Do not trust scientific evidence portrayed in daily newspapers. Wait for the final publications in scientific reviews" (Figure 2.1).

Figure 2.1 Science journalists surround planetary scientist Lawrence Soderblom during Voyager 1's encounter with Saturn, Jet Propulsion Laboratory, Pasadena, California, November 1980.

Source: Jet Propulsion Laboratory, courtesy Xaviant Ford.

Scientists are a special breed of intellectual: Every day, they struggle to exact a new insight from nature. They might spend decades exploring a single type of cell or particle. By contrast, we newspaper writers are condemned to write for an audience of laypeople: Most are non-intellectuals, even anti-intellectuals. In high school they struggled through science, and have disliked it ever since. (Some of them become newspaper editors.) Every day, I face this challenge: How can I, as a science writer, attract and hold such readers' attention? As they flip from the sports page to the comics, how do I convince them to stop and peruse my account on, say, the absence of dark matter in a few elliptical galaxies?

One way is to sugar-coat the pill—that is, to "dress up" the story with catchy phrases and melodramatic anecdotes. This process of sugar-coating begins when I propose the story to the editor: Without his/her approval, it will not see print. Many years ago, while working at the *Los Angeles Times*, I tried to persuade an editor to run a story about a San Diego mathematician's groundbreaking research on the fourth-dimensional Poincaré conjecture. The editor's eyes glazed over. Desperately, I mentioned that the mathematician relaxed by climbing the sides of campus buildings, like a human fly. The editor's eyes lit up. "Go for it!" he ordered. Our published story included a photo of the mathematician scaling a building. Later he won the Fields Medal, which is sometimes called the Nobel Prize of math.

In the two decades since, working at three different newspapers, I have written numerous math stories. Almost all of them attract positive mail. Yet, my "math" stories are ultimately no more about math than the movie "A Beautiful Mind" was "ultimately" about John Nash's work on the isometric embeddability of abstract Riemannian manifolds in Euclidean spaces and the application of game theory to economics. In reality, the movie was about a guy (who just happens to be a mathematician) who goes crazy and is saved by the love of a beautiful woman: It is pure Hollywood. If Nash had not lost his mind, no producer could have cared less about his Nobel Prize-winning work, and with good reason: The audiences would have stayed away in droves. Sadly, in the United States today, that is how the managers of all capitalistic, profit-making media—be they movies, TV, or newspapers—must "package" their very rare ventures into the depiction of scientific life: by turning complex intellectual sagas into formulaic soap operas.

Of course, this is true of almost all American newspaper coverage of the intellectual world, not just the sciences. Consider deconstruction and Heideggerian philosophy, which have had enormous impact on American academics over the last few decades. Yet American newspapers—whose snooty-but-middlebrow coverage of art and opera is rarely more thrilling than any page in *Being and Time*—have ignored these intellectual maelstroms, save on one type of occasion: when there is a new fuss over the "Nazi angle."

I have learned how to sell math stories by dressing them up with little melodramas— for example, news of mathematicians who are squabbling over some obscure proof, or who are caught committing silly goofs. Readers seem to love these stories. Time and again, they e-mail me: "I read your story and I realized, gosh! Scientists are real people!" I am always tempted to answer: "What did you think they were?"

The "scientists are real people" comment is such a persistent cliché, though, that it is worth a moment's reflection. Why would anyone think scientists are *not* "real people"? Consider Stephen Hawking. The journalistic mythologization of figures such as Hawking has reached quasi-religious proportions; I cannot read much of it without laughing. People are offended when I half-jokingly suggest that Hawking might be the Lord Kelvin of our time, a man whose extraordinary courage in the face of bodily decay is no guarantee that his paradigms are not as mortal as the Maxwellian æther. Hawking has a good sense of humor and I am sure he would agree with me.

Such irreverence is rare, though (and is never, ever printed in daily newspapers). Whatever we think in private, in print we science writers come across as science worshippers. Science reporters love their jobs more than any other reporters, save perhaps those who fanatically cover sports and fashion. In my experience, political reporters despise politicians; many business reporters mock Bill Gates, and some of them nurse socialist sentiments; film and art critics grow sick of the mediocre tripe they must review day after day. A few crime reporters relish their grisly "beat"—the *Miami Herald*'s Edna Buchanan was a poet of the sordid—but many more develop tics and ulcers. By contrast, science writing is an endless joy. Science is about nature, and who does not love nature? Science is one of the most amazing cultural innovations since the first Devonian fish crawled from a swamp; much of it is so exciting that it "sells itself." The Big Bang! The origin of life! Thanks to science, humans have gained the eyes and powers of demigods. Who could cover it cynically?

Yet—and here, after much meandering, I come to my main point—we must begin to do so. We science journalists must begin to cover science far more critically, even harshly, before it destroys much of the civilization it has dazzled since Galileo gazed awestruck at the phases of Venus.

This is a hard thing for me to say. In saying it—again, at age 52—I feel as if I am repudiating much of my career as an underpaid science propagandist. Well, maybe

I am: Though we pretend to be jaded cynics (Cary Grant and Rosalind Russell in *His Gal Friday*), most journalists remain idealists inside. In our hearts, we are still Robert Redford and Dustin Hoffman—I mean, Bob Woodward and Carl Bernstein— hurling Late Editions through the plateglass windows of The System. And at present, The System is busy bankrolling billions of dollars of new technologies that might well make this world a much worse place to live. You would hardly know it, though, judging by the generally optimistic, upbeat science coverage of the research upon which those new technologies are based. Our coverage of biotech and nanotech- nology, in particular, often reads like rewrites of press releases from corporate marketeers.

One way to stiffen our spines—to make us less vulnerable to the marketeers' hype—is by learning more about the history of science.

This brings me to Carl Sagan and Thomas S. Kuhn, two men whose names are not commonly associated. When it comes to science writing, I owe my earliest ideals to these men. Both influenced my youthful attitudes toward science's status and power in a democratic, pluralistic society that has thrived partly because of its traditional suspicions of authority and elites.

Yet Sagan and Kuhn influenced me in radically contradictory ways. I discovered Sagan's early writings in the 1960s. That was long before he became a Pulitzer Prize- winning author, the emcee of TV's science series *Cosmos*, and an anti-nuclear activist who was arrested in desert protests. As an adolescent Saganite, I was captivated by his wit, his optimism, his lucid explanations of difficult science, his vision of cosmic possibilities untainted by religious superstitions. For a time he convinced me, the junior atheist, that science was the royal road to truth—the last, best hope for humanity.

But that conviction wavered with the coming of the seventies. Powerful men (they were all men, in those days) were perverting science for evil ends: the napalming of Vietnam; the mass automation that left thousands jobless; the industrial despoliation of land, sea, and sky; the seemingly unstoppable growth of nuclear weapon arsenals. I became a political radical, less outwardly than inwardly. (I was too timid for protest marches, although I briefly co-ran a campus committee to impeach Nixon.)

One day in the early seventies, in a used bookstore in Atlanta, I stumbled across Kuhn's *The Copernican Revolution*. I thrilled to its epic depiction of the fall of a seem- ingly impregnable scientific worldview (Ptolemaic cosmology). By that time, a real-life political revolution was sweeping the streets of the world, most vividly in Chicago and Paris in the summer of 1968. As we later learned from FBI records, the US government viewed SDS, the Black Panthers, and the Weather Underground as real threats to national security. And strange though it seems now, some of us Baby Boomers who were scientifically inclined and politically disaffected—but were too innocent for Marx and too shy for serious street activism—found solace in Kuhn. True, he seemed totally apolitical—he wrote about scientific revolutions, not political ones. Yet we were reassured by his implicit point: Nothing collapses faster and harder than an apparently rock-solid System, be it a scientific ideology or the American political-military machine that we despised. In time the war ended and the nation's political passions cooled, but I continued to view science as

Kuhn did in his 1962 classic *The Structure of Scientific Revolutions*—as an icon ripe for demythologization via a reassessment of its history.

True, Kuhn did not welcome what he viewed as the radicals' co-opting of his ideas. An insecure careerist, he had contradictory goals: He wanted to say radical, disillusioning things about traditional scientific epistemology while maintaining the respect of conservative scientists and their fellow travelers within academia. This was no easy task, especially for a loner like Kuhn: He lacked the "people skills" and taste for self-promotion that helps one to build a sturdy power base within academia. (He bounced from Harvard to Berkeley to Princeton to MIT.)

Still, despite his post-*Structure* denials and obfuscations, Kuhn's implicit message remains a radical one. The radicalism of that message has been somewhat forgotten in recent years, amidst the hubbub over the "science wars." (The hubbub peaked—if that is the word—with the Sokal hoax of mid-1996, about the time of Kuhn's death.)[4] To me, that radical message is: If history is any guide, much of the seeming permanence of scientific "truth" is an illusion. Likewise, scientific "progress" is a far more problematic notion than we are raised to believe, especially in American society (which treats the deeply pessimistic less as shrewd counsels than as candidates for Prozac). Thus, science must not be trusted too much.

Again, that was not Kuhn's *explicit* message, yet it is what I have "learned" from him. Balzac was no radical, either; yet as Engels informed Margaret Harkness, he and Marx learned more from the novelist than from any number of social theorists.

Kuhn and Sagan's differences owed much, I suspect, to their different readings of the history of science. Both men had strong interests in the subject. Trained as a physicist at Harvard, Kuhn became the best-known historian of science since George Sarton. Trained in physics and astronomy at the University of Chicago, Sagan wrote many excellent (albeit Whiggish) short articles on famous scientists, dramatized Kepler's career in one episode of TV's *Cosmos*, and served on an advisory panel to the History of Science Society. On the whole, Sagan viewed science history as a saga of heroes vs. intellectual bigots and progress vs. superstition; his novel-turned-film *Contact* includes a surprisingly dated conception of the intellectual dialectic between science and faith. One might describe his historical writings as a mixture of Sarton and Andrew Dickson White, plus a pinch of Arthur Koestler.

Kuhn, by contrast, despised Sarton and Sartonism—especially its Whiggishness, its hero worship, its half-apologetic traces of Comtean positivism. He attended Sarton's Harvard classes after the war and found them "turgid and dull" (quite contrary to the perception of some other students, such as Robert K. Merton and I. B. Cohen, who later recalled Sarton's messianic showmanship). A new age was dawning for science historians—an age when Newton's alchemy would no longer be swept under the carpet, when Millikan's shifty handling of data would raise new questions about the process and meaning of "discovery," when the notion of "scientific method" would be beaten almost into dust, when famous episodes of scientific insight would be reassessed in terms of their sociocultural-economic-political contexts, when the scientific paper would be subjected to the rhetorician's scrutiny, when scientific "knowledge" would be interpreted less in epistemological terms than in terms of Power and Gender.

Of course, Kuhn did not launch this historiographical revolution alone. Fragments of his theories had been anticipated by others (Polanyi, Quine, Hanson, Bachelard, etc.). Still, thanks to the bestseller status of *Structure* (almost a million copies have been sold), Kuhn "popularized" the revolution's early ideas like no one else. For a decade or two, he became the prime lightning rod for its critics (Shapere, Scheffler, and others) until Feyerabend, Latour, and the Strong Programme *et al.*, became more attractive targets.[5]

Kuhn said little about science popularization, but what he said was unflattering. The opening sentence of *Structure*'s first chapter proposes that a revised view of science history "could produce a decisive transformation in the image of science by which we are now possessed." (Note the subtle sneer behind the word "possessed," as if Kuhn, an atheist, were describing an evangelical who is "possessed" by religious delusion.) Obviously, then, Kuhn was not being complimentary when he briefly alluded (on p. 139) to "the generally unhistorical air of science writing."

Such hints—which I barely heeded as a young, first-time reader of Kuhn—anticipated my insight during my late thirties: Most science journalists, myself included, are actually science propagandists. Wittingly or otherwise, we are the court stenographers of scientism, a pervasive and dangerous—yet rarely acknowledged—ideology. No less than my peers, I am guilty of what the late Dorothy Nelkin called "selling" science.

One reason we are so gullible, I suspect, is that we do not know much about the history of science—I mean, the real history of science, not the Whiggish, Sartonish variety marketed in so many of those "pop" science history books that have flourished during the last decade. Most science reporters lack a sophisticated knowledge of the history of science and technology. Not all, mind you; many educate themselves in the subject by haunting university libraries (as I have), or obtain degrees in the subject (e.g. Bill Broad of the *New York Times* has a master's in history of science from the University of Wisconsin). The historical ignorance is especially rife among those who are younger and less jaded by experiences such as the interferon "miracle," the nuclear industry's lies, the Pentagon's costly techno-gullibility, the irreproducibility of some much-hyped neuroscience research, the insidious problem of laboratory fraud, and the boundless mendacity of polluters, drug companies, and nuclear weapons labs. These, the young, are likeliest to be gulled by the latest alleged "discoveries"—for example, the reports of "gay genes" and "gay brains" that make front-page news, yet are never successfully replicated.

We could do readers a real service by reminding them how many great "discoveries" fall flat: The list is long. Remember the Mars rock? Lacking a historical consciousness, we lack memory of the ways in which huckster scientists fool us over and over again by selling us their "gay genes," "cancer drugs," and other flummeries. We are—incredibly, considering most reporters' liberal-to-left political leanings—especially susceptible to "discoveries" that reinforce social prejudices. Consider the claim, which attracted mass coverage a decade ago, that gay men have "different" brains from non-gays. So far, to my knowledge, there is no published research upholding it. When I reported this fact a few years ago, it was universally ignored by other media.[6]

Of course, the failure to "follow up" is a disease of journalists in general, not just science journalists; it is always more fun to write an exciting story than to follow up with the sad news that it is all hooey. But such failure to follow up is especially dangerous in science, I believe, because it tends to give a "scientific" aura to prejudices about gays, women, minorities, and other historically oppressed groups. The best cure for naïve science journalism, I believe, is the regular reading of history of science—not just the "pop" history of science books that have flourished in recent years but also the shrewder, more socially oriented academic history of science that has thrived since Kuhn's book appeared.

In short, as a science writer, I have migrated intellectually from Sagan to Kuhn. As a biographer of both men, mine has been a painful journey. Sagan "was a hero of my childhood and youth," as I wrote in my 1999 biography of the planetary scientist, *Carl Sagan: A Life*, published by John Wiley & Sons.[7] "But a childhood hero is a dangerous thing to have, because one eventually outgrows childhood." Mind you, I still deeply admire Sagan in many ways; I remained moved by his joyous spirit, his lucid (if occasionally purple) prose, his witty campaign against pseudoscience, his fight against nuclear weaponeers. But in an important sense, I have outgrown Sagan; I have come to suspect something that he could never quite admit to himself—that science, far from being the royal road to truth, may actually be the road to Hell.

I am not alone. Twice, I have interviewed the Astronomer Royal, Sir Martin Rees. Despite his cheerful air ("Call me Marty!"), he recently told me he gives humanity only a 50–50 chance of surviving the coming century. He bases his pessimism on the new or promised means of global self-destruction (micro-"nukes," biotech "bugs," nanotechnology, etc.) that are emerging from the laboratories. Personally, I think he is too upbeat.[8]

True, Kuhn was no Lewis Mumford or Jacques Ellul; his prime interests were epistemological rather than moral-political. Yet *Structure* is important for those of us who are concerned about the moral and political abuses of science. That is because Kuhn challenged the foundation of science's fame—its claim to being epistemically privileged. Mind you, he did not insist that other ways of knowing (art, religion, etc.) are *more* epistemically privileged, or even *equally* privileged. In his later years, he expressed revulsion for the most extreme social-constructivist views of science, which treat "nature" and "external reality" as socially negotiated concepts. Of course, Kuhn's own work, especially his notion of incommensurability, invited equally radical interpretations; but this is not the place to rehash *that* old debate.

The bottom line is this: Whatever Kuhn's real intentions, and however confused his subsequent "modifications" of his own ideas, his main point stands: Science is less epistemically privileged than it pretends to be, as the rubble of ruined paradigms reminds us. And by deflating science's epistemic pretensions—by exposing the emperor as half-clothed—Kuhn unwittingly gave moral support to those of us who fear handing too much epistemic authority over to scientists, "experts," or other synonyms for whatever group is the present epistemic aristocracy (as physicists have been since the Second World War). Hence Kuhn is, or should be, an unwitting hero for those science journalists who want to be more than cheerleaders for science, who wish not to wreck but to *reform* science and its societal masters. These science journalists want to make science more socially responsible, to inspire science to live up to its own professed ideals: Do not kowtow to authority; question everything; continually re-check and reassess your own ideas.

Looking back, how do I assess these two figures, Sagan and Kuhn, who influenced me so much? Other than Isaac Asimov, Sagan was the most effective American popularizer of science during the Cold War era. He left behind a legion of (mostly) adoring graduate students who still carry his torch. A few even try to imitate his career (although so far none have matched Sagan's remarkable combination of showmanship, literary prowess, sparkling personality, intellectual breadth, dizzy imagination, and progressive politics). How "great" a scientist was Sagan? Hard to say. An interviewer once asked me that question, and I joked (alluding to one of Sagan's odder ideas) that if we eventually find "balloon animals" in the atmosphere of Jupiter, then we will be able to say Sagan was a great scientist. Otherwise, I suspect his most colorful scientific research will be remembered as a period piece, that is, as a relic of what we might call Late Scientism—as a spinoff of the naïve optimism he acquired from the pulpy science-fiction magazines of his youth. His hyper-optimistic calculations of the possible abundance of organic molecules in the environs of other worlds (even the Moon!) are the epitome of what one might call "back-of-the-envelope" self-delusion. They reflect an underlying assumption that the universe is somehow "friendly" to the "chemistry of life," as if the existence of vast clouds of carbon molecules across the galaxy, or of stinking ponds of organo-molecular muck on Titan, are equivalent to "life." This premise is central to NASA's present "astrobiology" sideshow, which the media has covered almost totally uncritically although it is based mostly on wishful thinking. Sagan is dead, but his clones live on throughout the NASA bureaucracy and its grantees in academia and private industry.

By contrast, Kuhn (another atheist) is Mr Gloom. When it comes to the life of the mind, he envisions no Theories of Everything, no Final Truth, no End of History; all he expects, rather, is History Without End (history as a kind of Brownian motion, one might say). Kuhn offers no final truth, no announcement that science (or the rest of society) is on the right track (whatever that is) to wherever we would like to go. Kuhn's *Structure* is infamous for its insinuation that scientific "progress" is an illusion, at least in the ontological sense.

* * *

A common complaint against the news media is its lack of historical consciousness. American press, in particular, cover crises in the Middle East, Bosnia, and Kashmir as if they are sports events, as if our daily reports of "body counts" and officials' "sound bites" convey meaningful information to readers. Yet to those who are fighting and dying, the deep past looms large; the average day's bloodshed is no more "urgent" than the Balfour Declaration, or of the partition of '47. Likewise, the history of science and technology casts long shadows over present-day debates involving (say) stem cells or nuclear proliferation or space exploration or the Strategic Defense Initiative, lessons about the oft-ironic consequences of research and the hazards of seeking simplistic technical "fixes" for what are, at bottom, complex social, political, economic, diplomatic, and cultural woes. I have learned many such lessons in my quarter-century as a science writer. If I "published" these lessons inside a bag of fortune cookies, they would include homilies and worthy clichés such as: "If it sounds too good to be true, it probably is." "Laugh in the morning, cry in the evening: Hype almost always ends in disappointment." " 'Extraordinary claims require extraordinary evidence.' 'A blueprint is not the cathedral.' " "Trust your own (educated) common sense: The 'experts' are often hired guns for special interests." "The map is not the territory: Reality is always more complex than theories and meta-narratives." "Catchy phrases such as 'theory of everything' are sales pitches, not hypotheses." "If it isn't broken, don't try to fix it: Don't proffer a 'solution' that must look for a problem." " 'Progress' is sometimes a synonym for moneymaking schemes that benefit a small economic elite and disrupt or wreck the lives of millions." "The heart has reasons that the reason never knows: What passes for 'rationalism' is sometimes a thicket of unexamined presuppositions." "There are no true 'eureka' moments: Even so-called 'overnight' breakthroughs usually require many years of preparation, refinement and mopping-up work before they transform science and society—if ever." If science writers had heeded these sober lessons from the history of science and technology, they might have covered less gullibly topics such as interferon, all-electric cars, and nuclear power, where premature high hopes have (so far) foundered in disillusionment. Being steeped in such history, I can not help viewing present excitement over stem cells, nanotechnology, and Mars terraforming more warily than most of my colleagues do. Perhaps such high-tech dreams will pan out and perhaps they will not; as reporters, our responsibility is to take seriously both possible outcomes. Unfortunately, we usually cover only the former possibility. We are human; we are as gullible as the next guy. How, then, can we journalists expect our readers to make intelligent judgments—to think and vote like adults—based on such historically shallow news coverage? Could they understand the plot of a movie based on a single frame of film?

Ditto for science reporters: Judging by their output, few have a deep sense of history. What history they "know" tends to be Sartonian, that is, a saga of "heroes," "Eureka events," and highly mythologized "moments of truth" (such as the Arthur Stanley Eddington eclipse expedition of 1919, which has been ably deconstructed in recent years by scholars such as John Earman, Clark Glymour, Alistair Sponsel, Lewis Elton, and others). Eddington hoped to detect evidence of stellar light-bending due to the gravitational pull of the Sun, as predicted by Einstein's theory. Popular folklore and science journalists (including myself, I regret to say) have treated Eddington's expedition as a classic case of "scientific method" and hypothetico-deduction, as a clear cut instance in which a single empirical "fact" (Eddington's photographs) neatly and decisively verified Einstein's theory and falsified those of his predecessors. If only science were so simple! Even a distinguished philosopher like Karl Popper appears to have built much of his life's work on the emotional impact of Eddington's expedition, which occurred when he was young and impressionable; to him, the eclipse expedition was the epitome of Popperian "falsification." As historians now know, though, the results of the expedition were more ambiguous than is generally appreciated; its attendant publicity owed as much to the promotional skills of certain scientists as it did to the persuasiveness of Eddington's photos. Not until later years was the reality of stellar light-bending confirmed beyond what most physicists would regard as reasonable doubt. Yet many science reporters (especially the younger ones) tend to cover science as if it were an atemporal, asocial enterprise, whose practitioners exist in a cultural-socioeconomic void—as if (say) NASA "science news" or biotech "discoveries" or anti-missile "tests" result from "objective" research, untainted by institutional or corporate priorities. Oblivious to the oft-disillusioning lessons of history, they lack the skeptical, questioning sensibility—the wariness of sales pitches and self-promoters, be they scientists or Pentagon publicists—that a journalist needs to be anything more than a stenographer.

Obviously there are exceptions. Daily science reporters such as the *New York Times*'s William J. Broad (he of the master's degree in history of science from the University of Wisconsin) are tough-minded, take-no-B.S. role models for us all. Unfortunately, there are not nearly enough Bill Broads out there. They stand almost alone against the ever-swelling P. R. machinery of scientific institutions. NASA has hundreds of flacks who work night and day to sell the space agency's multibillion-dollar agenda to taxpayers, although much of it is based on debatable science.[9]

Curiously, when a science reporter dares to criticize NASA, he or she risks being accused of lacking "objectivity"—which, translated, means: "You're objective if you venerate the status quo and you lack objectivity if you question it." A deeper knowledge of history, I think, would convince my media colleagues that our "objectivity" ethic is too often a form of self-paralysis, a way of numbing our consciences—of kowtowing to the powerful without losing a wink of sleep. We are not "objective" when we cover child molestation, political corruption, or corporate embezzlement; our consciences tell us that these are awful things, and that we have a moral responsibility to expose them. So why should we science reporters be ashamed of calling attention to questionable science, be it creationism, Laetrile, "gay genes," or NASA's peculiar notions of "scientific method"? With luck, we will be able to quote some "expert" who shares our concern. Unfortunately, much of the time, the "experts" are only brave enough to criticize their colleagues *off* the record (this happens all the time); hence, sometimes, we journalists must take the lead in calling a spade a spade. Not being "experts" ourselves, we must do so with great caution, to be sure that we do not commit elementary technical gaffes. At the same time, we should remember that the "experts" have a lousy track record, too, and that "expertise" often camouflages narrow special interests. Remember the chemists who blasted Rachel Carson?

Okay, then: To transcend mere stenography, what historical works should today's science journalists be reading? Kuhn's *Structure* is a good start, I think, but I am prejudiced; and besides, it is too theoretical for many readers' tastes. For absolute beginners, you can not beat Kuhn's *The Copernican Revolution* (1957, but still in print), a very readable, even moving account of everyone's favorite paradigm shift. Paul K. Feyerabend's *Against Method* (1974) is a controversial book that has been much criticized, yet, however flawed, it remains a delightful introduction to related and broader historical topics. Reading it is like tossing a hand grenade into your head.

At the least, I urge science reporters to become familiar with the history of their own little profession, in addition to the history of science per se. To be blunt, there is no better way to hone one's skeptical sensibility than by studying the gaffes and gullibilities of our professional ancestors. I say that as someone who recently spent several days pawing through the private filing cabinets of men, now dead, who were legendary science writers of the early-to-mid-twentieth century. While reading some of their old articles, I occasionally burst out laughing. Yet with every laugh, I knew: "Someday, someone will mock *my* published prose, too." All "knowledge" is potentially malleable, and we should quit telling our readers otherwise.

An outstanding early review of the historical study of popular science is Roger Cooter and Stephen Pomfrey's "Separate Spheres and Public Places: Reflections on the History of Science Popularization and Science in Popular Culture"; it appeared in the journal *History of Science* in 1994, but is still essential introductory reading. (The journal is available at any decent university library near the many rows of *Isis* and *Social Studies of Science*; if your neighborhood campus does not carry any of these, protest vehemently.) At the time, Cooter and Pomfrey lamented that

> surprisingly little has been written on science generally in popular culture, past or present. Still shrouded in obscurity are the effects of even the most obvious mechanisms for the transmission of scientific knowledge and culture: the popular press, radio and television, to say nothing of science texts, museums, school curricula, and the overtly propagandist productions of the science lobby itself. From coffee houses to comic books and chemistry sets, from pulpits to pubs and picture palaces, from amateur clubs to advertising companies, from Science Parks to Jurassic Park, our ignorance both of the low drama and the high art of science's diffusion and modes of popular production and reproduction is staggering

they wrote.

Fortunately, there has been some real progress since 1994. An example is James Secord's *Victorian Sensation* (2001), an account of journalist Robert Chambers's *Vestiges of Creation* and its impact on pre-*Origin of Species* England. Secord's book dramatizes an important point, one also made by other historians of "popular science": The "popularization" of science is not a simple trickle-down effect, that is, a process in which elite "knowledge" gradually seeps down (via press vulgarizations) to the down-trodden masses. Rather, the influences extend in multiple directions—up, down, and sideways. Thus "public" interests and agendas sometimes set the stage for, and even influence the content of, science itself. (Contrary to myth, Kuhn didn't totally deny such "external" influences; still, they didn't interest him much after the 1950s). This seemingly radical idea is, in fact, obvious to anyone who has covered topics such as global warming, AIDS, cancer research, the search for extraterrestrial life, and inves-tigations of stem cells. No one can competently assess trends in such fields without taking into account their larger societal contexts; behind the technical-sounding debates over "facts," there loom grander, murkier clashes over value systems, ideologies, lifestyle choices, etc.

Other articles (among many) that have influenced my thinking about the prehistory of modern science popularization:

Lewis Elton's 1986 "Einstein, General Relativity, and the German Press, 1919–1920," and Alistair Sponsel's 2003 "Constructing a 'Revolution in Science': The Campaign to Promote a Favourable Reception for the 1919 Solar Eclipse Experiments."[10] These and other articles detail the personal and institutional machinations, and epistemic ambiguities, behind one of history's greatest so-called "Eureka events."

Of course, there is no longer any significant doubt that Einstein's theory was right. But thanks to these and other historians, there is also little doubt that Eddington and George Dyson prematurely oversold the theory to the press, and for reasons that were not entirely "scientific." Their media savvy foreshadowed the mass-marketing of much more dubious enterprises including commercial nuclear power, the space station, and the "war on cancer." These articles offer a classic cautionary tale for today's science journalists.

A somewhat related topic appears in Marshall Missner's 1985 "Why Einstein Became Famous in America."[11] It *was not* because the average citizen was stunned to learn that $E = mc^2$.

Steven Shapin and Barry Barnes, "Science, Nature and Control: Interpreting Mechanics' Institutes."[12] When does science "popularization" become a covert mode of social control? This study is worth the attention of any science journalist or science educator who, at times, wonders if he is an unwitting pawn in someone's institutional agenda. What good is "science literacy" if people remain powerless against bureaucracies that camouflage their empire-building and class-related agendas as "public understanding of science," a.k.a. PUS?

The latter crack in no way intends to mock the superb, Cornell University-based journal *Public Understanding of Science* (available online). It offers scholarly studies of the many different forms of popular science, past and present, that appear in venues as diverse as newspapers, TV, museums, classrooms, and popular movies.

* * *

In closing...I sometimes wonder: What would Sagan have become if he had lived many years past age sixty-two, rather than dying from complications of a blood disease? Would he have retained the boyish hopefulness of his young adulthood? Perhaps I am reading too much into the writings of his last years, but I sense an uncharacteristic somberness in *Shadows of Forgotten Ancestors*, the superb (and underrated) 1992 book that he co-authored with his third wife, Ann Druyan. For decades, Sagan had championed the effort to detect radio signals from extraterrestrial life. By the 1990s, though, the search was entering its fourth unsuccessful decade. The sky was silent. Where were the aliens? An ominous possibility began to suggest itself. Sagan had already spent years fighting with the nuclear weaponeers; he feared they would reduce the planet to radioactive ashes covered in the snows of "nuclear winter." Sagan began to muse about whether the celestial silence contained a warning—a hint of the fate of all cosmic intelligences. Do all technical civilizations tend to self-destruct?

Is intelligence, far from being a miraculous evolutionary adaptation, in fact a perverse mutation destined to follow the dodo?

Like Sagan and Sir Martin Rees, I increasingly ask myself such questions. The questions are easier to ask because, thanks to Kuhn, I am less deluded by science's epistemic pretensions than I once was. What bothers me is this: Very, very few of my fellow science journalists are asking the same questions—in print, anyway. By contrast, our newspaper predecessors knew well what was at stake: They lived through the dawn of poison gas and atomic bombings, and had no illusions about the misuses of science. Remember Gobind Behari Lal, the Indian science writer I mentioned? This charming journalist-scholar—who (I read in his 1982 obituary) had campaigned for India's independence, personally known Mahatma Gandhi, taught the philosophy of non-violence, and strongly advocated nuclear disarmament—once observed: "My job is to create a public taste for science. We must make science accessible to the people and for the people. Otherwise, it's dangerous."

Let us suppose, for argument's sake, that science is really paving the road to Hell. If so, then what purpose do we science writers—we science propagandists—serve other than to imitate the scribes and orators and polemicists who, century after century, have hailed the madmen who lead armies and nations to doom?

Notes

1 Stephen G. Brush, "Should the History of Science Be Rated X"? *Science* 183 (1974): 1164–1172.

2 Jacques Loeb, "Sensational Newspaper Reports as to Physiological Action of Common Salt," *Nature* 60 (February 14, 1901): 372.

3 A medical writer, Rorvik wrote a book, *In His Image: The Cloning of a Man* (New York: Pocket Books, 1978), which claimed that scientists had cloned the first human with funding from a millionaire. For details, see William J. Broad, "Court Affirms: Boy Clone Saga Is a Hoax," *Science* 213 (July 3, 1981): 118–119; and Broad, "Publisher Settles Suit, Says Clone Book Is a Fake," *Science* 216 (April 23, 1982): 391.

4 The "Sokal hoax" refers to the publication of physicist Alan Sokal's infamous essay "Transgressing Boundaries: Toward a Transformative Hermeneutics of Quantum Gravity"—a sophisticated mockery filled with spurious references to postmodernist theory—in the journal *Social Text* 46/47, vol. 14 (Spring/Summer 1996): 217–252, and his own disclosure of the episode in *Lingua Franca* (Alan Sokal, "A Physicist Experiments with Cultural Studies," *Lingua Franca* [May/June 1996]: 62–64) that same year.

5 Dudley Shapere, "The Structure of Scientific Revolutions," *Philosophical Review* 73 (1964): 383–394; Israel Scheffler, *Science and Subjectivity* (Indianapolis: Bobbs-Merrill, 1967).

6 Keay Davidson, "No Easy Link Between Genes, Behavior," *San Francisco Chronicle*, February 13, 2001, sec. A, p. 3.

7 For reviews, see David A. Hollinger, "Star Power," *New York Times Book Review*, (November 28, 1999): 15; David H. DeVorkin, "Carl Sagan: A Life," *Physics Today* 53, no. 9 (September 2000): 64; and "Carl Sagan, Carl Sagan: Biographies Echo an Extraordinary Life," *American Scientist* 88, no. 1 (January–February 2000): 74.

8 Keay Davidson, "Saving the Universe by Restricting Research; Astrophysicist says Technology has Potential to Annihilate," *San Francisco Chronicle*, April 14, 2003, sec. A, p. 6; Martin Rees, *Our Final Hour* (New York: Basic Books, 2003).

9 These comments are in no way intended to cast doubts on the very fine, even historic work by many NASA staffers and NASA-funded scientists; their colleagues and peer reviewers

know who they are. In addition, my remarks certainly do not apply to one of NASA's admirable branches, its history office. (See also David Cantor, this volume.)

10 Lewis Elton, "Einstein, General Relativity, and the German Press, 1919–1920," *Isis* 77 (1986): 95–103; Alistair Sponsel, "Constructing a 'Revolution in Science': The Campaign to Promote a Favourable Reception for the 1919 Solar Eclipse Experiments," *British Journal for the History of Science* 35, no. 4 (2003): 439–467.

11 Marshall Missner, "Why Einstein Became Famous in America," *Social Studies of Science* 15 (1985): 267–291.

12 Steven Shapin and Barry Barnes, "Science, Nature and Control: Interpreting Mechanics' Institutes," *Social Studies of Science* 7 (1977): 31–74.

3   The history of now

Reflections on being a "contemporary archivist"

Bruce V. Lewenstein

Although formally a historian of science, I work on very contemporary issues involving the public understanding of science. My interest in science journalism takes me each year to the annual meeting of the American Association for the Advancement of Science, where science journalists in the United States gather. When journalists realize that I am not there to write stories emerging from that year's meeting, they invariably ask: "Why are you here?" To which I reply: "I'm watching you—you're my lab rats."

In writing these notes on my experiences creating archives of contemporary science (including those for cold fusion and Y2K), I feel a bit like the science journalists I normally study. Over the years, I have not been especially reflective about my activities as an archivist of "science as it happens" or "science and technology in the making." I can tell you what I do, but I cannot tell you much about the theory or deeper meaning of how I approach archiving. For you, the reader, I am a lab rat— although, as you will see, a particularly opinionated rat who cares how the maze of contemporary history is organized.

What follows is based on my experiences in the last fifteen years of contributing to various archives on issues or events in science and technology as they happened. After describing the experiences, I will address two types of issues: practical and conceptual. But before I proceed, another comparison to journalism: There is a famous saying, that journalism is the first draft of history. The same is true for the contemporary archivist: he or she is creating the first draft of what historians will have access to later. I know that the questions below of comprehensiveness, of the value of particular records, about the challenge of identifying materials that might not be in the most obvious place, etc., have been dealt with many times before by other historians and archivists. But one of the effects of my focus on documenting contemporary issues has been that I have much less time to review the literature about historiography, archiving, and so on. (Perhaps I should not admit this in a chapter to appear in a book co-edited by Thomas Söderqvist, but I have not even read carefully some of the earlier works that he and others have produced that address issues relevant to contemporary archiving.[1]) I am too busy trying to create the archives themselves. So I apologize in advance for what will seem like beginner's mistakes.

My experiences

Cold fusion

In March 1989, I was just a year past my PhD in the social history of science, teaching a new course at Cornell University on "how to study science in the media." As the media frenzy developed around the announcement that two chemists in Utah, B. Stanley Pons and Martin Fleischmann, had discovered a way to produce nuclear fusion at room temperatures, in apparatus little more complicated than that found in a high-school chemistry laboratory, my students and I began collecting the media stories. I called my science journalist friends, asking them to tell me about how they were covering the story—just so I could report back to the class. It happened that an experienced sociologist of science, Thomas Gieryn of Indiana University, was spending a year at Cornell as a visiting professor. With a graduate student in management, he had started downloading and saving the messages about cold fusion appearing on several Internet-based bulletin boards. In 1989, such bulletin boards were essentially unknown in the science studies community, and Gieryn was interested in how scientists used them.

One day, a few weeks after the initial announcement of cold fusion—and well before it was clear what the "right" answer was about the existence of cold fusion— we realized that by combining our collections we might have a unique set of materials documenting "science as it happens." Our material might provide a window into the ephemeral judgments made by people involved in establishing stable scientific knowledge. We knew that, in a few years, people would look back and say "Oh, I always knew it was true," or "Of course, it was obvious it was ridiculous" (whichever turned out to be the primary judgment), when in fact the judgments had been much more muddled or changeable in the first days and weeks after the initial announcements.[2] Many of those processes of coming to judgment would be documented only in conversations, drafts of papers, personal e-mails, quotes in media stories, and so on. If someone were to try to write a history of cold fusion five or ten or fifty years after the fact, interviews would yield only retrospec- tive, reconstructed memories. The materials that might document the process of coming to judgment, especially those documenting a *communal* judgment as opposed to an individual's thought processes in the contemporary moment, would probably have been lost to historians—they were too ephemeral, too spread out among indi- viduals, too fragmentary.[3] Thus perhaps our collection of ephemera would have historical value, especially if cold fusion turned out to be "true." Being far more experienced than me at grantsmanship, Gieryn immediately called colleagues at the US National Science Foundation, who agreed that creating an archive on cold fusion made sense. Within a month after the initial announcement, we had submitted our proposal, asking for $15,000, mostly in travel money so that we could go interview participants in the controversy. We have always joked (though I have no idea if it is true) that the money was awarded out of the "earthquake fund"—that is, "natural disasters."

Over the next year, we collected more than 5,000 e-mail messages; more than 1,000 media stories in print, radio, and television; nearly 100 taped interviews ranging in length from 15 minutes to several hours; printouts, photocopies, and reprints of e-mails, reports, published articles, notebooks, and other documents; and a small collection of material culture (mostly t-shirts, mugs, and other items of humor associated with cold fusion). Until 1994, we continued to sporadically collect some material, mostly as people sent it to us or it fell into our laps. Finally, when we realized that cold fusion would not "end"—a small community of researchers continues to work in the field as of 2004—we declared the collection "closed." By then, the collection had been deposited in the Cornell University Library's Division of Rare & Manuscript Collections, where its 17 cubic feet of material remains available to interested researchers.[4] Gieryn and I had written a number of articles or chapters drawing on our work.[5] At least 25 other researchers used the collection in the 10 years after it was created, and several books and television shows—skeptical, supportive, and scholarly—drew on its contents.[6] At least one defendant accused of libel by Pons and Fleischmann used the archive to prepare for the trial.

The O. J. Simpson trial

By early 1994, I had finally put cold fusion behind me, declaring the archive closed in March five years after the original announcement, having in press several articles drawing on my work, ready to return to study of popular science. Then, in June 1994, US football legend O. J. Simpson was accused of murdering his estranged wife and a friend of hers. Much of the evidence involved the (then) relatively new use of DNA-typing or "DNA fingerprinting" on blood samples found at the scene of the crime and at Simpson's home.

At the time, my Cornell colleague Sheila Jasanoff was directing an NSF-funded study on the uses of DNA evidence in courtrooms. She asked if we could build an archive on the O. J. Simpson trial similar to the one we had created on cold fusion. I agreed to direct the project, and NSF supplemented her grant so that we could send some graduate students out to interview relevant experts and so that we could pay an undergraduate to organize the materials. (One thing I had learned from the first project: budget for labor.) We also had funds to purchase videotapes from *Court TV*, which was planning to broadcast the trial in its entirety.

For the next year, we scanned newspapers and watched the trial. We were especially interested in May 1995, when evidence about the DNA fingerprinting was presented in court (and when, according to TV ratings services, viewership of the trial dropped dramatically). The graduate students were also conducting general research on the use of DNA evidence in courtrooms, collecting transcripts and records of trials that helped establish precedents in the field, interviewing lawyers and scientists active in the forensic DNA field, and so on. In the end, we purchased 235 *Court TV* tapes and deposited them along with hundreds of newspaper articles in the Cornell archives.[7] More importantly, we deposited a number of the records of other trials, including the only publicly available transcript of *New York State v. Castro*, the trial that established the US precedent for allowing DNA evidence into a trial. The archive occupies more than 13 cubic feet.

Although I had taken some friendly insults for the cold fusion archive ("why are you studying something that isn't really *science?*" was what many scientists asked), the O. J. Simpson trial led to many more jokes. It was a highly publicized murder case in which even the lawyers, judge, and ancillary witnesses became celebrities. One witness in particular, Simpson's houseguest Kato Kaelin, seemed to embody the stereotypical California blond. When Cornell issued a press release about the archiving project, I was interviewed by a number of media outlets. One morning drive-time radio disc-jockey was particularly sarcastic, wondering why the US government was funding an academic study of the O. J. Simpson trial. All I could do was play along, laughingly noting that "I'm only archiving the DNA stuff—I don't archive Kato." I was quoted in the *New York Times*, defending academic interest in celebrity trials: "Intellectuals are people, too," I said. "We need excitement just like everybody else."[8]

The scholars and students involved in producing the DNA archive produced a special issue of the journal *Social Studies of Science* drawing on the material, and one of the students published an award winning book on the general issue of the history of fingerprinting as a tool of identification.[9] Fewer outside researchers have come to use the collection than the cold fusion collection, but it may have more lasting value because of the access it provides to court transcripts that are otherwise difficult to find (the Castro transcript was retrieved from the file cabinets of a private law firm). Once, while interviewing an academic job candidate, I was reminded that we had met several years earlier, when he had come to Cornell to use the DNA archive for his dissertation research. So while the jokes about the "O. J. archive" continue, I believe it serves an important purpose—a topic to which I will return.

A final important note about the DNA archive: a second lesson that I had learned from the cold fusion case was the need to establish a clear end-point for an archive on "science-and-technology-in-the-making." We had originally said that we would keep the cold fusion archive going "until we know if it's true." But, as noted above, while the scientific mainstream has clearly decided that cold fusion was an aberration, a fraud, perhaps even "pathological," a clear subcommunity of researchers continues (as of 2004) to pursue cold fusion research. There has been no "end." (In retrospect, that is one of the lessons of the cold fusion case—the difficulty of absolute closure in cases of scientific controversy, what Bart Simon, one of the first users of the archive, has called "undead science.")[10] In the case of the O. J. Simpson trial, we were more careful: we explicitly noted that "the Simpson case has an identifiable beginning (the murders of Nicole Brown Simpson and Ronald Goldman), a middle (the trial), and an end (the initial verdict) [so that] the task of identifying material will have relatively clear bounds." We expected (as did many observers) that Simpson would be convicted and we did not want to keep the archive open through multiple appeals. In fact, of course, Simpson was acquitted. Although a subsequent civil trial found him liable for the deaths of his wife and Goldman, we did not collect materials on that trial.

The Y2K bug

Once again I left archiving behind, and spent the late 1990s working on the history of the AAAS and on other non-historical projects.[11] Then, in January 1999, one of my graduate students, Joshua Greenberg, stopped by after returning from the winter holidays. I asked him how his break had been; "fine," he said, except that he had returned to Cornell to find his apartment broken into and his computer stolen. Knowing that Greenberg was a highly technologically attuned student, I commiserated. It was not a problem, he said—it had been an old machine and he had all of his data backed up. Insurance would pay for a new one. In fact, he had been investigating new machines and had found that he could get a card that would let the computer watch TV. "I can even scan the close-captioning text," he said. "I could tell the computer to capture, say, every mention of the words 'Y2K' or 'millenium bug', and we could create an archive of popular culture references to the Y2K bug."

I stared at him. That is *fundable!* I said. It was brilliant. By now, I had learned about grantsmanship and made my own call to NSF. Yes, they would be willing to fund us for a year to document the upcoming response to the Y2K bug (the belief that many software programs might break down at midnight on December 31, 1999, unable to cope with the third millennium).[12] By July 1, Greenberg had three computers set up in his apartment (we could not get cable TV in the office), watching TV 24 hours a day. The system he programmed required that he change the recordable CDs between 3:00 and 4:00 in the morning; this did not involve any change in his normal sleeping patterns.

Again we budgeted for labor (Greenberg's tuition and stipend for a year) and specified an end-point: January 31, 2000. On New Year's Eve, several of us set our VCRs to record across the midnight hour, to catch whatever happened in case Greenberg's computers themselves developed a Y2K problem. (He had successfully debugged his own software, however, and the system worked smoothly.) In the end, we deposited 400 CDs of data in the Cornell archives, along with a small collection of print media stories and, as in the cold fusion case, joke items and other examples of popular culture.[13] Greenberg has presented a few papers on the archive, and we briefly considered patenting the technology, but otherwise the archive has remained essentially unused.[14]

The voting technology archive

While I had developed some expertise in the archiving of science-and-technology-as-it-happens, I had also by 2000 begun to question the value of such archives. As I will discuss below, they raised questions of historiography and effort. So in January 2001, when Jasanoff (now at Harvard) and another Cornell colleague, sociologist Steve Hilgartner, asked whether I thought an archive devoted to the role of technology in the disputed US Presidential election of 2000 would be useful, I counseled against it. I questioned whether we, as historians and sociologists of science, could bring anything unique to the *archival* role. The National Archives, I was sure, would be collecting material on the election, including examples of the technologies responsible for "hanging chads" and "butterfly ballots." Precisely because the election was such an important historical event, I believed that traditional archives would preserve the ephemera that for me were a prime motivation for creating a contemporary archive. Analysts of science and technology could more usefully concentrate on thinking about the role of technology (as they did, for example, in contributions to a special issue of *Social Studies of Science*).[15]

My arguments were not sufficient (another example of the fleetingness of expertise), and the Voting Technology Archive was duly created and eventually deposited in the Cornell archives in 2004. Although I provided some suggestions to the students working on the archive, I chose not to participate in this project. (I do have to admit that it was fun to have one of the Palm Beach, Florida, voting machines in our offices, purchased on eBay.[16] But I still suspect that the butterfly ballot technology and other aspects of the technology of the election have been preserved elsewhere, as well.)

Nanotechnology

My determination to avoid archiving, however, soon faltered. In 2003, I became involved in various projects exploring the social and ethical issues associated with nanotechnology. As I helped put together a nationwide team to create an infrastructure for studying those issues, it was clear that documenting the emerging history of nanotechnology was a fundamental need. All of the arguments for documenting science-and-technology-in-the-making applied: a crucial "next big thing," a problem of rapid growth in scientific and technological knowledge, a diffuse research community, frequent discussions in meetings and online settings, all in a much more electronically based scientific community than had been true even a decade earlier. Unlike cold fusion, I felt confident that nanotechnology was not simply a particular research blip, but a major area of scientific research that deserved documentation.

That argument made sense to others, and the "social and ethical issues" component of the National Nanotechnology Infrastructure Network (http://www.nnin.org) includes a commitment to collecting and preserving documents and other materials from the history of nanotechnology. The leader for this effort was Timothy Lenoir, a historian of science who has been among the pioneers in online documentation with Sloan Foundation funded projects on the history of the mouse and the history of bioinformatics (http://hpslab.stanford.edu/). It is too soon to know the outcome of this project, but we are excited by the possibilities for using new technologies in novel ways.[17]

What have I learned?

Much of what I have learned about archiving falls into the "practical" category. As noted above, funding must cover the cost of labor to collect and organize the collection, not simply the direct costs of travel and copying. Unless major unrestricted funding is available, projects need definite boundaries, both in scope and in time. I have also learned the importance of keeping good—nay, excellent—records of what has been collected, what has been identified but not yet collected, how the collection has been organized, and various identifying information about the collected material (dates, provenance, etc.). Creating an archive is a team endeavor (unlike doing historical research, which is often an individual exercise), and I have learned the importance of team meetings, memos documenting decisions, regular communication with contributors and advisors, and so on. I have learned to work with Cornell's public information office to get publicity about the projects, which helps us identify people with collections that we can raid or materials that we can copy. (An amazing amount just comes in through the mail.)

But perhaps more important are the inchoate thoughts I have had about the relationship between archiving and history. Here I attempt to put those thoughts into some coherent order.

Historians depend on archives

We cannot write about things that are not documented (at least, not if we want to call our writing "history"). We all struggle with the need to find relevant archives with material suited to the questions we bring. When students ask us for topics, we often suggest: "first find the archive, then write about what you find there." We know that is imperfect, since the archives often do not have everything we need. But it is one of the truths about writing history.

Those of us who work on contemporary topics think we have an advantage: we can interview some of the participants in our stories, recording their memories of events, probing their interpretations of data, looking for the story-behind-the-story—what *really* happened at that meeting that the minutes describe so blandly? Even when we use interviews, we are still dealing with archives. Whether we rely on informal discussions or formal oral histories, we are simply using a different kind of document. For many contemporary historians, or those who rely on oral histories, the key methodological problem is "memory" and especially the conflict between contemporaneous documentation and *post hoc* oral history.[18]

Deliberately formed archives can actually create history

The "problem" of memory depends on *having* the documentation, both contemporaneous and *post hoc*. We tend to think of contemporary history as data-rich. But it still depends on what is saved and what is accessible. Contemporary topics are often not "public record." The drafts, the memos, the e-mails, the peer-review reports, and so on, are in people's private files. Those files might become available, but more likely they will be thrown out (unless the creator is a major figure). A particular historian working on a topic may gain access to the private records, but other historians coming later will not have access, and so will not be able to double-check the findings or interpretations of the first historian, or draw on the material for other purposes.

The key advantage of the prospective, consciously created archives I have described above is that they can actively seek out ephemeral material, items from people's files, material from the peripheral actors, things that the actors think of as unimportant (such as drafts of papers) but that the trained historian recognizes as a treasure. In the cold fusion archive, we have peer-review reports on some of the crucial manuscripts. In the DNA archive, we have trial transcripts retrieved from private filing cabinets.

Figure 3.1 Creating a contemporary archive offers opportunity to preserve cultural ephemera that might otherwise be lost. The stuffed toy closest to the computer, if dropped on the floor, plays a recording of breaking glass—the sound of a "Y2K bug crash."

Source: Photo by Bruce V. Lewenstein.

In the Y2K archive, we have not only news reports about preparations for the Y2K moment but also the LifeSavers® advertisement showing how Y2K had become a cultural trope in the months leading up to midnight on December 31, 1999. (Figure 3.1). Google® and the Internet have given us access to many more documents and media stories than ever before, but they still depend on those documents having been saved and scanned.

What topics are worth creating archives about?

This key advantage of saving documents, however, is also the source of one of the most crucial criticisms that can be made of prospective archiving. Precisely because it *is not* history, because it is the "first draft," how can we know if the topic is important? Was cold fusion an important moment in the history of science in the late twentieth century, or was it simply an aberration, of no more importance than hundreds of other brief claims that did not happen to create frenzies of media attention? And if it *was* only an aberration, has the action of creating an archive created a historiographic "bump," a focus on a minor incident in the history of science that people are writing about simply because the records *are* available? In the case of the O. J. Simpson trial, the verdict showed very little about how DNA evidence is used in a trial; it showed only that the technical questions about DNA evidence are secondary to more basic questions about the "chain of evidence," about proof that evidence has not been altered or planted. Will future historians use this archive as though the Simpson trial was an important marker in the use of DNA evidence? They might, only because they have easy access to the hundreds of hours of court time that we kept.

Unintended value

I have noted several times, however, that an ancillary value of the DNA archive is the preservation of trial transcripts. In the cold fusion archive, we have examples of peer review, a fundamental element of modern science that is notoriously hard to document. These materials were not the specific ones we hoped to preserve, and they do not necessarily address some of the research questions that shaped our initial collecting. That raises another issue: to what extent is a prospective archive limited by the questions and interests of its creators? The traditional archivist can work only with the materials that survive. But in contemporary archiving, you, the archivist, are the one deciding what survives. That means that you have some obligation to be as comprehensive as possible. This is distinctly different than being a contemporary *historian*, where you need only collect the materials that seem relevant to your project. If you are an archivist you have an obligation to the people who come after you. Some of that obligation, some of those limitations on what you collect, can be handled through explicit statements about what you collected and what you did not collect. But nonetheless, you have some obligation to constantly be surveying the surroundings, to be completely up-to-date on as much of the material as possible, to ensure that the collection is as broad as it can be.

Imposing order on complexity

A similar problem, of how the creators' interests shape the archive, is evident in the process of creating order in the material. The issues of series, folders, documents, and finding aids—these are the day-to-day practical issues that the contemporary archivist faces (I usually dealt with them by creating computer databases). Actually, of course, all archivists face those issues, and standard texts in the field devote time to dealing with organizational questions. But for the contemporary archivist, who is imposing order on material as it comes in, these issues are both daily decisions and conceptual decisions. They are the place where one defines what the topic is, what its boundaries are, what its constraints are. Again, the archivist must be explicit about what decisions are made, and why. In the cold fusion finding aid, for example, I created multiple indexes, to allow users to find material by name, by media, by date, by topic.

Responsibility to your sources

Another set of issues concerns the role that participants in the current science should have in the archiving. Anthropologists are very clear about this issue. They have an obligation, an ethical obligation, not only to treat their subjects with respect, but also to return something to their subjects. There should be some value to their subjects for having spent time with the anthropologist. What is our obligation as historians? Should we protect the anonymity of sources who give us access to private records? Should we allow sources to re-tell the story in their own words, even when those words contradict other sources of information? Many years ago, I had a heated argument with a sociologist at a meeting about documenting the Human Genome Project. He argued that his interviews with scientists were explicitly private and could not be identified by name. This was both a practical issue for him—he had promised his sources anonymity—and a methodological one: he was not interested in the particulars of individuals, but in the social patterns that their interactions revealed. For me, as a historian, anonymous interviews were essentially useless. I needed to be able to say who said what, when, to whom. If possible, I would like to say why, too.[19]

Yet, as a contemporary historian, I have agreed to allow my sources to control their information to some degree. Without that approach, they may not even talk to me. During one of my projects, I was interviewing a key player. At one point, I described an event that I had heard about from many sources, as a way of leading into a new topic. To my surprise, the person I was interviewing denied that the event I mentioned had occurred. I was startled for a moment, but decided to move on to another topic, as I wished to retain my rapport with the individual. At the end of the interview, I asked the individual to sign the release form allowing us to deposit the tape in the archive. The person asked for some time to read the form and I readily agreed. An hour or two later, I was called back to the individual's office. "I can't sign this form," the person said. "I lied to you. When you asked me about [the event], you caught me off-guard and I lied. The event did occur. But now I don't want anyone to hear that tape, because it has me lying on it." After a discussion, we came to an agreement. I had the tape with me. We found the spot on the tape, and I erased my question and the person's answer. There is now no evidence about the individual's statement to me, and the person signed the release form.

Conclusion

In the end, I agreed to let the individual control the material because I wanted the other information from the tape in the archive. For me, that is the crucial test of an archive of science-and-technology-in-the-making: will it yield data that would not otherwise be available, that some other historians can use to answer their questions? Yes, the information will be shaped by the questions and interests that I bring to the creation of the archive. But constructing my research in such a way as to save information for others means that I can be much richer in what I retrieve from contemporary history, which serves both my own research and the research of others to follow.

Acknowledgments

Much of what I have learned about archiving has come through the patience of the professional archivists in Cornell's library system. Elaine Engst, Director of the Division of Rare and Manuscript Collections and University Archivist, and

H. Thomas Hickerson, Associate University Librarian for Information Technology and Special Collections, have both spent many hours with me discussing particular details of archiving, making suggestions for how to handle specific problems or opportunities. Much of what I list in the section on "lessons learned" are things that I could have learned more quickly and less painfully had I paid closer attention to what they had to say. My work on contemporary archives has also been supported generously in both material and intellectual ways by my colleagues in Cornell's Departments of Communication and of Science and Technology Studies, and by the Chemical Heritage Foundation. Much of the work has been supported by various grants from the US National Science Foundation.

Notes

1 I am thinking particularly of Thomas Söderqvist, ed., *The Historiography of Contemporary Science and Technology* (London: Harwood Academic Publishers, 1997), but also of historiographically self-reflective works such as Frederic L. Holmes, Jürgen Renn, and Hans-Jörg Rheinberger, eds, *Reworking the Bench: Research Notebooks in the History of Science* (Dordrecht/Boston, MA: Kluwer Academic Publishers, 2003) and books on oral history generally, such as David K. Dunaway and Willa K. Baum, *Oral History: An Interdisciplinary Anthology* (Nashville, TN: American Association for State and Local History, 1984) and the Oral History Association's "Guidelines for Oral History," (1995).

2 A similar case is William Glen's contemporary oral history study of the mass extinction controversy; see Glen, *The Mass Extinction Debates: How Science Works in a Crisis* (Stanford, CA: Stanford University Press, 1994).

3 On this theme see also Lillian Hoddeson, this volume.

4 Bruce V. Lewenstein, *Cornell Cold Fusion Archive (Collection #4451): Finding Aid*, 5th ed. (Ithaca, NY: Cornell University Library, 1994.)

5 Thomas F. Gieryn, "The Ballad of Pons and Fleischmann: Experiment and Narrative in the (Un)Making of Cold Fusion," in *The Social Dimensions of Science*, ed. E. McMullin (Notre Dame, IN: University of Notre Dame Press, 1992); Thomas F. Gieryn, *Cultural Boundaries of Science: Credibility on the Line* (Chicago, IL,: University of Chicago Press, 1999); Bruce V. Lewenstein, "Cold Fusion and Hot History," *Osiris*, 2d ser., 7 (1992): 135–163; Bruce V. Lewenstein, "Do Public Electronic Bulletin Boards Help Create Scientific Knowledge?: The Cold Fusion Case," *Science, Technology, & Human Values* 29, no. 2 (1995): 123–149; Bruce V. Lewenstein, "From Fax to Facts: Communication in the Cold Fusion Saga," *Social Studies of Science* 25, no. 3 (1995): 403–436.

6 John Huizenga, *Cold Fusion: The Scientific Fiasco of the Century* (Rochester, NY: University of Rochester Press, 1992); Charles G. Beaudette, *Excess Heat: Why Cold Fusion Research Prevailed*, 2nd ed. (South Bristol, ME: Oak Grove Press, 2002); Bart Simon, *Undead Science: Science Studies and the Afterlife of Cold Fusion* (New Brunswick, NJ: Rutgers University Press, 2002).

7 *Guide to the O. J. Simpson Murder Trial and DNA Typing Archive, 1988–1996* (Collection 53-12-3037), Cornell University Library, Division of Rare and Manuscript Collections 2002, http://rmc.library.cornell.edu/EAD/htmldocs/RMA03037.html (August 30, 2004).

8 Janny Scott, "The Joy of Deconstructing O. J.; At a Symposium, Deep Thoughts and Cheap Thrills," *New York Times*, September 25, 1996, sec. B, p. 1.

9 Michael Lynch and Sheila Jasanoff, eds, "Contested Identities: Science, Law, and Forensic Practice," special issue of *Social Studies of Science* 28, no. 5–6 (1998): 675–686; and Simon Cole, *Suspect Identities: A History of Fingerprinting and Criminal Identification* (Cambridge: Harvard University Press, 2001).

10 Simon, *Undead Science.*

11 Sally Gregory Kohlstedt, Michael Sokal, and Bruce V. Lewenstein, *The Establishment of Science in America: 150 Years of the American Association for the Advancement of Science* (New Brunswick, NJ: Rutgers University Press, 1999); Bruce V. Lewenstein, "Advocacy vs. Objective Reporting—a Historical Perspective" in *Risikoberichterstattung und Wissenshcaftsjournalismus (Risk Communication and Science Reporting)*, ed. Winifred Gopfert and Renate Bader (Stuttgart: Schattauer Verlag, for Robert Bosch Foundation/Free University of Berlin, 1998), 179–190; Bruce V. Lewenstein, "Editorial: Reflections on Visiting a Science Center" *Public Understanding of Science* 7, no. 4 (1998): 267–269; Bruce V. Lewenstein, "Why the 'Public Understanding of Science' Field Is Beginning to Listen to the Audience" in *Transforming Practice*, ed. Joanne Hirsch and Lois Silverman (Washington, DC: Museum Education Roundtable, 2000), 240–249; Bruce V. Lewenstein and Steven W. Allison-Bunnell, "Creating Knowledge in Science Museums: Serving Both Public and Scientific Communities," in *Science Centers for This Century*, ed. Bernard Schiele and Emlyn H. Koster (St Foy, Quebec: Editions Multimondes, 2000).

12 No good history of the Y2K bug has yet appeared; for background, see http://en.wikipedia.org/wiki/Y2k_bug (November 29, 2004).

13 Joshua Greenberg, "Finding Aid," in Joshua Geenberg, ed., *Public Perceptions and Construction of the Y2K Problem Archive, 1999–2000* (Collection 53-12-3225) (Ithaca, NY: Cornell University Library, Division of Rare & Manuscript Collections).

14 Joshua Greenberg, "Constructing Y2K: Representations of Experts on U.S. Television" (paper presented at the annual meeting of the Society for Social Studies of Science, Vienna, Austria, September 2000); Joshua Greenberg and Bruce Lewenstein, "Using Closed Captioning for Content-Based Television Archiving: A Case Study" (paper presented at the annual conference of the International Communication Association, Washington, DC, May 28, 2001). The fascination with technology served Greenberg well; after completing in 2004 his dissertation on the history of the VCR, he took up a post-doctoral position at George Mason University's Center for History and New Media, http://chnm.gmu.edu (accessed on 30 June 2005).

15 Michael Lynch, ed., "Pandora's Ballot Box: Comments on the 2000 US Presidential Election," special section of *Social Studies of Science* 31, no. 3 (2001): 417–467.

16 The outcome of the US Presidential election in 2000 was determined by only a few hundred votes of the millions cast in the state of Florida. Many ballots were incompletely punched through (leaving what the media called "hanging chads.") Other ballots were designed so that they opened like a butterfly's wings, but the name of the candidate and the place to mark the vote did not always align properly, so that many voters cast ballots for candidates they did not support; in one district, for example, as many as 5,000 votes were apparently cast for the wrong candidate. For further information, see Political Staff of the Washington Post, *Deadlock: The Inside Story of America's Closest Election* (New York: PublicAffairs, 2001); and Jeffrey Toobin, *Too Close to Call: The Thirty-Six-Day Battle to Decide the 2000 Election* (New York: Random House, 2001).

17 See Hessenbruch, this volume.

18 See Hoddeson, this volume.

19 The question of how much control interview subjects should have has now become contentious in the United States, with some institutional review boards insisting that all interviews should be anonymous. For background, see John E. Sieber, Stuart Platter, and Philip Rubin, "How (Not) to Regulate Social and Behavioral Research," *Professional Ethics Report*, XV, no. 2 (Spring 2002): 1–3; Jonathan T. Church, Linda Shopes, and Margaret A. Blanchard, "Should All Disciplines Be Subject to the Common Rule?" *Academe* 88, no. 3 (May–June 2002): 62–69; and Joseph B. Walther, "Research Ethics in Internet-Enabled Research: Human Subjects Issues and Methodological Myopia," *Ethics and Information Technology* 4 (2002): 205–221.

Part II

Whose history? Ethics, lawsuits, national security, and the writing of contemporary history

4 The politics of commissioned histories (revisited)

David Cantor

Academic journals are often considered a sleepy backwater of litigation. Legal eyes rarely flicker over their pages; scholarly controversies may excite little more than a lawyer's yawn. No place here for the legal quarrels associated with the publication of, say, material labeled critical to national security, or magazine articles that disparage public figures. Or so it may seem. In fact, lawsuits against historians are far more common than might be thought, as I was to discover in the late 1980s and early 1990s.[1] For, in writing a review article on commissioned medical histories, called "Contracting Cancer," my publishers and I were threatened with lawsuits. Indeed, I was involved in an intense controversy that threatened to reach the higher levels of British government.

I will not keep you in suspense over one matter: Contracting Cancer *was* published, and the legal threats turned out to be bluffs. (You will find the article, as it originally appeared, beginning on p. 48 in this book.) I did not end up in court, nor did my publishers. But at the time no one could be sure that this would be the outcome. The mere threat of legal action served to intimidate, and for about two years it was quite unclear whether the article would be published. The "Cantor Affair"—as one historian called it—is thus a cautionary tale about the danger that legal action can pose to scholarly debate. My article nearly fell foul of threats of litigation, and so became part of the issue it sought to address. An essay on censorship was almost censored itself.

Contracting Cancer began life as a short book review for *Social History of Medicine* (*SHM*) of Joan Austoker's history of the Imperial Cancer Research Fund (ICRF). *SHM* was then and remains the journal of the Society for the Social History of Medicine (SSHM), the leading professional organization for historians of medicine in Britain.[2] Austoker's book had been commissioned by the ICRF, and it seemed to me to exemplify some of the worst features of a growing trend towards contract histories in the 1980s. So, on a fateful train journey from Manchester to London in 1988 or 1989, Roger Cooter (then the book reviews editor of *SHM*) and I agreed that I should expand the review into a longer essay that used the ICRF history to comment on the broader problems of commissioned histories.[3] The expanded review was refereed and revised, and the page proofs were on my desk.

Then the problems started.

Figure 4.1 Author David Cantor, himself a contract historian.

Source: David Cantor.

Austoker had received a copy of the article from the editors of the journal. She responded with a threat of legal action for defamation if publication went ahead.[4] Consternation and months of wrangling followed, for British law had (and still has) a very broad definition of defamation. In my case, the situation was further complicated by threats of a separate legal action from Charles Webster, then director of the Wellcome Unit for the History of Medicine in Oxford and the official historian of the British National Health Service (Webster is mentioned in a footnote on the first page). Webster also threatened to involve the Cabinet Office, the British Government department that supports the Prime Minister and the Cabinet, and which administered the NHS history.[5]

Part of the difficulty was that under British law, any critical review could be technically defamatory. *Social History of Medicine*, founded 1988, was (and still is) published by Oxford University Press (OUP), and the Press sought legal advice. OUP's lawyer explained defamation as involving the publication of a statement that reflected on a person's reputation and tended to lower him or her in the estimation of "right-thinking members of society generally," or tended to make them shun or avoid him or her. The phrase "right-thinking members of society generally" was important to this lawyer. He explained that it made little difference legally that the review was to be published in an academic journal where critical reviews might be expected—even demanded—by the readership. Academic publications were to be judged by what "reasonable people in general" might think of an article, and not by the opinions of any special class of persons, such as historians of medicine or science.

The real question for the lawyer was whether there was a legal defense. According to him, there was. In his view, the article could be defended on the grounds of "fair comment" on a matter of public interest, and he recommended a few minor changes.[6]

I revised the article to meet the lawyer's concerns, but the legal technicalities were only part of the story. First, the article became embroiled in political controversy within the Society for the Social History of Medicine, with people on both sides of the debate resigning from the Society and its committees as the fight swung in favor of publication or against. Among the casualties on my side were Bill Luckin, Mary Fissell, and Roger Cooter, who all resigned from SSHM/*SHM* positions in protest at what seemed a decision not to publish the article. In addition, the cultural historian Ludmilla Jordanova turned down an invitation to become editor of the journal, partly because of the problems over my essay.[7] Webster was an influential figure within the organization, and he retained considerable loyalty among some members of its committees. Despite the bloodletting, significant opposition to publication continued up to the moment the article appeared in print.

Second, the Society, the journal, the publishers, and myself and my then wife, Mary Fissell, became concerned about the financial implications of the threatened legal action. The lawyer's claim that the review could be defended on the grounds of fair comment was hedged with qualification.[8] There was always the possibility that Austoker or Webster would succeed in their legal actions—and that OUP, the Society, the Journal, or myself would face huge legal bills. So it was that, at various times in this depressing saga, financial uncertainty seemed to stymie any opportunity for publication. At one point, the late Roy Porter—perhaps the leading medical historian of his generation—generously offered to publish the article in *History of Science*, but his publishers were understandably chary of a paper that had already attracted threats of litigation. OUP had already urged the editorial board of *SHM* not to publish.

And then, to my delight, one day in April 1992, the article quietly appeared. By that point, Mary and I had moved from Manchester to Baltimore, but SSHM supporters across the Atlantic kept us informed of events. The editorial board had reversed its previous decision to reject, opposition had somehow weakened, and the threats of litigation melted away. Austoker and Webster chose not to pursue their lawsuits. Nor, apparently, did the Cabinet Office get involved. But the mere threat of such actions had opened up a host of fearful possibilities. For a moment in the late 1980s and early 1990s, the pressures against publishing this article meant that academic journals had to take the possibility of legal challenges to critical reviews more seriously. Scholars worried that this was part of a broader trend towards litigation in academia; that publishers might be cowed into conservatism; and that a valuable resource for scholarship might be harmed.[9] My critics could have engaged in public debate. Instead, they sought to close it down. In so doing, they opened up a controversy that could have undermined one of the leading British medical history journals, and perhaps the possibility of academic criticism itself.

And now the centerpiece of the controversy itself: Contracting Cancer, as originally published in *Social History of Medicine*, 5, 1992, pp. 131–142. Afterwards, I will comment on the state of contract history today.

DISCUSSION POINT

*This is the third discussion point to be published in the journal and it is concerned
with an important set of issues to do with twentieth-century medical history.
The opinions of the author of this discussion point are not necessarily those
of the editors or editorial board. Rejoinders to discussion points, in the
form of brief articles or correspondence, would be welcome.*

Contracting Cancer? The Politics of
Commissioned Histories

By DAVID CANTOR*

Is the history of medicine in Britain for sale? In the past decade, contract work
has blossomed. Histories have been commissioned by the Government (for
the National Health Service), the Wellcome Trust, St Mark's Hospital, the
Oxford Department of Anaesthetics and the Imperial Cancer Research Fund,
and at least a further fifteen are in print or preparation, not counting four
business histories of pharmaceutical companies.[1] Such histories must be set

---

* The Institute of the History of Medicine, The Johns Hopkins University, 1900 E. Monument Street,
Baltimore, Maryland 21205, U.S.A.

[1] J. Austoker, *A History of the Imperial Cancer Research Fund, 1902–1986* (Oxford, 1988); C.
Webster, *Peacetime History. The Health Services Since the War. Volume I. Problems of Health Care.
The National Health Service Before 1957* (London, 1988); J. Beinart, *A History of the Nuffield
Department of Anaesthetics, Oxford. 1937–1987.* (Oxford, 1987); L. Granshaw, *St. Mark's Hospital,
London. A Social History of a Specialist Institution* (London, 1985); A. R. Hall and B. A. Bembridge,
Physic and Philanthropy. A History of the Wellcome Trust, 1936–1986 (Cambridge, 1986); G.
Tweedale, *At the Sign of the Plough. 275 Years of Allen and Hanburys and the British
Pharmaceutical Industry 1715–1990* (London, 1990); C. Maggs, *A Century of Change: The Story of
the Royal National Pensions Fund for Nurses* (London, 1987); V. Berridge, *The Society for the Study
of Addiction 1884–1988* (Carfax, 1990); J. Slinn, *A History of May and Baker, 1834–1984*
(Cambridge, 1984); J. Pickstone, *Medicine and Industrial Society. A History of Hospital
Development in Manchester and Its Region, 1752–1946* (Manchester, 1985); P. Bartrip, *Mirror of
Medicine. A History of the British Medical Journal* (Oxford, 1990); Jean Barclay, *'The First Epileptic
Home in England'. A Centenary History of the Maghull Homes 1888–1988* (Glasgow?, 1990); Jean
Barclay, *Langho Colony/Langho Centre. 1906–1984: A Contextual Study of Manchester's Public
Institution for People With Epilepsy* (Manchester, 1990); S. Holloway, *Royal Pharmaceutical Society
of Great Britain 1841–1991. A Political and Social History,* (London, 1991).
   Commissioned histories in preparation or prospect include, the British Medical Association
(Peter Bartrip), the Arthritis and Rheumatism Council for Research (David Cantor), the Royal
College of Obstetricians and Gynaecologists (Ornella Moscucci), the Cancer Research Campaign
(Joan Austoker), the King's Fund (Frank Prochaska), Smith and Nephew (L. Dallmeyer), Chalfont
Epileptic Colony (Jean Barclay), and the Westminster Hospital (Gillian Cronjé). In addition, the
pharmaceutical company, Glaxo, has commissioned a history of itself, and the tobacco industry
has financed historical work on tobacco use (D. Harley).
   In this essay 'contract work' includes commissioned histories, even where the latter are not directly
financed by the institution concerned. Access to institutional and private archives and to the recollections of historical actors may impose similar obligations and limitations on commissioned histories as
on those directly financed by a sponsor. Also, it should be noted that the financial arrangements of
'contract work' proper may differ. Monies for some contract work may be channelled through academic medical and business history units, others through academic units of the sponsoring organization, or may be given directly to the contracted historian. Such varied financial arrangements provide
individual contract historians with different degrees of immunity from their sponsors' demands.

Social History of Medicine, 5, 1992, pp. 131–142

against a growing concern about the effects of the recent resurgence of sponsorship in British universities.[2] While the peculiar funding structure of medical history may provide some insulation from government cut-backs in spending, the recent popularity of contract work ensures that significant numbers of British historians, especially those interested in the twentieth century, seem doomed to sponsorship by the very groups they write about. Critical distance is difficult enough to achieve when writing twentieth-century history, a distance shortened by the purse-strings and obligations of contract work.

It is not that sponsors necessarily have a clear idea of what history should be about—my experience, like that of other contract historians, is that they have difficulty explaining why they want a history. Some see contract histories as celebrations of survival, or as vindications of the foresight of founders and patrons, while others lament freely to their contracted historian about the neglect of some obscure innovation, fact, or individual.[3] Such views inevitably inform historical studies, whether they are imposed upon, or internalized by contract workers. Consequently, the writing of such histories merely raises in more acute form the problem of all histories—of institutional and ideological control by the dominant culture, what Roy Porter has termed the 'intellectual treason' of turning history into a service industry.[4] To make matters worse, a number of sponsors seem impelled to complete the historian's work for him/her, often arrogantly asserting the priority of science over history. Thus, Jennifer Beinart's history of the Oxford Department of Anaesthetics is completed by the present professor of Anaesthetics, M. K. Sykes,[5] while the historian Rupert Hall is joined in writing the history of the Wellcome Trust by the ophthalmologist B. A. Bembridge (from the Trust's offices). Other Wellcome scientists also contribute to round off that history.

Not that I am claiming any priority with this insight; Roger Cooter, for example, has complained that Lindsay Granshaw's history of St Mark's Hospital has a progressivist tone that is comforting to its patrons, but which will hardly satisfy the historical constituency she also hopes to reach.[6] Richard Davenport-Hines has suggested that Rupert Hall's account of the Wellcome Trust is too mild in its characterization of Henry Wellcome's altruism, and too deferential in its treatment of the onetime Chairman of the Trustees, Lord Franks.[7] Finally, Christopher Lawrence has noted that in her recent history of

---

[2] On the effect of sponsorship on social science research see G. Clare Wenger, *The Research Relationship: Practice and Politics in Social Policy Research* (London, 1987). Donald Coleman, 'The uses and abuses of business history', *Business History*, 29 (1987), 141–56. Special Issue on Enterprise, Management and Innovation in British Business, 1914–1980, *Business History*, 29 (1987). See also, 'And here are more dons from our sponsors...', *Observer*, 28 January 1990. For an American perspective on privately-sponsored history see P. Novick, *That Noble Dream. The 'Objectivity Question' and the American Historical Profession*, (Cambridge, 1988), pp. 512–21.

[3] See Sir Robert Mackintosh's forward to Beinart, *Nuffield Department of Anaesthetics*.

[4] Roy Porter, 'What is the History of Science', in Juliet Gardiner (ed.), *What is History Today?* (Basingstoke, 1988), 71.

[5] Beinart, *Nuffield Department of Anaesthetics*.

[6] Roger Cooter, review of Granshaw, *St. Mark's* in *History and Philosophy of the Life Sciences*, 12 (1990), 141–3.

[7] R. P. T. Davenport-Hines, review of Hall and Bembridge, *Physic and Philanthropy* in Times *Literary Supplement*, 20 February 1987, p. 192.

the Imperial Cancer Research Fund (ICRF) Joan Austoker seems 'timorous about biting the hand which has fed her.'[8]

As Lawrence's comments suggest, Austoker's history of the ICRF—Britain's oldest cancer charity, founded in 1902—illustrates some of the problems of commissioned histories. As with Beinart's and Hall's histories, Austoker's is brought 'up-to-date' by a leading scientist, Walter Bodmer, the present director of the Fund—a point already criticized by one reviewer.[9] As with other contract histories, it is the story of a 'successful' institution (where survival equals success), suffused with a naive progressivism, from its chapter 1, 'The origins of experimental cancer research (1802–1902)' to its chapter 9, 'The creation of a "centre of excellence" (1968–1979)'. Yet here the parallels with most other contract histories stop, for the ICRF has bought a very bad bargain indeed. A full commissioned history might have explored the growth of medical interest in cancer, the social history of the disease, and the history of philanthropy from the donor's perspective, For example, Beinart's study of the Oxford Anaesthetics Department provides a nuanced account of an institution which allows for complexity. Granshaw's account of St Mark's Hospital engages with the broader problem of medical specialization. Bartrip's analysis of the BMA uses the format of a contract history to address a broad range of medical historical themes.[10] In contrast, Austoker's narrow focus limits the range of historical inquiry, and fails to interweave its intellectual and institutional accounts successfully. Despite—or perhaps because—of these problems, Austoker's book exemplifies many of the complex dilemmas facing contract history. In addition, the politics of cancer research itself have clearly shaped this work, making the interpenetration of science and history very apparent.

Such issues are, of course, as much a problem of general history as of contract work. It is rarely clear whether writers consciously adopt the agendas of their patrons; the reader cannot distinguish between self-censorship and that imposed externally. Yet, as Jim Merod has noted in a different context, there is a marked reluctance among some academics to explore the political choices involved in their work.[11] Contract work highlights the need to make such choices explicit, not least because sponsors may wish to re-shape the past in ways that legitimate their present activities, and commissioned historians are inevitably drawn into that process. This is not to argue that history could or should be apolitical, but rather to explore the implications of the choices we make—the themes we include, and those we exclude. If we learn less than we might about the past from Austoker's book, we can learn something of the politics of present-day science. More importantly, we can gain some insight into the impact of such politics on our own field.

[8] Christopher Lawrence, review of Austoker, *History of ICRF*, in *Times Higher Education Supplement*, 5 May 1989, p. 24.
[9] John Galloway, 'The Imperial Theme', *Nature*, 12 January 1989, 125–6.
[10] Beinart, *Nuffield Department of Anaesthetics*. Granshaw, *St. Marks*; Bartrip, *Mirror of Medicine*.
[11] Jim Merod, *The Political Responsibility of the Critic*, (Ithaca and London, 1987).

I

First, however, it is worth pointing out how such history can mediate the values and interests of science. Partly this is accomplished by blending the voice of the historian with those of the scientists whom the story is about. Such blending is easy to understand. As I have mentioned, often the contract historian is faced with sponsors who wish to ensure that their version of events becomes the 'official' history. But, it is the historian's responsibility to maintain boundaries between his or her story and those of historical actors. Obviously, such distinctions become increasingly problematic as historians approach the present: even non-contract historians may feel the pressure. As Evelleen Richards notes, historians of contemporary science and medicine may be drawn into struggles over scientific or technical knowledge claims: 'the goal of neutral social [historical] analysis', she concludes, 'is as mythical in actual practice as the scientist's goal of neutral evaluation of competing therapies.'[12]

Austoker's book illustrates the problems generated by an unwillingness to make such distinctions; her prose blends the voices of historical actors with her own. This technique is well illustrated in her comment that the state of cancer research at the turn of the century consisted of, 'Diverging opinions and conflicting statements of fact [which] contributed to a chaotic state, compounded by a lack of any sound criteria of treatment' (p. 32). This is strong stuff, and one might have expected some references to illustrate the point. But her one source for this statement is a paper by E. F. Bashford, the first director of the ICRF. Thus, what appears in the text as a statement of historical interpretation also seems to be Bashford's own view—the opinion, that is, of one of the participants in the early twentieth-century conflicts between bio-medical scientists and clinicians. Actor and author, historian and scientist, witness and participant, flow into each other, so that any sense of causal agency is avoided. But not only does Austoker unconsciously conflate her own and her actors' voices, she also blends in the voices of other historians. Thus John Harley Warner's views on the multiple meanings of science are accommodated without trouble next to Bashford's rhetorical pleas for his own status. Alas, equal weight cannot be accorded to all, and so it is that Austoker generally favours the accounts of scientists and doctors. A typical instance is her statement that the epidemiological study that Richard Doll carried out with Bradford Hill in the 1950s on smoking and cancer 'provided the basis of what remains to this day the single most reliably established and practicable means of reducing the proportion of deaths from cancer' (p. 197), which has for its source a couple of articles written by Richard Doll himself.

This blurring of the boundaries of history and science hides the political voices of scientists, while at the same time enabling them to assert their own interpretations of past events under the guise of history, and behind the mask of a professional historian. There are particular reasons why this is significant today, to which I refer in a moment. But, staying at the level of textual analysis, it is important to note that such problems are not confined to contract history.

---

[12] Evelleen Richards, *Vitamin C and Cancer: Medicine or Politics?* (Basingstoke and London, 1991), p. 12.

Austoker illustrates some of the standard syntactical devices for establishing what J. B. Thompson (following the political theorist, Claude Lefort) has called the 'dimension of society "without history" at the heart of historical society';[13] devices which obscure ideological positions. Take, for example, the following quotation from Austoker's account of the early policy formation of the ICRF, which follows a discussion of the confused state of cancer research at the turn of the century:

By virtue of the experimental work conducted in the latter part of the nineteenth century, cancer had become recognised not only as a pathological problem, but also as a comparative biological one requiring systematic and purposeful experimental observations across the entire animal kingdom. The investigation of a disease which had long haunted the imagination clearly required a rational restatement of the problems which had so far frustrated solution. There was therefore an urgent need to establish a firm foundation on which experimental cancer research could be built (p. 33).

Austoker's use of the passive voice serves to reify historical and social processes, and to represent power relations as 'rational' choices. For instance, 'cancer' does not 'become recognised'—people (Bashford? Austoker?) recognize it. Note also the specious consensus in 'investigations required', 'cancer had become recognized' and 'there was therefore'; and the use of the past tense—the verbs 'required', 'had', and 'was'—to refer to Austoker's present views; and, perhaps above all, the denial of alternative views.

By using these conventions to give scientists authority to interpret the past, Austoker, in turn, rests her own authority as much in science as in history. Hers is a peculiarly neutral science, in which politics never impinge on scientific knowledge itself: social and economic factors serve only to blind people to the obvious implications of scientific research, as her comments on the value of Doll's epidemiological research suggest. And it is not difficult to see why. As Dorothy Nelkin pointed out several years ago, conflict among experts may reduce their political impact, highlighting their fallibility, and demystifying their expertise by showing it to be subject to social concerns.[14] In Austoker's study, science too easily becomes either 'true' or 'false'—as in 'Genuine scientific enterprise in this area was rare' (p. 179)—as if these were eternal values. No place here for the stuff of the social history of medicine, let alone the sociology of scientific and medical knowledge. Instead, Austoker's discussion of the research side of the ICRF proceeds smoothly, like some huge unperturbable leviathan, deep beneath the troubled waters of the rivalries between the laboratory and the clinic, the MRC and the charities, the scientist and his/her sponsors. Rarely is this huge monster aware of the tumult above as it swims its unerring course towards the present.

---

[13] J. B. Thompson, 'Mass Communication and Modern Culture: Contribution to a Critical Theory of Ideology', *Sociology* 22, (1988), 359–83, esp p. 371. Claude LeFort, *The Political Forms of Modern Society. Bureaucracy, Democracy, Totalitarianism* (Oxford, 1986), Chapter 6. Thomson has recently developed these points in *Ideology and Modern Culture. Critical Social Theory in the Era of Mass Communication* (Cambridge, 1990).

[14] D. Nelkin, 'The Political Impact of Technical Expertise', *Social Studies of Science*, 5 (1975), 35–54, p. 54.

Yet Austoker appears to have some doubts about the narrative of progress she relates. On the penultimate page of her account we read that, 'the vast and expensive commitment to the problem of cancer has not yet been rewarded with the striking success that might have been hoped for' (p. 321). Indeed, the best part of her account—her discussion of the emergence of the biomedical scientist as a strong rival claimant to the clinician on health matters, of which the ICRF provides a microcosm—could be read as an indictment of bio-medical science. Whereas clinicians argued that progress could best be achieved through research in clinical settings carried out on patients or at least on material drawn from patients, bio-medical scientists took quite a different view. For them progress could only be achieved in the laboratory through animal experiments or research on cells cultivated *in vitro*. To these men and women, 'pure' science was science independent of clinical control. They argued, that clinicians were simply unfit to organize, carry out or direct research. Yet they did not win their case by demonstrating a better way to cure or prevent most cancers. One of the grim implications of this is that cancer researchers have succeeded in little more than elaborating the scientific complexities of cancer. In so doing they have created unending career opportunities for themselves and, after the Second World War, have beaten the old-fashioned clinical model of progress by creating research posts which only men and women of their ilk could fill. Only recently has the ICRF begun to turn its eyes towards clinical research again.

Austoker is at her best when delineating the back-stage battles between participants in her tale. She highlights the significance of Walter Fletcher, the secretary of the state-financed Medical Research Council, to the power struggles between bio-medical scientists and clinicians, and indicates how Fletcher turned the ICRF, and its younger sibling the British Empire Cancer Campaign (founded in 1923), into a focus for such disputes. These disputes were not simply about how the Fund's or Campaign's resources should be spent, but about the nature of medical science itself. Fletcher's victory, she argues, laid the foundations for a national bio-medical research policy.[15] How unfortunate, then that she never explores her own doubts about the value of ICRF research. This silence is all the more telling as her narrative nears the present, since she shows that from the 1970s the cancer charities have provided the financial basis of oncology as a specialty in Britain, and have become responsible for the bulk of cancer research carried out in this country. In a climate such as this, a history which hymns the praises of a charitable research institute is inevitably constituted in political terms. Austoker's technique of fusing science and history, and her neglect of the charitable side of the ICRF merely plays into medicine's current attempts to defend its professional and cognitive authority.

II

The privileged epistemological status of medical knowledge has taken a battering in recent years. Anthropologists, sociologists and historians have

---

[15] Austoker has developed these points subsequently in 'Walter Morley Fletcher and the Origins of a Basic Biomedical Research Policy', in Joan Austoker and Linda Bryder (eds.), *Historical Perspectives on the Role of the MRC. Essays in the History of the Medical Research Council of the United Kingdom and its Predecessor the Medical Research Committee, 1913–1953* (Oxford, 1989), 23–33.

highlighted the cultural contingency of medical knowledge, the ways in which medicine acts as an institution of social control, and how its knowledge is the result of professional struggles for power and control.[16] Doctors and scientists have become increasingly sensitive to such criticisms, which coincide with a Thatcherite and populist (if not popular?) assault on professionalism, and with a growing sense of crisis in health care.[17] Cancer, along with other chronic and degenerative diseases and their disabling consequences, has been implicated as an important component in this crisis, highlighting medicine's ineffectiveness, the dehumanizing way in which it is delivered and its escalating costs.[18] Orthodox medicine has been challenged further by renewed interest in holistic alternatives, which has forced a reluctant American medical establishment to test the purported anti-cancer drugs, Laetrile and Vitamin C.[19] Furthermore, the supremacy of medical over lay knowledge has been subverted by renewed attention to experiences of illness, and to the ways in which medicine expresses its power over the body.[20] Recent criticisms of randomized clinical trials have even forced medical advocates of such trials to defend themselves in the national press against accusations that they abuse their patients.[21]

---

[16] P. Wright and A. Treacher, *The Problem of Medical Knowledge. Examining the Social Construction of Medicine* (Edinburgh, 1982); K. Figlio, 'Chlorosis and Chronic Disease in Nineteenth Century Britain: The Social Constitution of Somatic Illness in a Capitalist Society', *Social History*, 3 (1978), 167–97; E. Friedson, *Profession of Medicine: A Study of the Sociology of Applied Knowledge* (New York & London, 1970); *idem, Professional Powers: A Study of the Institutionalization of Formal Knowledge* (Chicago & London, 1986); T. J. Johnston, *Professions and Power* (London, 1972); E. E. Evans Pritchard, *Witchcraft, Oracles and Magic Among the Azande* (Oxford, 1937); P. Morley and R. Wallis (eds.). *Culture and Curing: Anthropological Perspectives on Traditional Medical Beliefs and Practices* (London, 1978); A. Kleinman, 'The Meaning Context of Illness and Care: Reflections of A Central Theme in the Anthropology of Medicine', in E. Mendelsohn and Y. Elkana (eds), *Sciences and Cultures* (Dordrecht, 1981) 161–76.

[17] H. Perkin, *The Rise of Professional Society. England Since 1880* (London and New York, 1989), Chapter 10. For an American study see M. Haug and B. Lavin, *Consumerism in Medicine: Challenging Physician Authority* (London, 1983).

[18] T. Mckeown, *The Role of Medicine* (Oxford, 1979); A. L. Cochrane, *Effectiveness and Efficiency: Random Reflections of the Health Services* (London, 1972); I. Illich, *Limits to Medicine* (Harmondsworth, 1977); I. Kennedy, *The Unmasking of Medicine* (London, 1981). On the links between disablement and the current crisis in western health care systems, see, Gareth H. Williams, 'Disablement and the Ideological Crisis in Health Care', *Social Science and Medicine*, 32, (1991), 517–24.

[19] J. C. Petersen and G. E. Markle, 'Politics and Science in the Laetrile Controversy', *Social Studies of Science*, 9 (1979), 139–66; *idem*, 'Expansion of Conflict in Cancer Controversies', in L. Kriesberg (ed.), *Research in Social Movements, Conflict and Change*, Vol. 4, (Greenwich, CT, 1981), 151–69; *idem*, 'The Laetrile Controversy', in D. Nelkin (ed.), *Controversy: Politics of Technical Decisions* (Beverly Hills, CA, 1979), 159–78; *idem*, 'Resolution of the Laetrile Controversy: Past Attempts and Future Prospects', in T. H. Englehardt and A. L. Caplan (eds), *Scientific Controversies; Cases Studies in the Resolution and Closure of Disputes in Science and Technology* (Cambridge, 1978). 315–32; E. Richards, 'The Politics of Therapeutic Evaluation: The Vitamin C and Cancer Controversy', *Social Studies of Science*, 18 (1988), 653–701; *idem. Vitamin C*. See also Philip H. N. Wood and Gareth H. Williams, 'The Attraction of Alternative Therapies', in B. A. Stoll (ed.), *Ethical Dilemmas in Cancer Care* (London, 1989), 117–25.

[20] P. Conrad, 'The Experience of Illness: Recent and New Directions', *Researches in the Sociology of Health Care*, 6 (1987), 1–31.

[21] M. Baum, 'I Didn't Abuse My Patients', *Observer*, 16 October 1988. Baum's letter was a response to 'How Doctor's Secret Trials Abused Me', *Observer*, 9 October 1988. See also, C. Faulder, *Whose Body Is It? The Troubling Issue of Informed Consent* (London, 1985).

For the ICRF—indeed for British cancer researchers generally—these attacks on the status of medical knowledge are linked to fears of a loss of professional prestige and power and income. As Austoker notes, in Britain more than in most countries, cancer research and oncology are particularly dependent on charitable support, and their dependence is growing as state support for cancer research diminishes.[22] Cancer researchers and oncologists have responded by asking the public to renew their trust in science, in the professionalism of scientists and in the questionable equation of health care with beneficence. As Bodmer puts it in his epilogue to the ICRF history:

As a publicly supported charity, we are particularly conscious of the trust that the public places in us to advance cancer research. We can only honour this trust if we strive for excellence in all respects. The quality and success of the work we do depends on the quality and dedication of our staff at all levels (p. 359).

But cancer researchers do not reciprocate the trust they expect of the public. As Austoker—subsequently director of the CRC's Science Policy Research Unit—puts it elsewhere: 'we should not allow the nation's medical research portfolio to be determined predominantly by the sympathies of the public or by commercial considerations.'[23] She makes this comment in the context of an appeal for more state funding for basic research. Her remark points to a growing fear among scientists that donors will eventually bring them to account for the large sums of charitable money being spent on scientific research, just as in the last thirty years they have increasingly been asked to demonstrate the relevance of state-financed research. As with the state, scientists seem to find it difficult to show their charitable paymasters the value of basic research, and they fear the public's patience may run out. 'The public naturally expects results quickly, and persistently ask whether the next "breakthrough" is just around the corner,' writes Bodmer in his epilogue (p. 361). 'They must be convinced that breakthrough is not part of the scientific vocabulary'—an astonishing withdrawal from a myth that has served science well! And of course the public is often far more wise and skeptical about research than Bodmer's élitist and technocratic comments suggest. The irony of such calls for public education and appeals for funds is that they may in fact encourage the very ideologies of individual self-interest and community responsibility that have facilitated the state's withdrawal from cancer research, and which in turn have increased medicine's reliance on philanthropy.

If the issue of reciprocity between donor and charity is complex, that amongst medical personnel, donors and charity is more so. As Bodmer's quote illustrates, what medicine feels donors might legitimately expect and what donors actually want may differ dramatically. Such conundrums are further complicated by the ambiguous role of the contract historian of medical charity. Often funded by the public's donations to medical science, the historian's work

---

[22] J. Austoker 'Cancer Research Left to Charity', *New Scientist*, 15 October 1987, 28–9.
[23] Austoker, 'Cancer Research Left to Charity', p. 29.

is shaped by at least two potentially conflicting demands.[24] Many charities see their histories as publicity ventures—narratives of success which will perpetuate public confidence and public beneficence. Some charities also think that a history might reveal how and why donors gave in the past, providing clues for future successful fund-raising campaigns. Yet it is not clear that either goal is what the individuals who give to charity intend. It is only with historical analysis of those who funded scientific research that an understanding of these relationships of reciprocity might be reached. (The more difficult problem of the contract historian's role as recipient of medical charity remains unsolved.) Within the history of science, studies of the funding of research by philanthropic foundations have begun to suggest the ways in which the interests of donors and administrators shaped the science they funded.[25] Indeed, Austoker's own work on the MRC can be placed in this historiographic tradition.[26] It is thus all the more unfortunate that her account of the ICRF fails to address the charitable dimension of the organization.

In fact, Austoker so rarely discusses the philanthropic side of the ICRF that one might forget that this is a charity. We are not provided with basic information, such as the annual income of the charity, let alone whether this has kept pace with inflation or the cost of living. One would not know from this book that the ICRF and its chief rival the Cancer Research Campaign (CRC) each currently pull in more that any other medical charity in Britain. Indeed, of the top four hundred fund-raising charities in 1988—not just medical—these two came 4th and 5th respectively, exceeded only by Oxfam, the National Trust and the Royal National Lifeboat Institution.[27] How the ICRF achieved this preeminence is never discussed. If it has to do with the inexorable rise in cancer mortality figures, or is a reflection of medicine's growing cultural authority, or a product of broader changes in philanthropy—we are not told. All we read in this connection is of debates over how donations to 'cancer research' should be divided between the CRC and ICRF, whether money should be invested for long-term research or spent within the year of receipt, and whether funds donated for cancer research should be used for basic research. The nearest we get to individual philanthropists is her account of how Thomas Rudd, a City businessman, launched the ICRF with a promise of financial support, and her account of the newspaper proprietors and other 'gentlemen' who in 1921/2 funded the ancestor of the CRC, the British Empire Cancer Campaign (BECC). Austoker does not pursue the connections between the ICRF and business and high society concerns. Yet as Pinell and Brossat have argued for France and Patterson for America, donors have exerted considerable influence on cancer charities.[28] Such an omission

---

[24] Historians sponsored by government may occupy a similarly ambiguous role, funded by taxpayers money but answerable to a government which may wish to manipulate the taxpayers voice to its own ends. Similarly, there may be conflicts for historians funded by business as to whether they are answerable to research staff, management, shareholders, or directors.

[25] Robert E. Kohler, 'The Management of Science: The Experience of Warren Weaver and the Rockefeller Foundation Programs in Molecular Biology', *Minerva*, 14, (1976), 279–306.

[26] Austoker, 'Walter Morley Fletcher'.

[27] Michael Brophy and Judith McQuillan, *Charity Trends, 12th Edition* (Charities Aid Foundation, 1989).

[28] James T. Patterson *The Dread Disease. Cancer and Modern American Culture* (Cambridge, Mass., & London, 1987); P. Pinell and S. Brossat, 'The Birth of Cancer Policies in France', *Sociology of Health and Illness*, 10 (1988), 579–607.

perhaps fits with Bodmer's view of public accountability in terms of professional regulation of the scientific quality of research that the charities organize. Had the ICRF history concentrated more on donors it could have raised awkward questions about the extent to which scientific research has been shaped by philanthropists, a politically sensitive issue at a time when the question of the extent of public participation in charities' research strategies has begun to be raised.

III

Such history reveals the ways in which contract histories can embody the interests of their sponsors. As I say, this is an extreme example, but it is one which illustrates many of the conflicts inherent in commissioned histories, highlighting the problems facing historians who want to raise issues their sponsors might feel uneasy about. Inevitably, to the extent that commissioned histories express the collective identities of scientists and other patrons, narratives which question the progress of knowledge or the success of an institution become problematic. Few would claim that the ICRF, the BMA, the Arthritis and Rheumatism Council, the King's Fund or the host of other sponsors advocate radical critiques of medicine. And the dominance of such organizations in sponsoring contract histories raises serious questions about the extent to which institutional histories may obscure the study of more diffuse social or cultural movements or of institutions, ideas, groups or individuals that fell by the wayside over time.

These problems are not confined to the history of medicine. G. Clare Wenger and others have shown the ways in which sponsors alter the framework, methods, conclusions and accessibility through publication of social science research.[29] Donald Coleman has also complained of the limitations of commissioned history, noting that the multitude of contract business histories have tended to focus on individual companies at the expense of issues of interest to historians.[30] Indeed, when the journal *Business History* commissioned a number of historians to write on certain themes using a stock of the best recent company histories, some of the contributors found these histories silent on their selected topics.[31] Such company histories generally adopted the ideology of the institution they studied. And with few exceptions, many of the medical histories noted at the beginning of this paper can be criticized to varying extents on precisely this points. A number are social histories only to the extent that the institutions are located in a social context, as Cooter complains of Granshaw's study of St Mark's.[32] Few of these studies use institutions as a way to an historical understanding of wider communities, and they so often lack a sociological perspective of any sort so that questions of social formation and of the ways in which historically constituted interests shape institutions are often neglected.

Of course, we as historians cannot be absolved of all responsibility: much depends on our own political perspectives, and our own awareness of the ways

[29] Wenger, *Research Relationship*.
[30] Coleman, 'The uses and abuses of business history'.
[31] Special Issue on Enterprise, Management and Innovation in British Business, 1914–1980, *Business History*.
[32] Cooter review of Granshaw, *St Marks*.

in which our work supports (or not) such views. There is ample evidence to show that non-contract historians have been influenced by similar concerns to those of Austoker. Thus, Christopher Lawrence recently mourned the neglect of twenty-five years of sociological and historical work on the social construction of knowledge in Ann Harrington's (non-contract) study of nineteenth-century neuro-science. He is able, however, only to hint at the connections between Harrington's assumption that deviant science is tainted and real science is pure, and what he loosely calls 'the politics of the late 1980s'.[33] With contract histories such connections are clearer, as I have indicated in the case of Austoker. The extent to which similar politics are fostered outside of contract work may be a sign of a desire among some historians to neglect the questions raised by anthropologists, by Foucault, Marxists or the Edinburgh School about the nature of scientific and medical knowledge. We have yet to persuade many social historians that modern scientific knowledge is just as open to social and economic forces as religion, politics and history itself. What a shame if we were co-opted into medicine's and science's attempts to defend their authority, and forced to abandon the treatment of beliefs symmetrically, whether conventionally 'true' or 'false', 'scientific' or 'ideological'. We might then only become another voice seeking to encourage what has come to be known as the 'public understanding of science' as if there were historically only one science, and the understanding of it could encourage the public to shower the contents of its purses over the scientists behind such a movement.[34]

A final question relates to the impact of contract history on history of medicine/science's relationship to mainstream history. As Bob Young has noted, the history of science (and medicine) 'sits in an uneasy niche as a cultural ornament to science or as a tiny sub-specialty within history';[35] but it ought to be treated by historians as the fundamental part of culture that it has long been in practice. If contract histories make it difficult to explore the social or cultural shaping of knowledge, if they focus on institutions without explaining their historical significance and if they ignore opposing voices they will send the history of medicine spinning into the orbit of science. At the same time it is doubtful if mainstream history will make much sense of the intermittent and garbled messages that cross the ever widening space between it and its straying sub-specialty.

Acknowledgements

Many thanks to Roger Cooter, David Edgerton, Mary Fissell, Bill Luckin, Mick Worboys and Gareth Williams who read and commented on this paper. All errors remain my own.

---

[33] Christopher Lawrence, 'Cognitive Issues: Having it Both Ways', *Social History of Medicine*, 2 (1989), 87–92, p. 87.

[34] In this connection it is worth noting that Bodmer was the chairman of the Royal Society's *ad hoc* group which reported in 1985 on the need for greater 'public understanding of science'. Royal Society, *The Public Understanding of Science*, (London, 1985).

[35] Robert Young, 'What is the History of Science?' in Gardiner, *History Today*, p. 77.

It has been more than a dozen years since my article appeared. How has the standing of contract histories changed since then?

Since the 1990s, medical historians, both in Britain and beyond, have been hired to write a new sort of commissioned history, one that is as much about *issues* as institutions. The subjects of these new histories are vast: they have included the role of German medical and biomedical organizations in the Nazi regime,[10] biomedical involvement in radiation and other human experiments,[11] and the health issues associated with the tobacco, asbestos, nuclear, and lead-paint industries.[12] Such issue-driven histories were not so prominent when Contracting Cancer was originally published, and they tend to attract a broader readership than the older institutional histories. Their recent emergence underscores my 1992 claim that commissioned historians may tailor their tales to fit the needs of their sponsors. Indeed, this now looms as a major issue for historians of science, technology, and medicine.

Contracting Cancer was not the first article to make such a claim about potential abuses of history. The problem is as old as commissioned history itself. The new brand of issue-driven commissions simply brings the problem to broader audiences. Of course, some of the medical histories I discussed in 1992 served to promote the agendas and interests of the organizations that sponsored them. But they tended to attract little attention outside of a small specialist readership. By contrast, these new issue-driven histories focus on much more politically and socially contentious topics. The attention they have garnered raises, once again, questions about the credibility of the entire genre of commissioned history.[13]

* * *

Many commissioned historians wish it were not so. Even before the current blossoming of issue-driven histories, such historians balked at the suggestion that critical distance might be undermined by contract work.[14] Public historians attacked Contracting Cancer as evidence of a broader denigration by the academic establishment of public history, so often reliant on commissions.[15] Academic historians worried that the article impugned the scholarly quality of any commissioned work they undertook. And a growing band of entrepreneurial historians feared such criticism as part of a broader attempt to undermine the market value of history.[16] With grants and university employment often difficult to come by, criticism of commissioned histories seemed to erode a welcome additional source of research funding for academics, and an alternative career path for those without a university appointment.[17] It is no small wonder then that contract historians have sometimes distanced themselves from the article.

But efforts to dismiss the issues raised by Contracting Cancer are unlikely to allay suspicions of commissioned history. It is problematic for journalists to be paid by those they report on, for junior faculty to review the work of members of their tenure committee, and for a physician to receive payment from the pharmaceutical company whose drugs he or she prescribes. Why then should it be any less problematic for a historian to be paid by an organization that has a direct interest in the subject of his or her research? Studies suggest that clinicians and scientists who accept gifts from corporations are more likely to choose their products, and that corporate sponsored research studies are likely to favor those sponsors.[18] Few believe historians to be any less susceptible to such influence. Attempts by commissioned historians to deny the significance of such influence can only serve to breed further suspicion of the genre.

Suspicion of commissioned history is thus part of a broader cultural concern about conflicts of interest facing individuals seeking an understanding of the recent past. Critics contend that commissioned historians—like others subject to conflicts of interest—have a financial or institutional motive in subverting disciplinary standards and integrity. Their fear is that historians working under such arrangements subordinate history to the interests of the organizations for which they work, or the voice of their marketing, communications, or legal departments. It is not that history funded in this way may not be competently researched. But critics repeatedly argue that the questions such historians are asked to address often preclude answers that might work against the interests of the sponsor, that commissioned historians sometimes conspire with such approaches, and that they often fail to adopt a sufficiently critical perspective.[19] The growth of issue-driven commissions has given these doubts added prominence and urgency.

* * *

Nowhere are concerns about commissioned histories more evident than in the recent involvement of medical historians as expert witnesses in litigation.[20] In many instances, contract historians have come to remarkably different interpretations of the historical evidence, raising questions of objectivity and ethical standards. Those paid by one side provide interpretations that support their case; those paid by another provide evidence in favor of the opposite interpretation. It is easy for commentators to point to the financial rewards of legal work as providing an incentive for historians to slant the story one way or another, especially as some historians had no research interest in their new subject before their hiring.[21] Consequently, one *New York Times* journalist has asked: ought this trend towards litigation mean that historians should declare whose wallet is behind their research?[22]

Critics have focused especially on industrial and commercial sponsorship. For example, the historian Robert Proctor has recently noted that the tobacco industry far outspends its opponents in supporting historical research on smoking and health.[23] According to Proctor, such spending is not the result of a disinterested desire by the industry to promote academic scholarship. On the contrary: He suggests it is part of a broader effort to create a body of scholars and scholarship that the industry might draw on in litigation.[24] Thus, the industry seeks to manipulate historical scholarship in order to further its own interests. The problem is that this has often been done in an underhanded way. Much of this research has been published in scholarly publications, but often without mention of the wallet behind the work. In his 2004 exposé of such doings, Proctor suggests that history journals might go further than requiring a declaration of funding. They might, he proposes, follow the lead of the American Medical Association (AMA) and consider rejecting publications funded by the tobacco industry.[25]

The tobacco industry is not alone in seeking to manipulate scholarship. The asbestos and lead-paint industries have also funded scholarly work related to litigation. In certain respects, this is a remarkable turn-around. In the past, company histories tended to downplay health issues associated with their industries. Now, however, these new histories tend to bring them to the fore, and often in ways that industry executives might find comforting. Not only do they provide a perspective sympathetic to industry on these contentious issues, the authors of some of these publications have also been accused of obscuring the source of their funding.[26] Proctor has indicated the substantial sums that consultants to the tobacco industry can earn.[27] We have yet to see what consultants for other industries may have earned.

But how different can their interpretations be to those of other (non-industry funded) historians? Compare, for example, Peter Bartrip's industry funded account of the British asbestos producer Turner and Newall (T&N) with Geoffrey Tweedale's history of the same organization. While Bartrip argues that T&N did all it could to address the problem of the health hazards of asbestos, Tweedale argues the opposite case: that T&N knew about the hazards of asbestos very early, and did less than it might.[28] Similarly compare Peter English's industry sponsored account of childhood lead-paint poisoning with that of Christian Warren. English portrays the lead-paint industry as a pioneer in prevention against the health hazards of lead, yet Warren and others argue quite the opposite: that the lead-paint industry did less than it might to promote awareness of the hazards of lead-paint.[29] It is such stark differences between industry sponsored and non-industry sponsored histories that encourage public distrust of commissioned histories.

* * *

In 1992 I concluded Contracting Cancer by highlighting what seemed to me a risk: that commissioned history might become irrelevant to mainstream history. Issue-driven histories confound that expectation. Because of their subject matter, such commissioned accounts have attracted critical attention from others, anxious to counteract misrepresentations, and to undermine the ability of organizations to use such history to legitimize their activities. And they may be succeeding, if the commentaries in the *New York Times*, *Nature*, the *Lancet*, and elsewhere are any indication of success in sensitizing readers to the social, political, and commercial interests behind commissions.[30] The silver lining on the cloud of issue-driven history may, therefore, be emergence of a skeptical, critical reader.[31]

It is here that I begin to question Robert Proctor's call for a ban on the publication of certain commissioned histories. While Proctor is correct to point to the ways in which the tobacco industry seeks to manipulate history to promote its own ends, his call has frightening implications. He tells us little about where the banning he proposes should stop, who should decide which history or sponsors should be proscribed, or what should be done with the many commissioned publications already in the public domain. Furthermore, his focus is almost exclusively on the ways in which such organizations seek to limit awareness and promote ignorance. His call for censorship betrays a pessimism about the readers of commissioned and other histories, as if they cannot be educated about the manipulations and attempted manipulations of those who sponsor history. Yet, paradoxically, this pessimism comes at the very moment that he provides such readers with the tools to read critically. His own articles highlight the interests behind histories commissioned by the tobacco and other industries, and the reluctance of some historians to admit to their funding. Thus, Proctor begins to teach his readers how to assess such accounts, only to deny them the opportunity to read the types of works he so carefully exposes.

Acknowledgment

The author would like to thank the editors of *Social History of Medicine* for permission to reprint this article, which originally appeared in volume 5 (1992): 131–142. *Social History of Medicine* is published by Oxford University Press for the Society for the Social History of Medicine.

Notes

1 Antoon De Baets, "Defamation Cases against Historians," *History and Theory* 41 (2002): 346–366; Antoon De Baets, *Censorship of Historical Thought: A World Guide, 1945–2000* (Westport, CT: Greenwood Press, 2002). For perhaps the most notorious recent example of attempts to censor history through the law see Richard Evans, *Lying about Hitler: History, Holocaust, and the David Irving Trial* (New York: Basic Books, 2001). See also Deborah E. Lipstadt, *History on Trial: My Day in Court with Holocaust Denier David Irving* (New York: Ecco, 2005) and also the legal attempts to censor David Rosner and Gerald Markowitz cited at Note 29.

2 http://www.sshm.org (accessed on November 5, 2004). The following account is based in part on a thick folder of correspondence that I compiled at the time. For a history of the SSHM see Dorothy Porter, "The Mission of Social History of Medicine: An Historical View," *Social History of Medicine* 8 (1995): 345–359.

3 Roger Cooter and I then both worked at Manchester University: Roger at the Wellcome Unit for the History of Medicine, and myself at the Arthritis and Rheumatism Council for Research's (ARC) Epidemiology Unit in the Medical School. In 1990, I moved to the Wellcome Unit, to an office across the corridor from Roger's, which I shared with two associates of the Unit who proved strong supporters of "Contracting Cancer" within the SSHM and *SHM*: Bill Luckin (from Bolton) and Mick Worboys (then, from Sheffield). The secretary of the SSHM, Mary Fissell, had an office on the same corridor. The "Cantor Affair," was sometimes cast in terms of a rivalry between the Manchester and Oxford Wellcome Units. That is not how it began.

4 Austoker also stated that she was considering legal action against me because I had distributed the unpublished version of the essay to colleagues. Like most scholars, I had sent the paper out to friends and colleagues for critical comment, and their names appeared in the acknowledgments.

5 An earlier controversy over the publication in *Social History of Medicine* of a critical review by Daniel Fox—of the first volume of Webster's NHS history—set the stage for Webster's involvement in the struggle over my article. It should also be noted that I was Webster's research assistant on the first NHS volume, and that Austoker was a member of the Oxford Wellcome Unit while she worked on the ICRF history. See Charles Webster, *The Health Services since the War: Vol. 1. Problems of Health Care: The National Health Service before 1957* (London: Her Majesty's Stationery Office, 1988); Charles Webster, *The Health Services since the War: Vol. 2. Government and Health Care: The National Health Service 1958–1979*

(London: Her Majesty's Stationery Office, 1996); Daniel M. Fox, "Anti-intellectual History?" *Social History of Medicine* 3 (1990): 101–104; Charles Webster, "Official History?" *Social History of Medicine* 3 (1990): 104–105.

6 The principle change was the removal of one line, which the lawyer felt was particularly vulnerable to an attack of defamation.

7 On the resignations see "Comings and Goings," *Society for the Social History of Medicine Gazette*, no. 2 (August 1991): 2. The front page of the *Gazette* dates the publication as August 1990. This is a mistype. A copy of Ludmilla Jordanova's letter declining the editorship of the Journal is in the file I compiled at the time.

8 The lawyer's main concern was Austoker: he was dismissive of Webster's letter, noting that the language employed by Webster suggested that he had not consulted a lawyer before making his threat.

9 The sociologist Gareth Williams, a colleague at the ARC's Epidemiology Unit, raised the matter at a meeting of the editorial board of *Sociology of Health and Illness*. His concern was that others should follow Austoker's lead in threatening legal action against critical reviews of their books, OUP's hesitancy about publishing my article might foreshadow similar hesitancy among other publishers.

10 Sigrid Stöckel, ed., *Die "rechte Nation" und ihr Verleger: Politik und Popularisierung im J. F. Lehmanns Verlag, 1890–1979* (Berlin: LOB.de-Lehmanns Media, 2002)—sponsored by J. F. Lehmanns Verlag. See also the program and publications of the Presidential Commission: "History of the Kaiser Wilhelm Society in the National Socialist Era," http://www.mpiwg-berlin.mpg.de/KWG/commission.htm (accessed on May 3, 2004)— sponsored by the Max Planck Society. Some of the results of this research appear in Carola Sasche and Mark Walker eds, *Politics and Science in Wartime: Comparative International Perspectives on Kaiser Wilhelm Institutes, Osiris* 20 (2005).

11 The US government established presidential commissions to examine human radiation experiments and the Tuskegee Syphilis studies. See *The Human Radiation Experiments: Final Report of the Advisory Committee* (New York: Oxford University Press, 1996); *Final Report of the Advisory Committee on Human Radiation Experiments. Supplemental Volume 1: Ancillary Materials* (Washington, DC: US Government Printing Office, 1995); *Final Report of the Advisory Committee on Human Radiation Experiments. Supplemental Volume 2: Sources and Documentation; Supplemental Volume 2a: Sources and Documentation—Appendices* (Washington, DC: US Government Printing Office, 1995); V. Gamble and J. Fletcher, *Final Report of the Tuskegee Syphilis Study Legacy Committee* (Atlanta, GA: Centers for Disease Control and Prevention, 1996).

One of the investigators on ACHRE, a permanent staffer at the Smithsonian (National Air and Space Museum), has written on the problems of obtaining information from government agencies: James E. David, *Conducting Post-World War II National Security Research in Executive Branch Records: A Comprehensive Guide* (Westport, CT: Greenwood Press, 2001), esp. 4–5, http://tis.eh.doe.gov/ohre/roadmap/achre/committee.html (accessed on May 3, 2004).

12 Peter Bartrip, *The Way From Dusty Death: Turner and Newall and the Regulation of Occupational Health in the British Asbestos Industry, 1890s–1970* (London: Athlone, 2001)— funded by Turner and Newall/Federal Mogul; J. Samuel Walker, *Permissible Dose: A History of Radiation Protection in the Twentieth Century* (Berkeley: University of California Press, 2000)—sponsored by the U.S. Nuclear Regulatory Commission; Barton C. Hacker, *The Dragon's Tail: Radiation Safety in the Manhattan Project, 1942–1946* (Berkeley: University of California Press, 1987)—sponsored by the Department of Energy's Nevada Operations Office; Barton C. Hacker, *Elements of Controversy: The Atomic Energy Commission and Radiation Safety in Nuclear Weapons Testing, 1947–1974* (Berkeley: University of California Press, 1994)—sponsored by the Department of Energy; Peter C. English, *Old Paint: A Medical History of Childhood Lead-Paint Poisoning in the United States to 1980* (New Brunswick, N. J. and London: Rutgers University Press, 2001)—research began in preparation of an affidavit submitted on behalf of the Lead Industries Association, Inc.

13 In addition to those listed in other footnotes, commissioned histories produced since the original publication of "Contracting Cancer" include: Louis Galambos, *Values & Visions: A Merck Century* (Rahway, NJ: Merck & Co., 1991); Louis Galambos with Jane Eliot Sewell,

*Networks of Innovation: Vaccine Development at Merck, Sharp & Dohme, and Mulford,
1895–1995* (Cambridge and New York: Cambridge University Press, 1995); Roslyn
Russell, *Building Strength Through Change: Twenty Years of the Health Insurance Commission*
(Tuggeranong, Australian Capital Territory: Health Insurance Commission, 1995);
Christopher J. Kauffman, *Ministry and Meaning: A Religious History of Catholic Health Care
in the United States* (New York: Crossroad, 1995); James Foreman-Peck, *Smith & Nephew in
the Health Care Industry* (Aldershot, England: Edward Elgar Publishing, Ltd., 1995);
W. D. Rubinstein and Hilary C. Rubinstein, *Menders of the Mind: A History of The Royal
Australian and New Zealand College of Psychiatrists 1946–1996* (Oxford: Oxford University
Press, 1997); Naomi Rogers, *An Alternative Path: The Making and Remaking of Hahnemann
Medical College and Hospital of Philadelphia* (New Brunswick, NJ: Rutgers University Press,
1998); Sally Wilde, *Joined Across the Water: A History of the Urological Society of Australasia*
(Melbourne: Hyland House, 1999); Catherine Waterhouse, *The Schizophrenia Fellowship of
Victoria, 1978–1999: A Brief History* (Melbourne: Schizophrenia Fellowship of Victoria,
1999); Paul Litt, *Isotopes and Innovation: MDS Nordion's First Fifty Years, 1946–1996*
(Montreal: McGill-Queen's University Press, 2000); Celia Davies and Abigail Beach,
*Interpreting Professional Self-Regulation: A History of the United Kingdom Central Council for
Nursing, Midwifery and Health Visiting* (London and New York: Routledge, 2000); Eileen
Magnello, *A Centenary History of the Christie Hospital Manchester* (Manchester: Christie
Hospital NHS Trust and Wellcome Unit for the History of Medicine, University of
Manchester, 2001); Edgar Jones, *The Business of Medicine: The Extraordinary History of Glaxo,
a Baby Food Producer, Which Became One of the World's Most Successful Pharmaceutical Companies*
(London: Profile Books, 2001); Carl M. Brauer, *Champions of Quality in Health Care:
A History of the Joint Commission on Accreditation of Healthcare Organizations* (Lyme, CT:
Greenwich Publishing Group, 2001); Thomas Schlich, *Surgery, Science, and Industry:
A Revolution in Fracture Care, 1950s–1990s* (Basingstoke, England, and New York: Palgrave
Macmillan, 2002); John O'Donnell, *Coriell: The Coriell Institute for Medical Research and a
Half Century of Science*, with a foreword by Jonathan E. Rhoads (Canton, MA: Science
History Publications, 2002); Chester R. Burns, *Saving Lives, Training Caregivers, Making
Discoveries: A Centennial History of the University of Texas Medical Branch at Galveston* (Austin:
Texas State Historical Association, 2003); P. Roy Vagelos and Louis Galambos, *Medicine,
Science, and Merck* (Cambridge, UK, and New York : Cambridge University Press, 2004).
14 Bronwyn Dalley, "Finding the Common Ground: New Zealand's Public History," in *Going
Public: The Changing Face of New Zealand History*, ed. Bronwyn Dalley and Jock Phillips
(Auckland: Auckland University Press, 2001), 16–29, on 25; Paul Ashton and Christopher
Keating, "Commissioned History," in *The Oxford Companion to Australian History*, ed.
Graeme Davison, John Hirst, and Stuart MacIntyre (Oxford University Press, 2001),
139–141, on 141; Peter Bartrip, *Themselves Writ Large: The British Medical Association
1832–1966* (London: BMJ Publishing Group, 1996), esp. xv–xviii.
15 Some historians have attempted to portray public history as promoting neglected voices
and constituting new social history rather than providing a further voice for the elite. Yet,
such histories are not common among the medical and biomedical commissions listed
above. See, for example, Dalley, "Common Ground." An exception may be Gamble and
Fletcher, *Tuskegee.*
16 In recent years, a small number of commercial history organizations have begun to
flourish. Those with a medical historical interest include: History Associates, based in
Rockville, Maryland, http://www.historyassociates.com/ (accessed on May 3, 2004), which
undertakes much work for the National Institutes of Health; The Vantage Center, founded
by John O'Donnell, which produced O'Donnell, *Coriell*; and HistorySmiths based in
Carleton, Australia, founded 1994, http://www.historysmiths.com.au/ (accessed on May 3,
2004), which published Waterhouse, *Schizophrenia.*
17 On public history as an alternative career for those displaced from academia, see Alexandra
M. Lord, "Historians and Other Careers," *Perspectives* 41 (February 2003): 10–11; Jeffrey
L. Sturchio, "Historians of Science and the 'Real World' " *History of Science Society Newsletter*
(October 1988): 1, 16–17 (updated version: http://www.hssonline.org/teach_res/

essays/sturchio.html (accessed on June 12, 2004)); see also http://www.beyondacademe.com (accessed on May 3, 2004). Also see David Darlington, *"Beyond Academe*: The Internet Gateway to Nonacademic Careers," *Perspectives* 43 (January 2005): 27–29; Alexandra Lord, "The View from Outside the Ivory Tower," *Perspectives* 43 (January 2005): 30–32. In the US, the National Institutes of Health (NIH), the Federal Drug Administration, and the US Public Health Service all have historical offices: http://history.nih.gov/ (accessed on May 3, 2004); http://www.fda.gov/oc/history/historyoffice.html (accessed on May 3, 2004); http://lhncbc.nlm.nih.gov/ apdb/phsHistory/ (accessed on May 3, 2004). In addition, various institutes within the NIH have independently hired historians to write histories related to their mission including David Cantor (National Cancer Institute), Mark Parascandola (National Cancer Institute), and Simon Baatz (National Library of Medicine). Peter Bartrip makes the point about commissioned history as an additional source of funding for academics. See, *Themselves Writ Large: The British Medical Association 1832–1966* (London: BMJ Publishing Group, 1996), esp. xv–xviii.

18 J. Avorn, M. Chen and R. Hartley, "Scientific versus Commercial Sources of Influence on the Prescribing Behavior of Physicians," *American Journal of Medicine* 73 (1982): 4–8; T. Shawn Caudill, Mitzi S. Johnson, Eugene C. Rich and W. Paul Mc Kinney, "Physicians, Pharmaceutical Sales Representatives, and the Cost of Prescribing," *Archives of Family Medicine* 5 (1996): 201–206; Lisa D. Chew, Theresa S. O'Young, Thomas K. Hazlet, Katharine A. Bradley, Charles Maynard, and Daniel S. Lessler, "A Physician Survey of the Effect of Drugs Sample Availability on Physicians' Behavior," *Journal of General Internal Medicine* 15 (2000): 478–483; M. M. Chren and C. S. Landefeld, "Physicians' Behavior and their Interaction with Drug Companies," *Journal of the American Medical Association* 271 (1994): 684–689; Joel Lexchin, Lisa A. Bero, Benjamin Djulbegovic, and Otavio Clark, "Pharmaceutical Industry Sponsorship and Research Outcome and Quality: Systemic Review," *British Medical Journal* 326 (2003): 1167–1170; Hans Melander, Jane Ahlqvist-Rastad, Gertie Meijer, and Björn Beermann, "Evidence B(i)ased Medicine—Selective Reporting From Studies Sponsored by Pharmaceutical Industry: Review of Studies in New Drug Applications," *British Medical Journal* 326 (2003): 1171–1173; J. P. Orlowksi and L. Wateska, "The Effects of Pharmaceutical Firm Enticements on Physician Prescribing Patterns," *Chest* 102 (1992): 270–273; M. Y. Peay and E. R. Peay, "The Role of Commercial Sources on the Adoption of a New Drug," *Social Science and Medicine* 26 (1988): 1183–1189.

19 Robert Martensen attacked the Final Report of the U.S. President's Advisory Committee on Radiation Experiments because it lacked critical historiographical distance from its funding agency, the Federal Government. Robert Martensen, "If Only it Were So: Medical Physics, U.S. Human Radiation Experiments, and the *Final Report* of the President's Advisory Committee (ACHRE)," *Medical Humanities Review* 11, no. 2 (Fall 1997): 21–36, esp. 24; Ruth Faden and Dan Guttman, "In Response: Speaking Truth to Historiography," *Medical Humanities Review* 11, no. 2 (Fall 1997): 37–43. See also Gareth Williams, review of Elizabeth W. Ethridge, *Sentinel for Health* (1992) in *Social History of Medicine* 6 (1993): 456–457; Jeff Hughes, "Whigs, Prigs and Politics: Problems in the Historiography of Contemporary Science," in *The Historiography of Contemporary Science and Technology*, ed. Thomas Söderqvist (Amsterdam: Harwood Academic Publishers, 1997), 19–37, esp. 30–31.

20 David J. Rothman, "Serving Clio and Client: The Historian as Expert Witness," *Bulletin of the History of Medicine* 77 (2003): 25–44; Robert N. Proctor, "Expert Witnesses Take to the Stand: Historians of Science Can Play an Important Role in US Public Health Litigation," *Nature* 407 (September 7, 2000): 15–16. See also Richard J. Evans, "History, Memory, and the Law: The Historian as Expert Witness," *History and Theory* 41 (2002): 326–345.

21 Laura Maggi, "Bearing Witness for Tobacco," *The American Prospect* 11, no. 10 (October 2000); John A. Neuenschwander, "Historians as Expert Witnesses: The View from the Bench," *OAH Newsletter* 30, no. 3 (August 2002): 1, 6; Brian W. Martin, "Working with Lawyers: A Historian's Perspective," *OAH Newsletter* 30, no. 2 (May 2002): 1, 4.

22 Patricia Cohen, "History for Hire in Industry Lawsuits," *New York Times*, June 14, 2003, sec. B, p. 7.

23 Robert N. Proctor, "Should Medical Historians Be Working for the Tobacco Industry?" *Lancet* 363 (April 10, 2004): 1174–1175. For responses to Proctor's article see John Burnham, "Medical Historians and the Tobacco Industry," *Lancet* 364 (September 4, 2004): 838; Robert N. Proctor, "Medical Historians and the Tobacco Industry," *Lancet* 364 (September 4, 2004): 838–839; David Rothman, "Medical Historians and the Tobacco Industry," *Lancet* 364 (September 4, 2004): 839.

24 In particular, Proctor highlights the efforts of Philip Morris to recruit historians for Project Cosmic (1987–1993), an effort to build an extensive network of scientists and historians across the world to address questions related to tobacco litigation.

25 "Tobacco-funded Research," Amednews.com, July 22, 1996, http://www.ama-assn.org/amednews/1996/amn_96/summ0722.htm (accessed on May 3, 2004).

26 Laurie Kazan-Allen (compiler), "T7N Exposed," *British Asbestos Newsletter* Issue 37 (Winter 1999/2000), http://www.lkaz.demon.co.uk/ban37.htm (accessed on May 3, 2004).

27 Proctor, "Should Medical Historians."

28 Bartrip, *Dusty Death*; Geoffrey Tweedale, with additional research by Philip Hansen, *Magic Mineral to Killer Dust: Turner & Newall and the Asbestos Hazard* (Oxford and New York: Oxford University Press, 2000). See also Jock McCulloch, *Asbestos Blues: Labour, Capital, Physicians & the State in South Africa* (Oxford: James Currey; Bloomington: Indiana University Press, 2002); Ronald Johnston and Arthur McIvor, *Lethal Work: A History of the Asbestos Tragedy in Scotland* (East Linton, Scotland: Tuckwell, 2000); P. W. J. Bartrip, "Irving John Selikoff and the Strange Case of the Missing Medical Degrees," *Journal of the History of Medicine and Allied Sciences* 58 (2003): 3–33; Morris Greenberg, "Letter to the Editor," *Journal of the History of Medicine and Allied Sciences* 59 (2004): 122–126; Robert Cooper, with Deb Chromow, "Letter to the Editor," *Journal of the History of Medicine and Allied Sciences* 59 (2004): 126–134; Peter Bartrip, "Letter to the Editor. Reply: Around the World in Nine Years: A Medical Education Revisited," *Journal of the History of Medicine and Allied Sciences* 59 (2004): 135–144; David Egilman, Geoffrey Tweedale, Jock McCulloch, William Kovarik, Barry Castleman, William Longo, Stephen Levin, Susanna Rankin Bohme, "P. W. J. Bartrip's Attack on Irving J. Selikoff," *American Journal of Industrial Medicine* 46, no. 2 (August 2004): 151–155.

29 English, *Old Paint*; Christian Warren, *Brush with Death: A Social History of Lead Poisoning* (Baltimore: Johns Hopkins University Press, 2000). See also Gerald Markowitz and David Rosner, *Deceit and Denial: The Deadly Politics of Industrial Pollution* (Berkeley: University of California Press; New York: Milbank Memorial Fund, 2002). Rosner and Markowitz's book has also recently been the subject of litigation by the chemical industry regarding their claims about the industry's response to concerns about the health hazards of vinyl chloride in the 1970s; Lila Guterman, "Hot Type," *The Chronicle of Higher Education* 51, Issue 13 (November 19, 2004), A20. See also Jon Weiner, "Cancer Chemicals and History," *The Nation*, 280, No. 5 (February 7, 2005), 19–22. Reprinted as Jon Wiener, "Why 2 Historians Now Have to Fear the Chemical Industry" in http://hnn.us/articles/9950.html (accessed on April 12, 2005). Rosner and Markowitz provide documents related to this controversy at http://www.deceitanddenial.org (accessed on April 12, 2005).

30 Cohen, "History for Hire"; Proctor, "Expert Witnesses"; Proctor, "Should Medical Historians"; Maggi, "Bearing Witness"; Kazan-Allen, "T7N Exposed."

31 For a historical account of the association of commissioned histories with the emergence of modern skepticism, see Brendan Dooley, *The Social History of Skepticism: Experience and Doubt in Early Modern Culture* (Baltimore and London: Johns Hopkins Press, 1999), esp. Chapter 3.

5   From behind the fence

Threading the Labyrinths of classified historical research

Anne Fitzpatrick

On one typically crystalline August afternoon at Los Alamos National Laboratory in New Mexico in 1994, I received a call from my secretary. She informed me that my long-awaited United States Department of Energy (DOE) security "Q" clearance had arrived, allowing me access to documents and conversations for a classified history project to which I had been assigned.[1] I could now actually visit my secretary, Mildred*, and my immediate supervisor, John*, whose office was until then in a building off-limits to me. John had instructed the secretary to tell me to pick up my new identification badge, then meet him at the entrance gate at the main Administration Building, where I was to move into a new office that was, as they said, "behind the fence."

I dropped the papers I was reading, and hurried over to the badge office, picked up my new, bright-blue, electronically encoded badge and walked next door to the gatehouse where the security officer looked at my identification and let me pass with a smile. I walked over to John at the main entrance and he enthusiastically shook my hand and said, "Welcome to the brethren!" Jokingly I reminded him I was a woman but thanked him anyway for his warm welcome to this apparently elite club. We both laughed.

John led me to his office on the building's top, fourth floor. Without a word he opened his combination-locked safe, and excitedly pulled out stacks of classified documents, saying "Take a look, you will need to review these materials for your research." I briefly scanned a few of the documents—labeled Secret-Restricted Data, or SRD for short—mostly nondescript and boring 1950s-vintage military policy committee meeting minutes redacted from the National Archives in Washington, DC. John put them back in his safe and locked it, and led me away around the suite of offices our group occupied. He pointed out where my new office was. Later we talked a little about progress on the project—an ambitious multi-author history of nuclear weapons development during the Cold War—that I had been hired for, and our work plans for the coming year.

The next morning I began moving in to my new office, but not before I attended a class that instructed me about some of the caveats of having a clearance, and some etiquette associated with it: for example, I was not to wear my badge when off laboratory grounds, such as in local restaurants, lest I encourage potential spies to approach me. The instructor also joked that we were not supposed to use our badges to the scrape the ice off our car windshields during the winter. After the class was over, I filled out several forms in order to move my then state-of-the-art Macintosh Power PC 7100 behind the fence: since it had a removable hard drive, it was ready to go to my new office.

After getting settled in my new space, with its a breathtaking view of the Jemez Mountains, I began to learn the rules and social norms of working in a secure space. There was no dress code: most of the scientific staff typically donned jeans, cowboy boots, and (sometimes unfortunately) shorts in the summer. The only serious dress requirement was that employees had to wear their badges above the waist at all times. Of the hundred-plus people in the wing I was now working in, I was one of less than a dozen graduate-trained women. Yet I was becoming an accepted member of the Q-cleared community.

Over the next month I learned the rules of conduct from our secretaries. Upon signing in every morning I fired up my Macintosh—now deemed my unclassified computer system, and placed on top of it a small flip-tent placard stating that I was operating in an open mode. I had also acquired a classified stand-alone (in other words, not connected to the Internet or any network) IBM computer whose hard drive had to be locked up in my safe when not in use. If I left the office even for a few minutes, I was to lock up any documents I may have taken out of my safe. As long as my safe was open, I could not leave the office. At the end of the day I had to check the lock and initial a paper check-list on my safe. Our secretaries, who knew the drill well, would then check the security of the lock on my safe again then initial the list too. One of our secretaries, Florence*, who started working at the Laboratory in 1944 and had frequently encountered the likes of John von Neumann and others who made this place famous, knew security measures so well that everyone always asked her for help in this area.

If I decided I wanted to work with classified documents on that or any day, I opened my safe, placed a bright red magnet on it that proclaimed "open," removed any documents I wanted, then took my classified computer's classified hard drive out of the safe. I removed the unclassified hard drive from the PC and set it aside, replacing it with its classified twin. Then I placed another red paper tent-placard on top of the machine, indicating that I was operating in classified mode.

Every page of every document I printed from that machine had to be marked SRD and stored in my safe. If I wanted to dispose of any printed matter, I took it to the shredder down the hall designated for disposal of such documents. One of the perks of the job was that photocopying was always free, but certain photocopiers were for classified documents, and others were not. I could not bring a camera to my office, or recording equipment not issued by the laboratory, or even a radio that had not been checked out and approved by a technical device security specialist. I never bothered to bring one in because by then Internet radio was becoming popular and there were no formal regulations against listening to that.

There seemed to be a certain pecking order among those who held clearances, but it had more to do with the type of work they were assigned to do than any deliber-ately constructed hierarchy. In addition to the Q, the DOE grants another type of clearance—an L, or limited variant. While a Q clearance grants a holder access to

SRD materials that frequently have to do with nuclear weapons information, an L clearance does not, but allows access to less sensitive levels of information.

If at any time a person with an L—sporting a yellow badge—entered my office to speak with me or simply to hang out, I had to place any SRD documents out of sight. Most of the time visitors had no interest in what I was doing; and moreover friends who were not cleared—in many cases visiting students and foreign post-doctoral fellows—could never see my workspace, which was mildly inconvenient. But other social aspects of the job were fun—I would sometimes join older scientist friends for lunch or coffee, and almost everyone I met was eager to talk about their research and teach me about nuclear weapons.

In his book about Lawrence Livermore National Laboratory—Los Alamos's sister institution in California, anthropologist Hugh Gusterson goes into vivid detail about scientists living the daily practice of secrecy, noting the social norms and communities formed within secret societies.[2] Gusterson asserts that the "rituals" of secrecy, much like what I performed in my daily safe lock-up as described above, lend a kind of air of importance to one's daily work and give scientists behind the fence a boosted sense of self-distinctiveness.[3] Although this may be the case for some workers whose principal function is intelligence-gathering, I would argue that this is not so for the average person who works behind the fence.

Indeed, the National Laboratories do have their own unique cultures that are at times family-like, bizarre, isolated, and unlike any other laboratories in the world as Gusterson has so argued. Describing a more day-to-day scenario, another anthropologist colleague of mine wrote nearly her entire PhD thesis based on classified and unclassified interviews she conducted at Los Alamos with nuclear weapons scientists, and she explained the complicated ins and outs of this environment.[4] In her description of the "geography" of secrecy, she rightly points out that classified information involves not only physical and virtual space (such as computing), but also a social network, and an elaborate compartmentalized system of human, electronic, and physical controls and access means.[5]

The social network is indeed real. The "brethren" that John had joked about did take on some meaning for me after a time, despite some of the drawbacks with becoming part of a somewhat closed community. No institution had ever treated me so well in my entire career. I felt looked after, protected, and that I belonged there in some way. And even if I departed—and I did for several years following graduate school—I would always be part of the security-cleared community. Whether I kept my clearance active for the rest of my life or not, my public discussions of my research were subject to the DOE's scrutiny. Having held a Q clearance for a decade now, many of my colleagues in the history and science studies professions have asked me, "Isn't having a clearance a terrible burden for you?"

The short answer is no. But having a clearance does make research and publishing more time consuming. Regardless of this, the tools of good scholarship are the same anywhere, either inside or outside of the fence.[6] Moreover, to truly understand and effectively communicate recent science to the public, a solid understanding of the science is imperative; a clearance can provide one such avenue towards this. Scientific content is important and counts towards accurate scholarship.[7]

Years ago when I signed my clearance application forms I considered carefully how my work-life would change, but have never regretted this. How did this affect my scholarly life generally? Most immediately, anything I wrote up from the results of my research had to be reviewed by the Laboratory's derivative classifiers—people whose sole jobs involved reviewing word-by-word all professional written or recorded materials intended for publication. This included any lecture, manuscript, book review, grant proposal, web page, or anything I authored on my workplace's time and payroll. Besides this, I could not discuss publicly certain aspects of my research—although usually very nitty-gritty technical details more akin to engineering specifics as opposed to scientific content. I could neither work on such material at home nor take it on business trips. But I could write whatever I wanted, and after review share my work with other scholars, publish, attend meetings, teach, give seminars, and hold email chats.

I could also write a doctoral dissertation on a subject I was thoroughly intrigued by and had become immersed in—the origins of thermonuclear weapons. I could not write such a dissertation and gain a true understanding of the workings of such devices at any other institution in the world, and had long wanted to become more familiar with this scientific subject area, a subdiscipline of nuclear physics that is often described by scientists as intellectually challenging, darkly seductive, and even fun. By the time I finished writing my dissertation I estimated that I had had to eliminate about 2000 words of sensitive material to allow my committee members at Virginia Tech to see it. It is available as a public document at the Los Alamos Laboratory library.[8]

Indeed, having a clearance gave me a means of research access unavailable to most others, but severe potential penalties come along with these privileges if they are abused. Taking classified materials out of secure areas, or downloading this sort of data to home computers, or other acts can result in felony charges, time in jail, and termination of not only the clearance but also employment. In the United States, a treason conviction can even result in the death penalty.

Importantly, having a particular type of clearance does not automatically award a scholar access to every classified document out there—a point I will explain later— or to purported vast government secrets and conspiracies. "The fence" is not some holdup where we smirk with the knowledge that there sit scores of documents that no one else can access. Fueled by popular television shows such as the now-defunct *X-Files* and *Roswell*, a common myth persists that the National Laboratories, other federal agencies, and military establishments such as the infamous Area 51 hold vast dark secrets of public interest that await Freedom of Information Act (FOIA) release.

These rumors persist. For example, after I completed my dissertation many of my academic and journalist colleagues pressed me for the so-called "secret" to the design of what is most commonly called the Teller-Ulam thermonuclear device. People offered me phony job opportunities to lure me to discuss technical details of this device; one person even got my home telephone number and called me, pressing me for information about thermonuclear weapon design. Yet no one believed me when I said there was no secret to this. On the rare occasion academic colleagues have snapped at me when I reply that I cannot discuss nuclear weapon engineering design details or cite certain data. One professor—whom I had never met in person—blasted me via email that never in his career had anyone been so unwilling to share information. I was only being honest, and did not wish to break the law.

Other colleagues of mine who hold security clearances have relayed similar stories. Academic historians have claimed at times that they would sue Los Alamos Laboratory's archivist for "withholding public information"; one person even threatened to burn his house down. Yet by law he cannot give away classified information, and is as subject to the DOE's policies as anyone else with a Q clearance. Besides this, and sadly, the DOE as a whole does not place archival materials' declassification high on its list of workplace priorities; Los Alamos Laboratory's archives cannot employ the number of staff necessary to review and declassify enough of its documents. In addition, inconsistent classification procedures within many government agencies are certainly shameful, and the DOE is among the biggest offenders in this regard.

* * *

Among the biggest joys of writing about contemporary science and technology are the numerous open archival and reading materials in libraries, archives, and now on-line, and the ability to interact with the characters one is writing about. Yet frustration comes when historians—especially those interested in theoretical, experimental and applied physics, high-tech weapons, and military science and technology and engineering—encounter the realm of science and technology for national security.[9] As is well known, the physical sciences in particular were among those receiving the most federal support up to the 1970s. The various subdisciplines of physics, particularly high-energy physics and nuclear studies, generated all kinds of scientific reports and other documentation in this period. Yet the very reason the physical sciences were so well supported affects current attempts to look at this massive amount of documentation. Much physical science research was conducted by universities, industry, and government laboratories for the military and based on the notion of supporting national security and the large network-like military defense complex, of which the DOE is one part.[10]

Scholarly inquiry into such areas may not be easy, but it is possible. And here I would like to point out that such work requires a broad perspective: researchers often—and rightly—complain about the sad state of declassification efforts by American government agencies, but forget that the corporate world can be even tighter about sharing proprietary information about inventions, biotechnology patents, computer technology, and other information. Even more difficult to access are state archives in many foreign countries; success can depend on whether or not one has the right personal connections, formal letters of introduction, or other, often unofficial, means.

Sarov, the formerly top-secret Russian nuclear weapons laboratory, previously known as Arzamas-16 and affectionately known as Los Arzamas, is still a closed city. It first appeared in the public literature as "The Installation" in Russian Nobel laureate Andrei Dmitrievich Sakharov's *Memoirs* (Figure 5.1).[11]

Sarov lies about 400 kilometers east of Moscow. As a closed city, one does not simply show up and walk in to Sarov, as the entire city and region around it are

Figure 5.1 Igor Kurchatov, the driving force behind the Soviet Union's crash program
   to develop atomic weapons [right], and Andrei Sakharov, father of the Soviet
   H-bomb project and later political dissident and human rights activist, in 1957.

Source: Emilio Segrè Visual Archives, Physics Today Collection, Niels Bohr Library/Center for History of
Physics, American Institute of Physics.

encompassed by three layers of barbed-wire fences. Visiting requires a special invitation
generated through the Russian Federal Atomic Energy Agency (ROSATOM), a
background check by the Russian foreign visitors ministry, the ability to obtain a
visa, and a special *propusk* to board the evening Sarov-bound train at Moscow's Kazan
station. If you are lucky, the guards might have your entry paperwork in order when
you arrive in Sarov the following morning. When there, you can talk to people, and
occasionally wander about by yourself, but you cannot access historical documents—
if they still exist at all. Although I visited Sarov several times and speak Russian, I
never expect to have archival access at that facility during my lifetime. Yet the mere
idea that foreigners can now travel to such places is rather thrilling, and a testament
that some aspects of state secrecy have been relaxed since the end of the Cold War.

   Despite its flaws and inefficiency, and considering the difficulties involved with
access to foreign sites such as that mentioned above, the American federal government's system of records-keeping and declassification is not the worst out there.
Various government agencies employ their own classification methods, levels, and
categories, and attempting to discuss all of them is beyond my experience and the
scope of this chapter. But generally it is possible to make some headway into the
classified labyrinth, or perhaps get around it somewhat.

   The classification of military-related documents is as old as the profession of war.
While many famous scientists and lawmakers have argued that classification of nuclear
weapons data and other military devices—funded by taxpayer dollars—serves no
purpose and may even escalate international tensions, this remains up for debate. The

late Edward Teller long advocated that there should be no nuclear secrets, yet he was also infamous for hiding behind the veil of secrecy when defending his political actions.[12] Sakharov also once supposedly said that there was no secret to the atomic bomb because it could be built, yet the secret of the Strategic Defense Initiative (SDI)—or Star Wars—program was that it could *not* be built. Others in the current American presidential administration maintain otherwise.[13] Yet at the heart of such debates is a grave and intractable communication problem fueled from both inside and outside the fence.

In some sense classification of documents in the nuclear weapons sector has not changed much since its inception in the 1940s, although policies have been altered somewhat over the years. Over twenty years ago former Department of Energy historian Richard Hewlett stated that, "Throughout its existence from 1946 to 1975, the United States Atomic Energy Commission (the AEC, predecessor to the DOE) consistently relied upon the 'born classified' concept in administering its statutory authority to control the dissemination of classified information."[14] Although this term, "born classified," was not an official AEC doctrine, it is a fitting description of the way that the AEC and its large network of laboratories and corporate contractors routinely categorized reports, letters, and correspondence under the SRD heading— a term that the United States Joint Committee on Atomic Energy created in 1946 to cover in a broad sense all aspects of nuclear weapons research and development.

If DOE documents are "born classified," then they have certainly tended to remain that way, partly because historians and journalists operating outside federal government agencies often didn't know of their existence to begin with, and there was no general effort made to declassify many of these materials. But even cleared individuals may not gain knowledge of the existence of some restricted materials because they are subject to the doctrine of "Need to Know," a written statement required to access documentary evidence and a compartmentalization measure used by federal government agencies.

For a cleared researcher, the "Need to Know" statement has to be quite broadly constructed in order to access more than one or two categories of documents. Moreover, access to data on some very highly classified topics requires, in DOE parlance, "Sigmas"—special categories that place various nuclear weapons and other data into fifteen levels of specific bodies of information. Some of the higher ones are practically impossible for *anyone* to obtain historical documentation about.

While various presidential administrations have tightened and relaxed declassification policies over the last decades, using the Freedom of Information Act remains the best way to obtain and have documents declassified. Signed by President Lyndon Johnson in 1966, this measure is a legal right to access federal information. FOIA procedures, exemptions, and other related information are readily available on the United States Department of Justice's website, and thus I will not review them here.[15]

I would like to emphasize, though, that the most important elements when redacting formerly classified material or making FOIA requests are time and patience, neither of which many scholars are always willing to employ. If they are, efforts can involve numerous letters and phone calls to, and personal visits with, the resident archivist or historian at the facility you are interested in, to correctly identify the specific subject area you want to research. The archivist might be able to point you towards the classified documents' titles (titles of such documents are not always secret) they hold. If they are unable or don't want to help, call another facility's historian or archivist and they might know more about the first laboratory's holdings than that facility's own employee. Even if you cannot name document titles verbatim, if you can be specific enough in the wording of your FOIA request then success might occur. The underlying notion behind doing all of the above-mentioned is to be *persistent* in your approach, while being aware that: (1) federal archival staffs continue to be cut and underfunded; (2) FOIA is not universally applicable; and (3) various mandates may bear on the timely progress of scholarly research. A notable recent example of such mandates is the October 12, 2001 United States Attorney General John Ashcroft memo emphasizing the protection of information pertaining to national security.[16]

Some researchers have been extremely successful in using FOIA to obtain once-classified materials for historical research. Among them, the late independent military historian Chuck Hansen stood alone, until his untimely death in 2003. Hansen wrote prolifically about American nuclear weapons systems, and was well known for his massive private document collection, assembled from over thirty years of requesting materials from the DOE, nuclear weapons laboratories, and military archives.

Hansen himself once stated that the DOE cannot distinguish between imaginary and real secrets, where for example a document might be classified at one facility, such as Livermore Laboratory, and a declassified version might be found at the DOE headquarters archive in the Washington, DC area. His work is proof that it not only pays to check around at different government research sites, but moreover, one declassified document might cite others that still await declassification.[17]

I can personally attest to the inconsistency in the DOE's declassification policies, having come across the exact same document at two different facilities. In 1995 I found copies of the same letter from Stanislaw Ulam to John von Neumann both at Los Alamos and at the National Archives in Washington. One was marked SRD, while the other was not. Could I cite it? No one in any position of authority seemed to know the answer to this. So I went to visit Ulam's and von Neumann's still-living friends and colleagues to see if they knew anything about this letter. No one with whom I spoke remembered the letter distinctively, but some knew something of the subject matter, which concerned the origins of the Teller-Ulam device. I did not discover anything new or profound about this topic, and in the end I was able to note the letter in my dissertation footnotes, because at that time DOE policy allowed its employees to cite the titles of some classified documents, although they could not necessarily cite from the documents' bodies.

* * *

The barriers to doing research in such areas as the history of nuclear weapons or defense technologies are not just physical in the sense that source materials may be housed at restricted facilities. Those cleared scientists who authored classified reports are of course incredibly valuable resources, and what they are allowed to discuss can be limited. Yet merely having the privilege to interact with such individuals is one of the most exciting aspects of writing about recent science, and any interview with a scientist-subject may provide all kinds of useful and sometimes entertaining information.

Interviews regarding work in classified subject areas are extremely important to complete a historical narrative, because the information obtained from the scientists may fill in the gaps in the declassified reports and provide leads towards obtaining new records and documents, new contacts, as well as reveal some firsthand participation in formerly secret work.

In order to prepare for a meeting with a scientific participant it is critical to read as many as possible of the interview subject's open publications on the particular topic of interest; this is why mastering scientific content is critical. An uncleared researcher may not be able to read his or her subject's classified reports and notes, but can often surmise much about that person's restricted work simply by reading their published reports and articles. In the history of high-energy physics, for example, journals such as the *Physical Review* series are good sources for tracing the science back to its classified origins. Here, this sort of "reverse engineering" can work.[18]

For example, it was no accident that many discoveries in nuclear physics and experimental mathematics grew out of fission and fusion weapons research. One such discovery—the Monte Carlo technique used for neutron diffusion simulations—was first developed by mathematicians Stanislaw Ulam and John von Neumann to handle fission and hydrogen bomb simulations.[19] While many other methods of modeling neutron diffusion were developed after the Second World War, Ulam and von Neumann developed this particular technique as part of an attempt to solve what historian of physics Peter Galison has described as the most complex scientific problem up through the postwar period: the thermonuclear "Super" problem.[20] Later, Monte Carlo found its way into all kinds of scientific calculations in many fields.

Much contemporary unclassified research has roots in formerly classified science. Ulam, along with Italian physicist Enrico Fermi, John Pasta, and Mary Tsingou, programmed the first unclassified simulations of nonlinear experimental mathematical problems on the Los Alamos MANIAC (Mathematical and Numeric Integrator and Calculator) computer in 1954 and 1955, constituting a significant beginning to the now hot field of nonlinear science.[21] The team's "heuristic work" in nonlinear physical systems, as they called it in one of their formal reports on the subject, was based on by then well-established hydrogen weapons computing efforts. Yet all of their nonlinear experimental mathematical work was unclassified, and anyone can view the final written report. In such documents, it is easy to see what had happened in the behind-the-fence weapons programs, in this case the development of ever-more complex techniques and codes for nuclear weapons explosive and effects simulations. Constructing interview questions from such kinds of materials can bring surprisingly pleasing results. For example, Richard Rhodes relied extensively on elaborate interviews for both his Pulitzer Prize-winning book on the atomic bomb, *The Making of the Atomic Bomb* (1986), and for *Dark Sun* (1995).[22] Rhodes, like Hansen, conducted his research and interviews without holding a clearance, and found interviews to be very revealing.

When one talks to and interviews scientists and engineers, and visits archivists, it is not only a matter of good manners, but also critical to cultivate good relationships with these people. It is *that* simple. In particular, historical and other scholarly researchers—including myself—tend to forget that librarians and archivists are generally underpaid and overstretched in their duties. If one treats the archivist as an enemy who keeps the sacred documents, chances are they will not welcome you. If one is eager and friendly and displays an obvious interest in the sought-after materials as well as in the archivist's knowledge of these items, one may get never-before-seen documents or perhaps learn a great deal.

Let me offer a personal example. The main archives of the Russian Academy of Sciences in Moscow at first appeared to me a dismal, un-user-friendly place. It failed to operate on regular, set hours, and provided little in the way of comfort to its users, causing a great deal of frustration.

But despite my poor initial impressions of this facility, I took some time to get to know the employees there. Sometimes I brought cakes and tea with me. Soon I noticed that I was warmly welcomed both to work and also to sit with the ladies who ran the archives to chat over tea about everything from politics to weather. I learned a great deal about things that at first did not make sense to me as a foreigner: sometimes these archivists did not get paid for months; too, they could not afford to heat the reading room in the winter, and thus all the users had to sit there bundled up in coats, hats, and gloves; finally, because getting around in Moscow via public transportation was so cumbersome, especially in the winter, the archives' opening hours varied according to when its employees could get there. Not only did gaining some understanding make the research go easier, but also getting to know folks at the Academy archives was well worth the effort. I would not have been able to even conceive of trying to sue them for not doing their job.

As with the archival staffs, conducting interviews with scientists can sometimes lead to the development of personal relationships with these people. There is no harm in this. Often they are glad to hear of historians writing about them and can be happy to help you get the story straight while also making additional contacts for you. They may even get involved working with you in various ways.[23]

Too often historians are taught by their academic mentors to work alone—as opposed to collaboratively—and to perform research in very specific ways such as only using formal archival documents as source materials. Browse any issue of *Science* or *Nature* where research articles normally sport anywhere between four and eight or more authors. Articles in *Isis* tend to have single authors. Historians of recent science might do well to think outside the black box of their training when it comes to research: recent science is far too complicated not to do this.

Traditional thinking on this difference in cultures harkens back to Paul Forman's well-written 1991 paper in *Isis*, in which he argues that historians of science need to be independent of scientists and their practice.[24] Indeed, solid, disciplined inquiry should never be compromised as a scholarly goal, but Forman is operating from the standpoint of a historical scholar working out of a traditional academic setting and historical training, a profession that he noted was about one generation old at the time of writing his article.[25] The reality for most academically-trained scholars today is quite different from that of this earlier generation. The difference lies both in the way that contemporary scientific practice—the way that science is done—is changing, and also in the employment situation that scholars of the current generation find themselves in. Both factors are contributing to the emergence of a new type of scholar who does not adhere solely to a distinct, rigid disciplinary method of research. Providing an accurate account of events is more important than any particular technique in or approach to studying recent science, classified or not. All scholars have a civic duty to provide the most accurate, reliable information that they can. Yet the study of science of the last fifty years or so requires generally a more interdisciplinary view than just traditional archival and documentary evidence. Towards this, historians in particular ought to look more towards the work of scholarly professionals from other disciplines in order to learn new tools from them.[26]

Horace Freeland Judson has often remarked that recent science is operative: it is ongoing, all the time. It has occurred in the near past and is occurring now constantly, more than ever before. Many of the individuals engaged in this work are still alive, and the kinds of interactions they have with others are different than in centuries past—they communicate by email as opposed to letters, and they less often maintain fastidious notebooks or diaries of their daily laboratory activities. Thus, these documents are often no longer available for historians to analyze.

Here is where oral interviews can be extremely valuable. As Ronald E. Doel stated in his 2003 article on oral history of American science, "historians of twentieth-century science are increasingly aware of the limitations of written sources. Many no longer automatically ascribe a higher epistemological value to newspaper accounts, meeting minutes, memoirs, and published reports."[27] This is not to diminish the value of such documents to good scholarship, but in the realm of recent science—some of which is classified—oral history may be the only information-gathering option. There is no one correct way to study scientific and technological activity. But given the nature of recent science—including government classification of certain subjects within that realm—the broadest analytical flexibility and reliability is found in the broadest approach to research: that which can be borrowed from historians, sociologists, anthropologists, and practitioners of various scientific techniques.[28]

Because it is operative, recent science is also strongly characterized by the astounding speeds at which it generates historical information in enormous volumes: the most successful historians of recent science will need to develop the ability to organize collaborative, cross-disciplinary research efforts on a particular topic to handle this otherwise overwhelming nature of current scientific and technological activity.

Some institutions have already realized the value of this. Notably, Los Alamos National Laboratory and other established scientific facilities, along with many large corporate research-oriented enterprises, have hired increasing numbers of social scientist teams in the last decade. This trend seems to indicate that scientific practice is facing challenges unique to the late twentieth and early twenty-first centuries. Social scientists employed in such nontraditional institutions (i.e. other than in a university social science department) are forced to innovate and thus take multi-disciplinary approaches to their research as a matter of course, given the demands of their employers.

A social scientist's—or any scientist's—possession of a security clearance is itself a phenomenon unique to recent science, with the advent of the military-industrial complex. Personally speaking, it has forced me to adopt novel or untraditional research techniques so I may say publicly what I need to say to meet standards of scholarship, and provide what physicist John Ziman deemed "reliable knowledge": the carefully reconnoitered landscape of consensually-reached, peer-reviewed findings.[29]

To return to a question posed earlier—is having a clearance a burden—the answer is mostly no. It did not hinder my getting an advanced degree, or prevent me from attending conferences or continuing public research. It did not affect my ability to compose this chapter, but did slow down the process slightly.

Having a clearance did alter my career choices. I found that, after three years behind the fence as a student, much as I tried, I could not go back to academia full-time, and eventually found myself back on staff in Los Alamos, working on a variety of projects. As with becoming part of any organization—academic, corporate, non-governmental, or other—one more often than not becomes one of their "brethren." Nor is there just one brotherhood of cleared people. Many agencies besides the nuclear weapons complex produce classified documentation. With the increasing concerns about biological weapons and cyberterrorism, scholars of recent science will need to be prepared to take on the massive amounts of virtual and hard-copy information these areas will generate. In the private sector, biotechnology companies are creating equally huge amounts of secret documentation that does not fall under traditional governmental classification policies, but is subject to all sorts of legal restrictions that historians will have to figure out how to deal with.

For better or for worse, it seems almost certain that the routine, "born classified" approach to sensitive areas in science and technology will pervade in the post-September 11, 2001-era. Yet what constitutes the actual "secret" parts of such materials is likely not so massive as is the whole procedure of classification itself. Moreover, secret details can often be trivial in the context of a larger, broadly focused historical perspective. Fortunately for historians and scholars of such current events, those scientists engaged in this work will likely be around for a while to elucidate it for us in our pursuit of reliable knowledge. Looking out at the view from behind the fence, Sakharov's supposed comment (noted earlier in this chapter) about secrecy can well be applied to the realm of recent science: routine bureaucratic classification of the petty details within scientific and technological innovation can never truly impede perseverant, well-researched, disciplined inquiry.

Acknowledgments

The author is grateful to the following individuals for their comments on and thoughtful critiques of this chapter: Horace Freeland Judson, Tomoko Steen, David Grier, Patrick McCray, Olivia Walling, Amy Crumpton, Steve Weiss, Nathaniel Comfort, Michele Garfinkel, Susan Garfinkel, Carl-Henry Geschwind, and Janet Abbate.

Notes

* Out of respect for their privacy I have used fictitious names for my office colleagues mentioned here.
1 United States Government agencies and military institutions all issue their own types of clearances. The Q is specific to the DOE. The Navy, Air Force, and Army, for example, issue "Secret" and "Top Secret" clearance levels.
2 Hugh Gusterson, *Nuclear Rites: A Weapons Laboratory at the End of the Cold War* (Berkeley, CA: University of California Press, 1996), 68–100.
3 Ibid., 88.

4 Laura Agnes McNamara, "Ways of Knowing about Weapons: The Cold War's End at the Los Alamos National Laboratory" (PhD dissertation, University of New Mexico, 2001).

5 McNamara, "Ways of Knowing," 60–74.

6 See Anne Fitzpatrick, "Coming to Terms with Recent Science," *Recent Science Newsletter*, 2, no. 1, p. 1, pp. 6–7 (spring 2000).

7 Alan E. Shapiro, "Historians of Science Must Again Master Scientific Substance," *The Chronicle of Higher Education* XLIV, no. 21 (February 20, 1998), sec. B, p. 4–5.

8 Anne Fitzpatrick, "Igniting the Light Elements: The Los Alamos Thermonuclear Weapon Project, 1942–1952" (Los Alamos, NM: Los Alamos National Laboratory report LA-13577-T, 1999). As of November 2003, this technical report is available in hard copy in the Los Alamos National Laboratory Study Center, which is open to the public daily depending on security conditions. An older version of the dissertation is also available from the Virginia Tech electronic thesis and dissertation library as a pdf document, via the Virginia Tech campus network, http://scholar.lib.vt.edu/theses/available/etd-121898-140317/ (accessed on November 24, 2003).

9 Paul Forman, "Behind Quantum Electronics: National Security as Basis for Physical Research in the United States, 1940–1960," *Historical Studies in the Physical and Biological Sciences*, 18 (1987): 149–229.

10 Forman, "Behind Quantum Electronics"; Daniel Kevles, *The Physicists: The History of a Scientific Community in Modern America* (Cambridge: Harvard University Press, 1995); R. W. Seidel, "A Home for Big Science: The Atomic Energy Commission's Laboratory System," *Historical Studies in the Physical and Biological Sciences* 16, no. 1 (1986): 135–175.

11 Andrei Sakharov, *Memoirs* (New York: Knopf, 1990).

12 Teller often lectured and debated about the Reagan administration's "Star Wars" program at Stanford University and the Stanford Linear Accelerator Center (SLAC). He often implied that he knew far more than others about weapons technologies: "If you knew what I know, you'd agree with me," he once stated in a debate with then-SLAC director Wolfgang Panofsky. See also Lillian Hoddeson's essay in this volume.

13 That is, as I write in 2003, the administration of George W. Bush supports the twenty-first century variant of SDI, the National Missile Defense Program. For more on the origins and history of the SDI program during the Ronald Reagan administration, see Francis FitzGerald, *Way Out there in the Blue: Reagan, Star Wars, and the End of the Cold War* (New York: Simon and Schuster, 2000).

14 Richard G. Hewlett, "Born Classified in the AEC: A Historian's View," *Bulletin of the Atomic Scientists* 37, no. 10 (December 1981): 20–27.

15 See http://www.usdoj.gov/04foia/foi-act.htm (accessed on November 24, 2003).

16 See http://www.usdoj.gov/oip/foiapost/2001foiapost19.htm, (accessed on November 24, 2003)

17 Chuck Hansen, "Open Secrets, Closed Minds: Classification Practices of the Department of Energy," *Bulletin of the Atomic Scientists* 51, no. 4 (July–August 1995): 16.

18 See also Tilli Tansey's chapter in this book.

19 Robert D. Richtmyer and John R. von Neumann, *Statistical Methods in Neutron Diffusion* (Los Alamos Scientific Laboratory, LAMS-551 April 9, 1947).

20 Peter Galison, *Image and Logic: A Material Culture of Microphysics* (Chicago: The University of Chicago Press, 1997), 693–694.

21 E. Fermi, L. Pasta, and S. Ulam, *Studies of Nonlinear Problems: I* (Los Alamos Scientific Laboratory, LA-1940, May 1955). Other published materials are easily found: many of Soviet physicist Yakov Zeldovich's published works were based on his nuclear weapons design experience. See Yakov B. Zeldovich and H. McNeil, eds, *The Mathematical Theory of Combustion and Explosion* (Kluwer, Dordrecht, 1985).

22 Richard Rhodes, *The Making of the Atomic Bomb* (New York: Simon and Schuster, 1986); Richard Rhodes, *Dark Sun: The Making of the Hydrogen Bomb* (New York: Simon and Schuster, 1995).

23 See Arne Hessenbruch's essay in this book.

24 Paul Forman, "Independence, Not Transcendence, for the Historian of Science," *Isis* 82 (1991): 71–86.
25 Ibid., 77.
26 For instance, I'm thinking of an anthropologist who goes off to live in New Guinea in order to study local culture and customs while doing her best not to go native; likewise, I believe there is no harm in a historian of recent science visiting and interacting with people in a biological weapons laboratory, for example.
27 Ronald E. Doel, "Oral History of American Science: A Forty-Year Review," *History of Science* 41 (2003): 349–378.
28 Fitzpatrick, "Coming to Terms."
29 John Ziman, *Reliable Knowledge: An Exploration of the Grounds for Scientific Belief* (Cambridge: Cambridge University Press, 1978).

6 On ethics, scientists, and democracy

Writing the history of eugenic sterilization

Lene Koch

In the summer of 1997, a media storm broke out in Sweden. The front-page headline on August 20 of the *Dagens Nyheter*, one of Sweden's most influential newspapers, grabbed the attention of the country: Sweden had sterilized about 60,000 of her citizens—many by force.[1] Even earlier, however, researchers had made the history of compulsory sterilization a topic of considerable interest. As the interest of the international media demonstrated, the practice of eugenics in Sweden and other Scandinavian countries has a history that casts long and compelling shadows.[2]

The use of compulsory sterilization, by contemporary Western standards, is usually considered appalling and revolting: a violation of basic human rights. Consequently it is often considered an anomaly that state-sponsored eugenics programs took place not only in Nazi Germany and the United States, but also in democratic welfare states such as Sweden, Norway, and Denmark. Compulsory sterilization seems incompatible with the ideals of democracy, and stamps its perpetrators as evil and criminal. It is a fact however, that eugenically motivated compulsory sterilization was generally encouraged by politicians, scientists, and the general public in Scandinavia, and eugenic legislation warranting compulsory sterilization was considered an important element in the welfare reform that was introduced in the 1930s (Figure 6.1).[3]

The contrast between past and present norms makes eugenic sterilization in Scandinavia a controversial topic in contemporary historiography. For several reasons, the issue is politically sensitive: in the Nordic countries, it links the leading Social Democratic parties with eugenics. It also associates eugenics with many well-known and respected scientists and physicians. Some of them are still alive, as are many of the people who were sterilized. The fact that democratic countries in Scandinavia supported eugenics is also a challenge to the generally accepted interpretation of the established political and ethical order, and blurs the clear demarcation line between fascism and democracy, between good and bad science, between good and evil.[4] Furthermore, the current governments of Sweden, Norway, and Denmark have all considered compensating the sterilized, which I will discuss below. All these circumstances make a study of the history of eugenics a difficult matter politically, scientifically and ethically—and as a consequence of this a theme within the recent history of medicine that deserves specific historiographical consideration.

How can historians write about one of the most wrenching issues of the present day? The following reflections originate from my work on the practice of

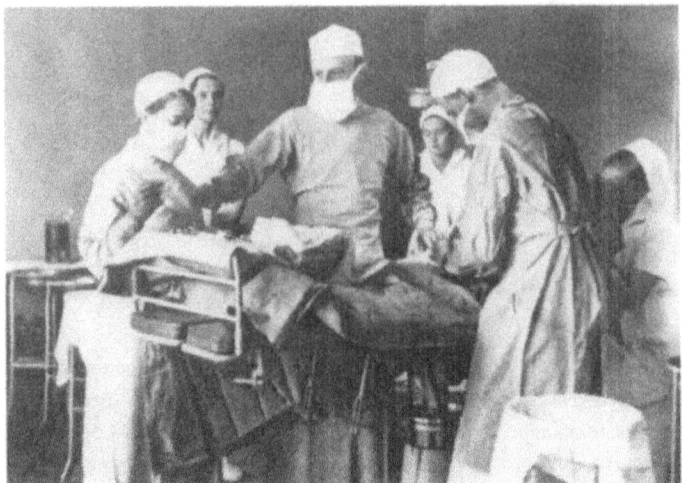

Figure 6.1 Operating scene in a Danish regional hospital, 1938. Perhaps a sterilization operation, and thus capturing the ultimate eugenic act: eliminating the capacity for reproduction. But it also illustrates what is judged as important in different historical periods. In 1938 this was a picture of the welfare state. Today it is a scary illustration of the surgical solution as a state-sanctioned violation of individual rights.

Source: Medical Museion, University of Copenhagen.

compulsory sterilization in Denmark in the period framed by the passage of the first sterilization act in 1929 to the abolishment of legal access to compulsory sterilization in 1967.[5]

The problem

There are many good reasons to be critical of compulsory sterilization as it was practiced in Scandinavia prior to the 1970s: many people were violated and prevented from living normal lives. During the media storm in 1997 and afterwards, newspapers, magazines, and televised documentaries reported heartbreaking, deeply personal stories of some of these people. Leading politicians proposed that the compulsorily sterilized should be rehabilitated. In 1999 the Swedish Sterilization Commission, founded two years earlier, recommended that some of these individuals be given economic compensation.[6] Norwegian and Danish researchers have also issued studies of compulsory sterilization. In these countries the discussion of public compensation has reached the ministerial level, although the issue has not yet been decided upon.[7]

Historians have been major contributors of the data used in the public debate and examination of our eugenic past. Thus it is difficult to avoid a discussion of the tasks and obligations of historians—to history itself, to present-day audiences, and last but not least to the survivors of compulsory sterilization. There are several options: one may decide upon a strategy of moralizing, one may impose judgment and choose sides, or one can stay neutral. One solution may initially seem surprising. Throughout the twentieth century, many historians have cast a critical eye at a key formulation by the German historian Leopold von Ranke (1795–1886). Ranke originated one of the best-known definitions of the purpose of historiography, to describe history *"wie es eigentlich gewesen"*—that is, as it really happened. The widespread, often reflexive dismissal of this view may be based on a misinterpretation. It seems to me likely that Ranke was not advocating the possibility of describing the past in an objective manner, but rather was distancing himself from another, more debatable perception of history. When this critical phrase (from Ranke's early 1824 work *Geschichten der romanishen und germanishen Völker*) is placed in context, it reads like this: "History has been assigned the task of judging the past in order to instruct the present for the benefit of posterity. The present attempt does not aim that high, it only wants to relate what really happened."[8] Here Ranke does not assert a naive belief in the possibility of objectivity, of which he is often accused, but rather represents a rejection of moralizing history, or—as we might now put it—of political correctness.

Ranke's concept provides a useful reminder in a discussion of how to write the history of eugenic sterilization. Nobody writes history without taking stands. We take stands in the choice of subject matter, research problem, theoretical starting points, and interpretation of source material. But the research process itself, in my opinion, ought aim at *understanding* the past rather than choosing the easy solution of *judging* it.

Those of us who grew up in the period after the Second World War have found it difficult to comprehend and accept that our grandparents and their contemporaries were governed by norms so fundamentally different from the ones we find are "right." Many have been tempted to characterize their actions as "wrong"—undemocratic, unethical and scientifically unjustified—as it has often happened in the public debate about compulsory sterilization. Such a condemnation of course has serious consequences, including a critical readjustment of our understanding of the welfare society. Eugenics as ideology and practice did not occur in a vacuum, but rather was an integral part of the welfare states that developed in Scandinavia in the 1930s. Leading eugenicists, politicians, and scientists considered the sterilization of the biologically "unfit" necessary so that the state could continue to finance the welfare of the unfortunate but biologically "fit" poor.[9] As the "unfit" were most often considered mentally deficient, the criteria to determine who was "unfit" was intelligence, most often measured by IQ tests. These less-valued citizens were sacrificed in order for others to benefit.

Such considerations certainly are not specific to the historiography of eugenics, but are relevant for the study of many other topics within the recent history of medicine. We may mention the use of lobotomy, scientific experiments on prisoners and the handicapped in asylums, and experimental use of atomic radiation in the clinic.[10] A thorough reflection on the relationship between the past and the present is urgent in a situation where the historian's work may be used to justify ethical-political decisions to pass judgment on the past—as has happened with the Swedish decision to compensate the victims of compulsory sterilization. The following sections attempt to illustrate several significant problems resulting from the research strategies most often chosen.

The historiography of indignation

The Swedish decision of 1999 to compensate victims of past eugenic practices was followed, as we have seen, by widespread media attention to the history of sterilization. These stories were closely associated with a particular historiography of eugenics, which might be called the historiography of indignation. One might have expected a historical approach aimed at understanding earlier sterilization practices to be the most obvious starting point for a research-based decision on whether to offer compensation. And an expectation that an enlightening historical study would be done to reveal *"wie es eigentlich gewesen"* arose from the Swedish state's public claim that it would further research the fate of those who were sterilized. Strangely enough, the assignment behind the 1997 Sterilization Commission in Sweden did not give a reconstruction of the eugenic past any priority. On the contrary, the political decision was clear even before the data had been scrutinized: those sterilized should be compensated. The historical study of the basis for compensation had the primary purpose of "providing supportive documentation for a discussion concerning the values of the Society and State actions during the years when the Sterilization Acts were in force."[11] Clearly no intention to do historical reconstruction was in play here. Instead, the political decision makers seem to have been guided by present-day political views of moral right and wrong.

In addition, media interest in the history of compulsory sterilization—what ultimately convinced Swedish politicians to take action—may be seen as an expression of a politicizing and moralizing attitude towards the past: an attempt to use the past to serve specific contemporary interests. Recent media stories on this issue might also be termed modernist or Whiggish, since the norms used in the judgmental historiographical process are those of the present, not those of earlier times. This approach is marked by an inclination to judge the past according to its compliance to what is considered the superior present-day social, legal, scientific, and ethical norms. When the Swedish Minister of Social Affairs, Margot Wallström, characterized the sterilizations as "barbaric," she was expressing judgments of just this sort.[12]

This is not a new tradition, but represents a historiographic orientation that has a lot in common with the social history of the 1960s and 1970s.[13] Within this tradition, the histories of oppressed groups such as workers, women, children, blacks, and the insane were written, and continue to be examined. The two leading historians of Swedish eugenics state their motive as follows: "What we wrote about had moral dimensions First of all we wanted to write the history of those who had not previously been heard."[14] Here the historian's task is to help restore justice—one result of which might be a payment of compensation for damages done to the survivors, as seen in the Swedish case.

Acting on unscientific grounds

Many historians of eugenics try to explain what they see as unethical practices of those responsible—particularly physicians, politicians and administrators—with the assumption that they were acting on unscientific grounds. To illustrate my point, I have chosen one of the most important books on this subject, *In the Name of Eugenics* by Daniel J. Kevles.[15]

Many positive reviews of this book have been written, most of which have called attention to its careful, meticulous research and broad scope.[16] Kevles is one of the most prominent American historians of twentieth century science. Yet I am troubled by his book. Kevles, guided by what I regard as a Whiggish approach to the subject, clearly announces his sympathies and antipathies. His attitude toward classical (so-called mainline) eugenics is negative. Throughout much of the book, he opposes a more positive account of recent reform eugenics. His primary criticisms of mainline eugenics are its social prejudices and unscientific nature. Kevles ironically characterizes eugenics as a "science"—putting the term in quotation marks to signify the false self-understanding of contemporary believers.[17] To Kevles, mainline eugenics is racist and class-conscious. Additional biases are revealed by the words he uses. Mainline eugenics is associated with religious beliefs, the opposite of true objective science. Indeed, the first four chapters in Kevles' book systematically describe mainline eugenics as a religious activity. The famous British mathematician Francis Galton is characterized as a "founder of the faith"—that is, eugenics. Galton's follower Karl Pearson, an internationally respected statistician, was a leading figure in the early days of eugenics. The chapter devoted to him is called "Saint Biometrica." Moreover, the important American geneticist and eugenicist Charles Davenport is described as a worshipper, and the chapter that deals with the growing popular support of eugenics in the Western world is entitled "The Gospel becomes Popular."[18] All in all, Kevles's account of mainline eugenics paints a picture of a movement which invites our ridicule and disbelief—in spite of the fact that a large section of the scientific community at the time supported eugenics, and in spite of the fact that eugenic ideas could not be called unscientific by the standards of the day.

To further substantiate his views, Kevles presents a number of contemporary figures who actually did consider mainline eugenics unscientific at the time. But these are rarely scientists speaking with any scientific authority at the time; rather they were at best marginal persons. An illustration may be found in a discussion of IQ testing as a means of separating the defective from the normal. Here journalist Walter Lippman's criticism of IQ testing is used to undermine the "basically flawed" views of Lewis Terman—an internationally acknowledged psychological authority.[19] In most cases, however, Kevles's own views suffice as the standard to which the eugenicists are compared. As Kevles distances himself from the Victorian and traditionalist attitude to race and class held by most scientists in the early twentieth century, Pearson and Davenport—the two leading characters in British and American eugenics in the 1920s and 1930s—are discarded as "blinded by eugenic prejudice."[20] Davenport's theories are characterized as downright "wrong."[21] While this is true for many theories of outstanding scientists working in the past, this characterization serves to discredit Davenport in general. In contrast, the political views of those Kevles considers the truly scientific reform eugenicists are rarely presented as influencing their scientific views in any inappropriate way. They held attitudes that Kevles presumably accepts: they were socialists, liberals, and feminists. A modern view of what constitutes science and acceptable political attitudes is used to explain what counts as false and true beliefs in the 1930s. The British reform eugenicists are the heroes of Kevles's account, and in spite of their own political prejudices they are presented as enlightened rationalists, and repeatedly ascribed the wish to eliminate "all the methodological sins and biases" of mainline eugenics.[22]

Kevles writes the history of the winners, and sees their victory as a result of their high political, scientific, and moral standards.[23] By doing so he fails to scrutinize his own political and ethical standards, with the result that adherents of mainline eugenics are seen as deficient scientifically, politically, and ethically. This is unfortunate, since genetics done with high scientific standards is obviously not immune to unethical applications. In the end, Kevles's presentation of the history of eugenics becomes a retrospective condemnation of the dominant theories and practices of the past. The subsequent idealization of the scientific norms of the present seems "blinded" by scientist prejudice—precisely what Ranke warned against. In addition to building upon a modern moral and political view of justice, Kevles's analysis privileges a modern understanding of science.

The past and the present

When the Swedish media storm arose, and sterilization became a controversial public issue on every front page of every newspaper in most of the world, it became clear to many informed observers the extent to which the historiography of eugenics is tied to the cognitive interests of the present. Almost all the facts reported in the media had been published by Swedish historians, in publicly accessible form, several years before the press "discovered" and presented them as something new and scandalous. Somehow the issue of eugenic sterilization was only "visible" to a small segment of the academic community until the late 1990s. In spite of the poor and oppressed conditions under which they lived and were prevented from reproducing, the "unfit" have not been considered a worthwhile historical topic. Interestingly, neither labor, feminist, social, or medical historians in Sweden had been tempted to dig into the history of sterilization and the sterilized until this time. The past is always examined through the eyes of the present, and the topics that historians regard as important reflect contemporary political and social problems.[24]

The most important factor determining contemporary interest in the history of eugenics has probably been the development of modern medical genetics. Here we have witnessed very rapid technological development that has greatly influenced social and human relations in modern society. Many informed citizens thus find the development of eugenics in the 1930s and 1940s of frightening interest—motivated perhaps by fear that history will repeat itself, indeed at a more technologically sophisticated level than the one eugenicists of the 1930s and 1940s had at their disposal. We may think of applications such as diagnostic screening of fetuses to determine birth defects and subsequent decisions for abortions. In this perspective, the present-day narration of the history of genetics is often related to a need to idealize the modern political order, representative democracy, and its commitment to the free market and individual autonomy. Within this democratic political order, proponents of the new genetics emphasize its value-free character and its independence from the eugenics of the past. Modern bioethics seeks to rescue human genetics by proposing a socially legitimate employment of its new technological powers—such as when it insists that the uses of genetic diagnosis of fetuses, children, or adults be voluntary and based on the individual's own decision.[25] Considering this association of modern genetics with individual self-determination, there is a strong interest among promoters of modern medical genetics in emphasizing the use of compulsion as the foremost and scariest aspect of past eugenics practices: the aspect that before anything else is to be denounced and labeled as a phenomenon of the past. At the same time, it seems that the issue of the scientifically correct use of genetics has disappeared from the agenda. It is tacitly assumed that the modern voluntary use of genetics—regardless the purpose—is scientifically well established. The important issue is voluntariness: the absence of compulsion.

Hence the question: how does the modern worry about the abuse of genetics influence the *historical* study of eugenics? In my study of Danish eugenics, I recognized this issue, and it has shaped my own approach to eugenics. Concern about eugenic abuses of the new genetics thus forms part of my own "pre-judices," as the hermeneuticians would phrase it.[26] An important part of this prejudice is the moral superiority and political primacy of individual self-determination and voluntariness. When we look at the conceptual opposites of compulsion and voluntariness, it is obvious that voluntariness is given priority in our society's norm system. Likewise, when we look at the relationship between individual and society, the interests of the individual are virtually always given priority.[27] This value system is dominant in most Western nation states. As a citizen of modern Denmark, these values are also constitutive for my "prejudices" and impossible to shed completely, no matter how hard I try to establish a critical view of my own society. They become apparent when I confront the archival sources and when I decide which problems I find important to study.

With the recent development of molecular genetics, an old dilemma has re-emerged—namely, the conflict between optimistic hope for positive uses of technology, and pessimistic fear of the abuses of technology.[28] Pessimistic fear of the abuses of genetics is often argued with reference to the historical past, to the history of eugenics. Pessimists usually refer to the State's abuse of genetics as it tried to identify the individuals with the poorest germ plasm and prevent these "unfit" individuals from reproducing—in the interest of the common good and of future generations.

I have been a strong critic of the development of the new genetics. Only gradually did I realize that I harbored an expectation to find the same basic conflicts in the archival sources as in the modern debate on human genetics—and that this expectation was part of my advance position or pre-judice. Since individual autonomy and voluntariness are the basic values of modern society, the risk of transferring these values to a study of the 1930s and 1940s is high. Such a transferral of norms may explain the well-known demonization of classical eugenics, leading to a condemnation of the past based on modern norms—be they scientific, moral, or political. The motives may be manifold, but an obvious one is the wish to repress the fear of abuse of genetics and relegate it to the past. To try to evaluate the history of eugenics in the 1930s and 1940s on the basis of modern ethical norms would be ahistorical, and would fail to do justice to the specific character of eugenics in that particular period.

The historiography of understanding

The autonomous individual is a crucial concept in modern politics. It is the standard according to which everything is compared, no matter if the issue is Chinese population policy ("how awful and unjust that they are not allowed to have all the children they want"), Swedish racial hygiene ("how awful and unjust to sterilize the mentally handicapped"), or West Indian slavery ("how awful and unjust that plantation owners used slave labor"). In spite of the fact that such phenomena emerged and developed in completely different historical contexts, characterized by political, scientific, and ethical norms differing strongly from our own, they all fail to live up to this standard and are today condemned as awful, unjust, or even criminal. The key question, however, is whether the choices we describe as voluntary can stand closer scrutiny. Voluntariness may be a more complex concept. From studies of the uses of modern reproductive and genetic technology, it often appears that choices to use such technology are, though not coerced, at least strongly shaped both by the mere availability of the technology and by prevailing social expectations.

This nuanced way of looking at the present may also be applied to the past. Rather than labeling the specific uses of the sterilization technologies available in the 1930s and 1940s as "awful and unjust," it seems urgent to study the history of eugenics and compulsory sterilization in a more congenial way. Inspired by the hermeneutic tradition, I propose the method of attempting a dialogue with the past in order to understand what it was that made the decisions we now find repulsive and disturbing politically and socially meaningful—and ethically acceptable—to decision-makers and to a large part of the public in the past. As an alternative to ridicule and indignation, I propose to take the actors of the past seriously, even if pro-eugenic politicians, scientists, and administrators may appear as "the bad guys" from a present-day point of view. In my study of the sterilization practice of the 1930s and 1940s, I have been interested to know what motivated physicians and administrators in this era, to understand what made them act as they did, to find out what at that time could explain and justify the use of compulsory sterilization on the weakest members of society, including the mentally retarded, prostitutes, tramps, and the very poor.

Facing major differences between our society and theirs in matters of science, politics, ethics, and reproductive concerns, I felt divided between competing world-views, trying to live in two worlds at once. The relationship between past and present may be seen as a meeting between old and new experiences, as a dialogue where new insight and understanding may be created. The full implications of this metaphor may not seem obvious, as actors of the more distant past are dead and don't make very eloquent partners in a conversation. But within the hermeneutical tradition, this problem is approached by using the concept of *"Wirkungsgeschichte,"* the effective history. This implies that part of history is still active and alive, and this active element remains a guiding force that influences our actions and consciousness. This process may also be described as a merging of two horizons, that of the historian and that of the historical period or text which she is trying to interpret. The spontaneous relationship between the present and the past may be compared to a feeling of alienation, where the past appears as strange and incomprehensible—structured by different principles and norms. The chance that this experience of alienation could be replaced by a feeling of familiarity is made possible by the fact that horizons may move as we change our point of observation. If we are lucky, a meeting takes place: the two horizons may intersect or even merge. This is the precondition for a meeting, a dialogue, between us and them, between now and then.

The first step in such a dialogue must naturally be taken by the historian, who directs her questions to the past—specifically to the sources, the historical texts. In meeting with the text, she may experience that the text in turn begins to question her, by challenging her norms and prejudices. The text may thus come alive. Since the end of the Second World War, the principles that guided sterilization practice have seemed to us beyond the pale, bordering on evil. Contact between or merging the horizons of then and now seemed impossible. The need to *create* a rupture with the past—politically, scientifically, and ethically—has been overwhelming for obvious political reasons associated with the proximity of German *Rassenhygiene* (racial hygiene). As the paradigm for relating to the past since the Nazi era has been one of rupture, interpretations stressing continuities have been avoided. An example is the widespread denial from the end of the Second World War to the mid-1980s that Danish eugenics shared any similarities with or contact with German *Rassenhygiene*.[29] As the postwar generation has grown to adulthood, this obstacle has at least to some extent broken down. But now, as the new molecular genetics has raced ahead, positive attitudes about the uses of human genetics have gradually moderated, raising for some new questions about whether individual self-determination and liberalism ought remain the central foundations for ethically acceptable and socially responsible control of genetic technology.

The next step in the interpretative process is to recognize the other view. When a historical text lends itself to interpretation, it asks a question of the interpreter. We may phrase it this way: we only understand a historical text if we understand the question that this text tried to answer.[30] The same is true of the actions of human beings in the historical past.

Let me provide an example. It turned out that the most important source material for understanding compulsory sterilization—that is, the patient files of the sterilized—systematically failed to give voice to the view that the reproductive freedom of the individual should be protected. Instead, the need to protect *society* from the unwanted and hereditarily tainted offspring of the mentally retarded and other unfit individuals was repeatedly articulated. Along with other evidence, this priority of interests may be interpreted as providing an answer to problems that Danish society in the 1930s and 1940s believed it faced. One of the dominant problems was the fear of degeneration of the population, including the threat of the mentally retarded, whose numbers were claimed to be growing exponentially.[31] In this pronounced class-conscious society, at a time when social and biological qualities were not clearly distinguished, mental deficiency was considered a serious threat to the public health. The risk that the biological quality of the population was deteriorating worried large segments of the Danish population. Parties from the whole political spectrum were united in this fear, which seemed firmly supported by scientific evidence. Established and respected scientists not only substantiated these fears but to a great extent took the initiative to articulate and document them. Eugenic sterilization legislation from the 1930s remains a tangible testimonial to the conviction of the politicians of the times. Specifically, the rights and duties of the individual were perceived and regulated quite differently than they are today: the notion of a right to reproduce or to control one's reproductive functions regardless of, or in opposition to, the public interest was not on the political agenda. Numerous official publications on reproductive issues from the eugenic period document this. For instance, in rejecting a more liberal abortion law, one Danish governmental commission from 1936 explicitly stated the primacy of the rights of the state vis-à-vis the rights of the individual: "Women's right to self determination goes precisely as far as the interests of society will allow it."[32] Furthermore, the individual had a duty to use her good sense in the public interest. This naturally included reproductive matters, and those who could not be relied upon to demonstrate such social responsibility had to be forced to do so by law. Hence I now find it understandable that the sterilization laws focused on the mentally retarded, a group that (besides its alleged sexual promiscuity) also demonstrated a poor understanding of the vital interests of society.

Seeking the question behind the answer: limits to reconstruction

The research strategy I have sketched here, to seek the question behind the answer, is not at all a straightforward one. To some extent it is based on the belief that the human mind is logical, and that the question-and-answer constitute a coherent whole. We all know that the actions of humans are not always bound by logic: irrationality and chance often play a role, and what is considered logical and coherent may differ from one period to another. When we try to interpret the actions of people of another historical period, we may have to apply other methods, including creative and subjective reasoning. Let me give an example. In my study of the reasons given to sterilize people who were considered mentally retarded, it seemed that eugenic and social reasons were interchangeable without any obvious explanation. The action taken often turned out to be the same, whether it could be documented that the individual in question had a hereditary load or whether there was no documentation for this. If a retarded person came from a poor family with alcoholism and prostitution, the decision makers—doctors and judges—considered this equivalent to a family history of mental retardation. At first glance, such a practice seemed contradictory— social and hereditary factors are exclusive in our world view—but as I dug deeper into the meaning of the concept of eugenics, it turned out that social and eugenic indications for sterilization at that time were considered two sides of the same coin. One was a sign of the other; they were seamlessly linked in a meaningful whole. Thus the much-investigated and criticized class bias of the eugenicists was not a separate issue that could be criticized as a particularly problematic aspect of mainline eugenics—as a sign of unscientific practice—but rather an integral element in the eugenic ideology as such. The abundance of social problems in such a family was seen as the phenotype of mental retardation.

This example may illustrate the difficulties in approaching the issue of compensating the victims of sterilization on the basis of past norms. In Sweden, Norway, and Denmark, many citizens expected that a historical study could show whether the sterilized were entitled to compensation. As I have shown, this is not as straightforward as it might have seemed to begin with. First of all, what should be the basis for compensation? The Swedish solution has been to provide restitution to all who had been sterilized against their will. But as we look at the patient files, it turns out to be impossible to determine which sterilizations were compulsory and which were voluntary.[33] In the Danish records, almost all affected individuals had signed consent forms to declare their voluntary participation. But such compliance does not demonstrate that all willingly went to the operating table. Rather, we know that the practice of compulsion was more complex. In a large number of cases, sterilization was the condition to be released from the asylum. In the face of life-long institutionalization, most mentally retarded preferred sterilization—and eventually signed the consent form.

If it could be shown that a sterilization had been carried out illegally, would compensation not be justified? But then again, which situations should be considered illegal? The text of the Danish law permitted compulsory sterilization of mentally retarded persons, so only those who were not mentally retarded could be considered illegally sterilized. Here our focus shifts to a new topic of discussion: who should be considered mentally retarded? This requires an understanding of the concept of mental retardation as it was then defined. The question was highly controversial in the 1930s and 1940s, and no clear consensus existed. Some argued that an IQ measurement could determine this question; others argued that the IQ was only one among several features to be taken into consideration before a proper diagnosis could be given. Thus the decision to sterilize was made in spite of this uncertainty: it was made pragmatically, as any medical decision was, and is. And in addition to the difficulties of determining what counted as legal, scientific, and ethical correctness in the past, we may add the difficulties in understanding what the eugenicists actually meant by mental retardation. As I have tried to illustrate above, hereditary features as well as social performance played a role for the diagnosis of those considered eugenically dangerous, and for the subsequent decision made by the medical authorities in the past. It is—to put it mildly—very, very difficult to balance all these factors and reach a clear conclusion on the issue of legality. It is exceedingly difficult to determine what should count as scientifically correct. From a historical perspective, it is a futile exercise to attempt to separate out the illegal or unscientific or unethical sterilizations from those considered legal, scientifically, and ethically correct.[34]

Still, the state may decide to compensate in cases of a violation of the law as it was understood and interpreted at the time, and solve the problems I have sketched by giving the sterilized the benefit of the doubt. This seems more or less what the Swedish Sterilization Commission has decided to do.[35] But in that case, compensation would not actually be given on the basis of a historical reconstruction but on the basis of *modern* ethical norms—that is, for moral and political reasons. It seems to me highly problematic however to issue compensation simply because the practices of the past are deemed ethically wrong in *our* time. Such action, while acknowledging the plight of the sterilized, simultaneously privileges the norms of the present. Good historical research may increase our understanding of practices that may seem morally and politically wrong—as many have suggested was the case with eugenic sterilization of the mentally retarded. But it seems very important to recognize the continuity between then and now, and to integrate this understanding of the practices of the past rather than judging it. As to the situation of the sterilized, it seems to me a more sensible solution to demonstrate to them that contemporary society understands their suffering and respects their grief—not by condemning the past, but by recognizing their present.

Conclusion

As I have been completing my study of the history of eugenic sterilization in Denmark, I have engaged in many silent conversations with the leading Danish eugenicists of the 1930s and 1940s. Sometimes I thought I recognized the familiar conflicts of modern genetics. But as I compared our contemporary conflicts with the original source materials, and reread the old eugenicists' arguments about the need to sterilize, I felt that my original attempts to interpret their actions had broken down. New aspects that were not originally visible appeared. Over and over again I had to revise original interpretations and my preconceived ideas of the leading men in Danish eugenics. Not that the "bad guys" turned out to be "heroes"—far from it. Yet the dialogical method made it possible for me to look at them not only as people from another culture: a culture which was different from mine, but which at the same time belonged in the same historical continuum. The historical texts—particularly the eugenicists' arguments to sterilize presented in the patient files—that had originally seemed incomprehensibly cruel and heartless to me, now appeared to hold hidden messages which only became clear to me as my own perspective—*my own horizon*—moved. This movement only became possible after a prolonged coexistence with the source material, after months and years of reading and studying and discussing their views. At a certain time their actions made sense and became understandable. I felt I understood their actions, which of course is not the same as condoning them.

It is an old truth that history has to be rewritten once in every generation, that new experiences reshape the relationship between the text (the sources) and its interpreter, just as these new experiences reshape the relationship between the past and the present. The experiences that shaped the postwar generation and its painful relationship with eugenics were the Second World War, the Nazi threat to democracy, and that state's persecution of the Jews, Gypsies, and other undesirables. My own generation also has its reasons to demonize eugenics, although these reasons are different.

As I have mentioned earlier, the present interest in the history of eugenics is highly determined by the success of the new genetics—the academic successor of eugenics—and our spontaneous disgust with the use of compulsory measures in eugenic practice is a product of the modern idealization of individual self-determination. In medicine, bioethics has become the institution to promote the norm of individual self-determination as the ethically right and legitimate way of handling the risks and chances of molecular genetics. This view may be considered the modern answer to the question that is raised by the "hereditary taint" of the new genetics and its close ties to eugenics. It has long been considered the best way to avoid the mistakes and crimes that were committed by eugenicists in the 1930s and 1940s and thus becomes a part of the *Wirkungsgeschichte* of eugenics.

In spite of the vocal support for individual self-determination in the uses of the new reproductive and genetic technologies, self-determination still has its limits. This is illustrated by the vehement criticism of these new technologies. It turned out that even those who spontaneously opposed compulsory sterilization when they looked back on the 1930s and 1940s reacted with a similarly spontaneous opposition to, for instance, granting gays and lesbians the right to artificial reproduction. One wonders: why should the principle of unlimited freedom to reproduce be granted to the socially deviant of the 1930s and 1940s but not our own? Such deviations from the seemingly universal support for individual self-determination in reproductive affairs represent the loci where the horizons of the past and the present merge. A study of the sterilization practice of the past may provide a much-needed commentary on our own difficult choices, and the divide between the idealization and reality of self-determination. Also, for those of us alive today, it may increase the understanding and respect for past norms.

Seen from a present-day point of view, a pure reconstruction of the past seems an impossible venture. One problem often cited is a lack of source material. While trying to pry open the past to give a meaningful description of an event, we often delude ourselves that if only this or that document were available the problem would be solved: then we would be able to write the true story. But the gap between the lived past and its narration makes the attempts at reconstruction a subjective venture no matter how rich the source material. As the American historian Simon Schama has written, "We are doomed to be forever hailing someone who has just gone around the corner and out of earshot."[36]

But even if we knew everything there was to know about the past, our own subjectivity, our own present, would still be of vital importance for the final result. As all historians know: any understanding of the past is guided by the perspective of a subjective historian living in her present. This modern context will highlight some aspects of the past and leave others in the dark. We also know more about eugenics than they did, and the historian is often much better informed about general political and cultural issues of the times than any individual could possibly have been then. We also know their posterity and—in contrast to them—we know the consequences of their actions. These in turn have influenced us, and contribute to the questions we ask the past.

The history we write thus will never be the same as the reality they experienced and created. Historiography is a meeting between two worlds rather than an attempt to reconstruct one world in another. The meaning of history and of the events of the past is constantly changing during this process of interpretation.

Notes

1   Maciej Zaremba published two articles on this issue in *Dagens Nyheter*, 20 and 21 August 1997. Later he published a book on the topic: see *De rena och de andra* [The Pure and the Others] (Stockholm: DN förlaget, 1999).
2   See for instance "Swedish Scandal," *New York Times*, August 30, 1997, 22.
3   See Gunnar Broberg and Mattias Tydèn, *Oönskade i Folkhemmet: Rashygien och Sterilisering i Sverige* (Stockholm: Gidlunds Bokförlag, 1991); Lene Koch, *Racehygiejne i Danmark 1920–1956* (Copenhagen: Gyldendal, 1996); Gunnar Broberg and Nils Roll-Hansen, eds, *Eugenics and the Welfare State: Sterilization Policy in Denmark, Sweden, Norway, and Finland* (East Lansing, MI: Michigan State University Press, 1996): Maija Runcis, *Steriliseringer i Folkhemmet* (Stockholm: Ordfront, 1998); and Diane Paul, *The Politics of Heredity: Essays on Eugenics, Biomedicine, and the Nature-Nurture Debate* (Albany, NY: State University of New York Press, 1998).
4   On this large theme, see for instance Robert Proctor, *Value-free Science?: Purity and Power in Modern Knowledge* (Cambridge: Harvard University Press, 1991) and Mark Walker, ed., *Science and Ideology: A Comparative History* (London and New York: Routledge, 2003).
5   Lene Koch, *Tvangssterilisation i Danmark 1929–67 [Compulsory sterilization in Denmark]* (Copenhagen: Gyldendal, 2000).
6   *Steriliseringsfrågan i Sverige 1935–75 [The question of sterilization in Sweden]*, Statens Offentliga Utredningar (Stockholm: Fakta Info Direkt, 1999), 2.
7   The reply of the Minister of Social Affairs to question nr S 3222 og S 3223 from Flemming Oppfeldt (V) to the Legal Secretariat of the Danish Parliament (Folketingets Lovsekretariat). Ministry of Social Affairs, Office of Handicapped People [handicapkon-toret] 7. kt. Jr. Nr. 503–1131, 9.9 1997.
8   "Man hat der Historie das Amt, die Vergangenheit zu richten, die Mitwelt zum Nutzen Zukünftiger Jahre zu belehren, beigemessen: so hoher Aemter unterwindet sich gegenwärtiger Versuch nicht: er will blos zeigen, wie es eigentlich gewesen" (in-text translation is mine). Leopold von Ranke, *Geschichten der Romanischen und Germanischen Völker von 1494 bis 1514* (Leipzig:Verlag von Dunker und Humblot, 1885 [1824]), vii. Hans-Georg Gadamer discusses this concept in *Truth and Method* (London: Sheed and Ward, 1989), 262; for a related discussion see Peter Novick, *That Noble Dream: The "Objectivity Question" and the American Historical Profession* (Cambridge: Cambridge University Press, 1998).
9   See for instance the major work of the pioneer of Danish eugenics, K. K. Steincke's *Fremtidens Forsørgelsesvæsen [The Social Security System of the Future]* (Copenhagen: Indenrigsministeriet, 1920). Steincke was Minister of Justice and later Minister of Social Affairs in the first Social Democratic government in Denmark and the driving force behind Denmark's sterilization acts.
10 Literature on these topics is growing; see for instance Gerald N. Grob, *Mental Illness and American Society, 1875–1940* (Princeton, NJ: Princeton University Press, 1983); Grob, *The Mad Among Us: A History of the Care of America's Mentally Ill* (New York: The Free Press, 1994); and Susan E. Lederer, *Subjected to Science: Human Experimentation in America Before the Second World War* (Baltimore, MD: Johns Hopkins University Press, 1997); on the deliberate release of radioactivity into the environment during the Cold War, see Ruth Faden, ed., *Final Report of the Advisory Committee on Human Radiation Experiments* (New York: Oxford University Press, 1996). Further reports appear in http://tis.eh.doe.gov/ohre/ roadmap/achre/index.html (accessed on October 4, 2004).
11 *Steriliseringsfrågan i Sverige*, 30.
12 Wallström's remark is quoted in Paul Gallagher, "The Man Who Told the Secret," *Columbia Journalism Review* 36 (January–February 1998): 155–172.
13 For an introduction to this literature, see Theodore Rabb and Robert I. Roberg, eds, *The New History: The 1980s and Beyond: Studies in Interdisciplinary History* (Princeton, NJ: Princeton University Press, 1982).
14 Gunnar Broberg and Mattias Tydén, "När svensk historia blev en världsnyhet [When Swedish History became World News]," *Tvärsnitt* 3 (1998): 2–5. See also Gunnar Broberg

and Mattias Tydén, "Introduction" [to special issue on Scandinavian eugenics], *Scandinavian Journal of History* 24 (1999): 141–143.

15 Daniel J. Kevles, *In the Name of Eugenics: Genetics and the Uses of Human Heredity* (Berkeley, CA: University of California Press, 1986).

16 Among them, Leon J. Kamin in the *New York Times*, 9 June 1985, sec. VII, p. 9; Donald K. Pickens, "In the Name of Eugenics: Genetics and the uses of Human Heredity," *American Historical Review* 91, no. 3 (June 1986): 632–633; and Horace Freeland Judson, "Gene Genie," *New Republic* (August 5, 1985): 28–34.

17 Kevles, *Eugenics*, 41.

18 Ibid., vii.

19 Ibid., 129.

20 Ibid., 49.

21 Ibid., 53 and passim.

22 Ibid., 199.

23 Ibid., 211.

24 This pattern has been noticed in a different context by Katherine McComas and James Shanahan; see their article "Telling Stories about Global Climate Change: Measuring the Impact of Narratives on Issue Cycles," *Communication Research* 26 (February 1999): 30–57.

25 One of the most outspoken advocates of this view is H. Tristam Engelhart, Jr; see for instance his *The Foundation of Bioethics* (New York: Oxford University Press, 1996). For a balanced criticism see J. F. Childress, "The Place of Autonomy in Bioethics," *Hastings Center Report* 20 (January–February 1990): 12–17.

26 See Gadamer, *Truth and Method* for a modern version of hermeneutics. Gadamer's concept is "Vorverständnis."

27 This is true for the whole corpus of bioethical codes governing the health care systems of the Western world. A fairly recent example is the Convention for the protection of human rights and dignity of the human being with regard to the application of biology and medicine (Council of Europe, Convention 164, April 4, 1997, http://conventions.coe.int/Treaty/EN/Treaties/ html/164.htm, accessed on October 4, 2004) ratified in a number of European countries in the late 1990s. In this document the interests of society are explicitly relegated to a secondary position if they conflict with health needs of individuals.

28 On this issue see David S. Landes, *The Unbound Prometheus: Technological Change and Industrial Development in Western Europe from 1750 to the Present* (Cambridge: Cambridge University Press, 1969).

29 An exception is the postwar work of the Danish professor of human genetics Tage Kemp, who corresponded with Nazi geneticists from the early 1930s until his death in 1964. See for instance T. Kemp, *Genetics and Disease* (Copenhagen: Munksgaard, 1951). For a broader overview see Nils Roll-Hansen, "Eugenics in Scandinavia After 1945," *Scandinavian Journal of History* 24 (1999): 199–213.

30 See Gadamer, *Truth and Method*, 370–373; and R. G. Collingwood, *An Autobiography* (Oxford: Oxford University Press, 1939).

31 Steincke, *Fremtidens Forsørgelsesvæsen* is perhaps the best illustration of the serious attention that the Danish political elite gave this threat.

32 Ibid.

33 The following argument is based on Koch, *Tvangssterilisation i Danmark 1929–67*.

34 This is, of course, one instance of a far larger issue: to assess what was understood about science-based hazards in the recent past by contemporary authorities. For an introduction to this emerging challenge for historians see John A. Neuenschwander, "Historians as Expert Witnesses: The View from the Bench," *Organization of American Historians Newsletter* (August 2002): 1, 6; and Patricia Cohen, "History for Hire in Industry Lawsuits," *New York Times*, June 14, 2003, sec. B, p. 7.

35 One member of the Swedish commission dissented on these grounds; see *Steriliseringsfrågan i Sverige 1935–75*, Svenska Offentliga Utredningar 1999, no. 2, 181.

36 Simon Schama, *Dead Certainties: Unwarranted Speculations* (New York: Vintage, 992), 320.

Part III

Witness to history

Issues in biography and ethics

7 What is the use of writing lives of recent scientists?

Thomas Söderqvist

Introduction

Few historians of science realize that scientific biography is a very old metascientific genre—one that goes much further back than the tradition of historiography.[1] While histories of science only came forth in the mid- and late eighteenth century,[2] the first *vitae* of astronomers and natural philosophers appeared already in the seventeenth century immediately after the rise of modern science.[3] Since then, about four to five thousand biographies of scientists have been published in the major European languages, not to mention obituaries, short biographical articles, and dictionary entries on individual scientists.[4] Cumulatively, scientific biography has been, and may still be, the best-selling and most widely read of all genres of writing about science's past. And, as long as the book review institution has existed, scientific lives have repeatedly matched the highest possible standards for scholarly writing.

This impressive quantitative and qualitative presence notwithstanding, the genre has not received much theoretical and methodological comment.[5] Whereas historiography (including the historiography of recent science) is a perpetual topic of reflection for historians of science, biography has gained limited attention. Biographies of recent scientists have received even less. In this chapter I will focus on some of the issues associated with writing lives of recent scientists (i.e. of scientists active within the life-span of the biographer). I will draw on my experiences of writing the biography of the Danish-British-Dutch immunologist Niels Jerne,[6] and the discussion will therefore at times be autobiographical. But the questions raised hopefully have a broader significance: what can be learned from writing biographies of scientists of the recent past? What can the genre be used for?

I: The road to the life of Niels Jerne

Let me first introduce my personal background for providing the typology of biographical sub genres in the following section of this paper. Niels Jerne is probably best-known today for his theories of antibody formation and for his theories of the self-regulation of the immune system—at least, this is what the Nobel Assembly at the Karolinska Institute in Stockholm cited him for back in 1984.[7] He was born in London in 1911 as the fourth child of an emigrant Danish bacon factory manager, but the family soon moved to the Netherlands where he attended school. After failing in his chemistry studies in Leiden he decided to go to Copenhagen for medical training. It took him thirteen years to get his MD, however, because he got married, had children, and spent a lot of time living a bohemian life among avant-garde artists during the Second World War. He was, as he later called himself, a *Spätzünder* (late-bloomer).

Jerne came into science by default. To support his growing family he found employment in the Danish State Serum Institute—first as a secretary, then to help with vaccine standardization work—and he continued there after completing his medical degree to do research on the antigen-antibody reaction for his PhD. It was this work that later led him, in 1954, to his major intellectual breakthrough: the selection theory of antibody formation. According to Jerne's theory, the immune system is not a *tabula rasa*; instead, the body has an inborn capacity to react to any foreign molecule.[8] Today we would say that this capacity is in the germ-line. But in the mid-1950s it was quite revolutionary among immunologists to suggest this.

A couple of years later, Sir Macfarlane Burnet modified Jerne's theory to the clonal selection theory.[9] Within a decade the selection principle gained the status of a kind of "central dogma" in immunology. Consequently Jerne rapidly earned fame as the leading theoretician of the new discipline. In 1960 he was asked by the World Health Organization (WHO) to establish an international office for immunology, and two years later he was hired as chairman of the microbiology department at University of Pittsburgh. There he created a new method for the detection of antibody producing cells (the hemolytic plaque assay). From the mid-1960s onward this became one of the standard methods in immunological laboratories around the world. Four years later he was asked to become director of the Paul-Ehrlich-Institut in Frankfurt. Finally, in 1968, he was recruited as the first director of the Basel Institute of Immunology. The Basel Institute soon became the world's leading immunological research institute, a mecca for immunologists. Here Jerne came up with yet another theory of the functions of the immune system: the idiotypic network theory, which held that all antibodies and lymphocyte receptors are best conceived as mutually independent parts of a steady-state system.[10] In Jerne's view the system is informatically closed; it regulates itself. Few today believe in this immune network, but it kept a lot of people very busy in their labs throughout most of the 1980s. And then, in September 1984, Jerne got the call from Stockholm.[11]

Serendipity: choosing a biographical subject

I will not say more about Niels Jerne here: I have always been more concerned with how to use any one scientist's life and work for biographical purposes than in him, or his work, or even in the history of immunology as such. I actually had been on the lookout for an interesting person (indeed any subject of some intellectual interest!) to write a biography about when Jerne showed up by accident. The circumstances deserve a short anecdotal detour, because they illustrate how serendipitous the road to a biographical subject may be.

In December 1984, a few days after the Stockholm ceremony, one of my history of science students suggested that I should attend Jerne's duplicate Nobel lecture in

Copenhagen. I went but gave it little attention. At the time I was too occupied with finishing my PhD. Two summers later, this student visited Jerne at his home in Castillon-du-Gard in Languedoc in southern France, where she interviewed him for her master's thesis on the origin of natural selection theory. When she went back for more extended interviews in the autumn, I took the opportunity to join her.

This trip to satisfy the curiosity of a thesis supervisor became a decisive turn in my academic career. Ignorant of the fact that France had just imposed new visa rules for non-EU citizens like myself (I carry a Swedish passport), I was stopped by the immigration authorities at Lyon airport and put in a prison cell (while my student, who carried a Danish EU passport, walked right through). I was to be sent back with the first morning flight. I spent a sleepless night under a naked lightbulb with a thin blanket and one of Jerne's scientific papers—about the analogy between the immune response and the Socratic theory of learning—as my only comforts.[12] Back in Copenhagen the next day I was met by three armed police officers who had just received a telegram about a "terroriste présomptif." This was before 9/11, and so, quickly realizing the absurdity of situation, they let me off to acquire the necessary documents.

My traveler's tale dominated the conversation around the lunch table in Castillon-du-Gard the following days. The fact that I had spent a long and chilly night reading and rereading his old musings on Socratic immunology apparently added to Jerne's accomodating attitude, because after lunch he invited us up to the attic. He had "saved quite a bit of paper," he declared. There, with the mistral howling through the window-frames, the dream of every incipient biographer came true: a 900 square feet room packed with tens of thousands of letters, notes, and manuscripts stored in hundreds of paper bags from the local supermarket. This was an exceptionally rich collection of documents from a late twentieth-century scientist. Since his adolescence, Jerne had collected almost everything that had passed through his hands. Drafts of scientific papers, lecture manuscripts, notes from scientific meetings, abundant laboratory records—all this in addition to many thousands of private letters, diary entries, scraps of paper (often undated) with passing thoughts, library book-loan receipts, movie ticket stubs, chess records, domestic bills, medical prescriptions, ledgers, and more. He had kept drafts or copies of almost all his outgoing correspondence, had reclaimed his own letters to parents and wives, and even saved little wrinkled scraps of paper (1 liter of milk/piece of bread/don't forget the cigarettes!) that his wife had written for him. Almost nothing of importance had been destroyed. He had lived a "biographical life," the life lived in apparent expectation of one's biographer.[13]

At that time, few collections for modern life scientists had been deposited in archival institutions. Most existing collections were either bowdlerized (more-or-less) by the donators or their families, or much poorer in content; few reflected a neurotic bent to save everything.[14] That Jerne's collection had not been read by any other person than himself (not to mention other biographers and historians), and that a substantial proportion of the documents remained *in situ* (some letters were still in their original envelopes, and many meeting and lecture notes remained attached to the original notepads) only made it more attractive. Ultimately I was also given the privilege to act as his literary executor and curator of the collection: to decide what to keep and what to discard. Few historians of recent science ever get such an opportunity. I was thrilled.

Rich, unspoiled document collections are an attraction in their own right, like virgin archaeological sites, and hardly need any further scholarly excuse to be examined. But there turned out to be other good reasons as well for choosing Niels Jerne as a biographical subject. First, within a few years the history of recent immunology became a highly profiled field within the historiography of recent science and medicine.[15] As the project unfolded, I could therefore draw on a rapidly expanding corpus of cutting-edge work in order to better contextualize my approach ("idiosyncratic," as three of my social history oriented peers called it; they apparently meant "biographical").[16] Jerne's stature as a top scientist in a prestigious scientific field made everything much easier. Indeed, his Nobel Prize no doubt helped pave the way for a generous grant from the Swedish Humanities Research Council to cover three years of travels and archival work. In addition, working on such a central and charismatic personality opened up for me many otherwise closed doors in the biomedical community. Piggybacking on his scientific status undoubtedly helped me later when I approached prospective publishers. Finally, Jerne's fairly small scientific production—only some eighty-five scientific papers altogether, many fairly non-technical—made it easy for someone like me, who knew very little immunology, to familiarize myself with his *oeuvre* without having to spend years deciphering a great many complicated experimental papers (the few he had written nevertheless consumed months of my time). As the project went on, there were thus many factors that corroborated my serendipitous choice of biographical subject.

Biographical use of the autobiographical voice

Jerne was not a passive subject. He was a man in charge of his own narrative. A year before we met, Sloan Foundation officials had asked him to write a volume in their new series of scientific autobiographies. He had considered the offer for some time but soon gave it up: "It's a large job," he commented. Nevertheless the idea of telling his life-story to others intrigued him; and me. He was only 75 when we started our collaboration, his mind was still sharp, and he was a charming and natural born conversationalist (he died eight years later, in 1994, at the age of 83).

Jerne's personal qualities provided me with yet another reason for engaging with him: I wanted to explore an idea about the construction of biographical narratives I had gotten after reading anthropologist Dale F. Eickelman's *Knowledge and Power in Morocco* (1985).[17] During his interviews with a local Islamic notable, Eickelman had realized that both of them were engaged in a mutual effort to construct stories—albeit for two widely different kinds of audiences. The Western anthropologist returned to Princeton with a book manuscript that constructed the picture of an Islamic Other for Western academic readers, while the "other" remained in his local town with an enriched set of oral stories about his life and culture (and probably some gossip about a curious man from a Western university).

My idea was to enter into in a similarly mutual biographical process, with the main difference that Jerne's story was supposed to appear in print as well (Eickelman very much regretted that the Morrocan notable did not publish his account). So we set out to discuss a variety of possible end products. My naïve enthusiasm over the prospect of creating a new kind of scientific biography was reflected in several premature sketches I made for the book cover. The most radical alternative was a book with parallel biographical (left) and autobiographical (right) pages, but this idea was discarded when we realized that several chapters would largely lack right pages, because he had very little to say about certain periods in his life. (It would have been fun to do, but probably impossible to convince a publisher to take it.) Less radical versions involved a mostly biographical text, with various amounts of interspersed autobiographical passages and/or transcripts from the interviews.[18]

During the next four years we became involved in a subtle continuous negotiation over the control of the story. It did not help that I had received substantial funding for the project; he wanted to see for himself that I was a *bona fide* biographer. Interestingly, his question of trust was not about how I would treat his troubled personal life. He was more worried about my lack of immunological training: would I understand his scientific work? After a first round of interviews in 1987, and especially after a week of intensive late-night sessions in Basel in early 1988, Jerne demanded to see a synopsis before continuing. I refused at first, claiming that one never asks a portrait artist for a sketch. But he stood his ground. My then wife asked me to swallow my pride and comply. In the following summer I produced a forty-page draft based on his scientific papers, the preliminary interviews, and a few published biographical data. A couple of days later he called me up, congratulated me on the accuracy of the sketch, and invited me down to Castillon-du-Gard to continue the project. For the next three years I visited him more than a dozen times, sometimes for a few days, sometimes for weeks in a stretch. I stayed in nearby hotels and came to his house almost daily where his wife treated me with coffee and pastries.

Even after his authorization, however, my access to the collection—and thus the extent to which I could construct my biographical account—remained a topic of subtle negotiation. In the beginning, Jerne brought selected documents down from the attic room for me to read. When he tired of doing so I was allowed to sit in the room and read alone. Later still I could bring papers to my hotel overnight. After a few months I was allowed to bring whole suitcases of documents back to Copenhagen for reading and copying. His wish to see what I wrote about him weakened too. In the beginning, I sent him chapter drafts to comment on (some of these remarks were later included in the book), but gradually he gave up the idea of having any kind of influence over the outcome.

Throughout these three years we were also engaged in a seemingly endless array of conversations, sometimes just for the fun of it, but often deliberately to elicit new autobiographical memories. Our talks were often prompted by some document I brought down from the attic and confronted him with in his study or at lunch. In all, I spent about 160 hours interviewing him, with perhaps as many hours of easy-going conversations over a bottle of wine.[19] The purpose of these interviews was not to produce information to supplement the abundant archival documents.

Indeed, Jerne's stories could rarely match the documents as historical sources. Even if he himself felt that he was telling the truth, his memory often deceived him, at least as far as I could tell from comparing his stories with the documents (and he was perfectly conscious of this possibility).[20] Furthermore, he may have found it in his interest to push me in one direction or another; it was often hard to decide whether his thoughts and attitudes represented factual experiences, exaggerations, or affectations; sometimes he amused himself with this uncertainty.[21] Finally, as the months and years passed by, he began to repeat the same stories over and over again—not necessarily because he grew older (he remained mentally sharp until a few months before his death) but probably because he only had a limited repertoire of stories. As a result, I found it more and more difficult to open up new autobiographical vistas.

I concluded that I could only use our discussions (some taped, others scribbled down in session) as source material in the case of events that could otherwise not be substantiated in the archival record. Instead I decided to use the interviews primarily to give Jerne his own narrative presence in my book. In other words, I wanted the reader to hear not only Jerne's voice of the past, as it issued from letters, notes, diaries, and his earlier autobiographical commentaries, but also Jerne as a present-day interlocutor. I wanted him to be heard as someone who interpreted and corrected, cleared up and rearranged his life-story in a way that he himself wished to see it—and for us to read and understand it. An autobiographical narrative (printed in *italics*) emerged that now complemented and, at times, contradicted my own.

Biography as a means for creating recent science archives

Another topic of great concern during the research phase was how Jerne's unique collection of documents could be utilized by future historians of science. Jerne occasionally threatened to follow the example of the young Freud—to burn it all. But when he was in a more temperate mood, we discussed several possible receiving institutions, including the Nobel archives in Stockholm and (because he was a British subject by birth) several archival institutions in the United Kingdom. Finally, however, he decided on the Royal Library in Copenhagen—partly because he had a Danish passport too, but primarily because he said he wanted his papers to rest together with those of his intellectual hero, the Danish existential philosopher Søren Kierkegaard. He also asked me to be his literary executor. He only kept some documents of strong emotional value, such as a handful of his father's documents and his first wife's farewell letter before she committed suicide. After the donation was made in the late summer of 1992, the library's manuscript department generously accepted me as an *ad hoc* curator to make the necessary decisions about what to keep and what to discard.

The establishment of the Jerne collection in the Royal Library thus was the result of a complex interaction between biographer and biographee. The long interview sessions and the gradual formation of a trusting relationship between us, in retrospect, both proved vital not only for any donation at all, but particularly its richness; his personal as well as his scientific papers were saved for posterity. I very much doubt that the collection would have been as complete as it is now (or even exist at all) if the estate had made the decision about what to do with the paper bags in the 900 square-foot attic room after Jerne's death. Most probably his private correspondence and notes would have been withheld or even destroyed.

Furthermore, although I did not start my biographical work with the intention to construct an archival collection, it turned out that writing a biography, and not a history of immunology, was precisely what made the eventual Jerne collection so fecund. My biographical collaboration with a subject with so many diverse interests and social connections resulted in a rich and varied archive, whereas a more narrow analytical approach—for example a study of the biographical origin of the network theory—would have resulted likely in a much poorer archival collection. My experience thus suggests that writing biographies of recent scientists is a major instrument for securing rich archival collections for later historical use.

In 1994, a few months before his death, Jerne somewhat reluctantly gave me unlimited rights to quote from the documents, a *sine qua non* for publication. Relieved, I spent another couple of years in the Royal Library reading room to write the final manuscript. In 1998, almost twelve years after we first met, *Hvilken Kamp for at Undslippe* (What Struggle to Escape) was published in Danish.[22] A thoroughly revised and abridged English edition, *Science as Autobiography: The Troubled Life of Niels Jerne*, followed in 2003.[23]

II: Seven uses of scientific biography

More than archival and curatorial lessons, however, can be drawn from my work on Jerne's biography, particularly concerning the overall value of scientific biography. Why should we write about the life and work of individual recent scientists?

Let me offer a number of approaches and motivations—a typology of seven biographical subgenres as it were—which I now consider meaningful and important in writing about scientists such as Jerne. I will differentiate between biography (1) as a method for writing contextual history of science, (2) as a means for understanding science-in-the-making, (3) as a genre for the popular understanding of science, (4) as belles-lettres, (5) as public commemoration (eulogy), (6) as labor of love, and finally (7) as research ethics.

This seven-fold typology is somewhat arbitrary: genres are not fixed species of literature.[24] But I have not pulled them out of the blue, either. Furthermore, the purpose of the typology is not to settle on these seven reasons for writing scientific biography, but to stimulate discussion of the many different uses of the genre. Even though I ended up favoring one of them, the point here is not to recommend this particular use of biography as better or more important than others. My aim, instead, is to clarify some of the decisions one has to make as a biographer of recent scientists.

1. Biography as an ancilla historiae

"Until recently," historian Elizabeth Garber wrote in 1990, "many of the more traditional approaches [to history of science] were pursued apologetically because of serious questions about their validity as history in light of historiographies derived from sociology and anthropology." She continued:

> No genre of history fell under more odium than that of biography. Belief in the priority of socially defined criteria necessarily destroys the importance of individual lives, especially of their peculiarities and eccentricities. Seeing the cognitive realm as derivative of social place, cultural values or political ideology undermines any concern with what makes an individual's contribution to the sciences or technology critical or crucial. Character and originality were lost to the crowd, suppressed as subjects for legitimate historical investigation.[25]

This attitude to biography among historians of science has changed in the past decades. Perhaps not so much because of a sudden interest in the individual scientist as such, but because biography has been coopted as yet another method for writing contextual history of science (where "context" almost invariably is understood as the larger social, cultural, or political context). Scientific biography has become an *ancilla historiae*, a hand-maiden of history of science.[26]

Seen in retrospect, a seminal event that changed the attitude of historians of science was an aptly titled article published by Thomas L. Hankins in the journal *History of Science* a quarter century ago: "In defence of biography."[27] Hankins's defense amounted to utilizing the individual as a smart way to get under the skin of a whole historical period or situation. Biography, he argued, "gives us a way to tie together the parallel currents of history at the level where the events and ideas occur.... We have, in the case of an individual, his scientific, philosophical, social and political ideas wrapped up in a single package."[28] Essentially, Hankins viewed a biography of a scientist as a kind of microcosm that could shed light on history of science at the macrolevel. Later, others have recognized that biography can be seen as a kind of microhistory taken to its logical extreme.[29]

Using biographies of scientists for historical purposes—especially for demonstrating the larger social, cultural, or political context of science—is probably the most commonly acknowledged aim there is for the genre today. Again, Elizabeth Garber stated: "Studies of individuals are proving invaluable in probing the values, behavior, and social life in complex societies. The idiosyncracies of the subject even help to shed light on the characteristics of the collective."[30] Justified by its usefulness for the cultural understanding of science—"by opening out [the] individual to social and cultural contexts," as Michael Shortland puts it,[31] or for understanding how historical actors fashion their cultural identities—scientific biography has become fully integrated into the historian's agenda. No serious historian of science rejects the genre out of hand today, at least so long as it contributes to a socially and culturally informed history of science. Adrian Desmond's two-volume life of Thomas Henry Huxley is a good example: "This is a story of Class, Power and Propaganda," he states, this is "a contribution to the new contextual history of science." And he continues with a series of rhetorical questions that reveal this new taken-for-granted status of the genre: "Isn't it the modern function of biography to carve a path through brambly contexts? To become a part of history? ... And isn't that our ultimate aim, to understand the making of our world?"[32]

As I began to appreciate the emerging interest in the history of recent immunology, I too became interested in using the story of Jerne's life and work for such contextual–historical purposes. His collection of documents provided rich materials for a contextualized history of the heroic decade of immunology. It was from the early 1960s to the early 1970s that the clonal selection theory became the unifying theory of the new discipline, the role of the thymus in the immune response was clarified, the molecular structure of antibodies were elucidated, and the B/T-cell-distinction was established. Jerne was continuously at the epicenter of these linked developments. He corresponded with almost everybody of significance, attended all the important scientific meetings, and took detailed notes (all of which survived) of the discussions. The Jerne collection thus functioned as a lens into these exciting years of the new cellular and molecular immunology; he was the single package that wrapped up these currents (as Hankins had advocated).

Part of the Jerne biography thus can be read as an *ancilla historiae immunologiae*. Some reviewers certainly have read it that way. More generally, since very little has yet been written about the history of recent science, biographies of near-contemporary scientists may prove one of the best methods for addressing what Arnold Thackray has called the "last frontier" of history of science.[33] There are several good reasons for this. One I have already mentioned: writing biography is an excellent way of securing rich archival collections for historical uses. Another good reason is Hankins's argument, that biographies give us a way to "tie together the parallel currents of history at the level where the events and ideas occur." Third, biography is a genre which—because it humanizes science—makes the technical complexity of recent science palatable to the general educated reader.[34]

Yet biography is not just history by other means. A servant is never totally subservient to its master. The distinction that Plutarch made two millennia ago, that *bios* and *historia* are two distinct ways of writing about the past, remains instructive in discussing the uses of scientific biography. *Historia* originally meant "an inquiry." Over time, such inquiries into the past by tradition have come to mean studies of phenomena like nations, classes, economic institutions, political movements, social interactions, cultural constructs, and the like. But *bios* meant—and still means—"a life" in the sense of "an individual life-course." Even if some historians today pretend that they are writing biographies of cities or countries or even diseases, most of us prefer to think about biography as the art of writing about individual human beings. While historiography by tradition deals with the collective phenomena of the past, biography by definition deals with individuals. One past, two genres.

Most historians today think of biography as a genre that plays a secondary role in assisting her more influential master. But this is not the whole story. Biography has other, and more independent, roles to play.

2.   *Biography as a means for understanding the construction of science*

Traditionally, historians of science also have used the genre of biography to understand the origin and construction of scientific results—such as experimental findings, concepts, theories, and innovations. Here the idea is that scientific results should be understood, or even explained, not primarily with reference to social or cultural circumstances (anything from the "social swamp," which L. Pearce Williams once called it with his usual sense for debasement)[35] but with reference to individual circumstances, such as motivations, ambitions, ideas, feelings, personal experiences, and individual experimental practices.[36]

One can find many examples of this use of biography from the past hundred years. Indeed, one of the major motivations for writing scientific biography instead of history of science has been to understand science as an individual achievement. This is not something particular to the historiography of science; it is a methodology for understanding cultural artifacts that historians of science have shared with literary historians, art historians, historians of music, and other scholars to the present. This was the kind of biographical history that the French literary critic Charles-Augustin Saint Beuve advocated, very successfully, in the mid-nineteenth century[37]—and which generations of scholars in literary criticism (first the New Criticism movement and then different brands of poststructuralists) have reacted so strongly against. But whatever accusations of naïve individualism can be leveled at it, it is a genre of writing that is still going strong, both in the historiography of science and other hyphenated historical disciplines (as anyone who opens the pages of a standard book review journal like the *Times Literary Supplement* can see).

A nice example of this approach to science-in-the-making is the late Frederic L. Holmes's two-volume, detailed, almost day-by-day, account of how biochemist Hans Krebs came to the understanding of the citric acid cycle in the 1930s. Relying primarily on Krebs's detailed laboratory notebooks, Holmes gives fascinating (almost microscopic) insight into the interaction between biochemical ideas and daily bench-work, with some personal details here and there to give the flavor of the man.[38] Holmes interacted closely with his subject to interpret the notebooks, although he never included Krebs's autobiographical stories into the narrative. Nevertheless, it is a masterful case-study of scientific creativity primarily based on an individualistic understanding of the construction of scientific results.

Holmes's two-volumes on Krebs came out when I was about to start writing the manuscript. I was tempted to emulate this painstakingly cognitive reconstruction of an experimental pathway. Jerne had created two influential theories (the selection theory and the idiotypic network theory) and a method (the hemolytic plaque assay) which soon acquired the status of a "citation classic."[39] These were three good examples of well-defined scientific constructs closely associated with the work of an individual scientist. They therefore fit an individualist approach. The document collection also turned out to be particularly well-suited to reconstructing the experimental and theoretical pathway that led to the selection theory. Hundreds of almost daily experimental protocols were preserved, along with letters to his peers where he tried to make sense of his experimental observations; there was a complete series of consecutive drafts to a final paper that made it possible to follow how he had changed his line of argument while working on that paper. In addition, he had already written a retrospective article in which he tried to explain the major steps in the construction of the theory. Finally we engaged in many hours of discussion to try to open up his memory (which largely failed, however!).[40] I thus had great expectations of doing for Jerne's selection theory similar to what Holmes had done for the Krebs cycle. After several months of work with the documents I was able to reconstruct almost every step in the interaction between experimental work and theoretical argument that went into the construction of the selection theory.[41]

Whatever merits this approach to the history of science has, however, it risks going against the gist of biography as a genre of life-writing. Too narrow a focus on the scientist's work waters down the rationale of the genre as a study of *bios* (a life-course). The Germans ingeniously coined the term "Ergographie" for a biographoid study of the work (*ergon*) of an individual, the work-counterpart of a biography; Holmes' study of Krebs thus is more ergography than biography.[42] To turn an ergography into a biography proper, one must relate the work to the life, i.e., contextualize the scientific work with reference to the life. A biographical contextualization of a scientist's work is a far more difficult task, however, than either reconstructing a cognitive pathway or solely narrating the life. The difficulty lies in the fact that few archives contain the appropriate documentary material. Most efforts to give a true biographical account of the scientific work have therefore been rather speculative—for instance, psychobiographical explanations.

As it turned out, Jerne's collection provided a unique opportunity for correlating *bios* and *ergon*. By keeping almost everything that passed through his hands, Jerne had lived a "biographical life." Yet it was only after intact donation of the entire collection to the Royal Library that I discovered the strength of some of these documents for making a true biographical explanation of his work, particularly his selection theory of antibody formation (1955). Among his papers were letters and notes which clearly showed how he understood his own social self: he characterized himself as a person who always had "a set of viewpoints in stock, which can be put to use on different occasions," or as a wrench that fits all kinds of bolts.[43] The picture emerging from these and other documents revealed a man who felt that he had a number of given mental states or conditions in stock to draw on in order to cope with influences from the outer world (see Figure 7.1).

Even more interesting, however, is that the selection theory of antibody formation appeared isomorphic with Jerne's view of himself. To summarize a long argument, I could show not only that there was a structural similarity between his self-understanding and the selection theory but also that the best explanation for the origin of the theory was that Jerne metaphorically projected this self-understanding on his experimental data in the local situation where the theory was conceived. Later the theory was communicated into a larger social world and was eventually transformed into the central dogma of immunology—but that is another story.

I was lucky to stumble upon documentary material that allowed this kind of biographical explanation of the origin of a scientific theory. Few archives contain such

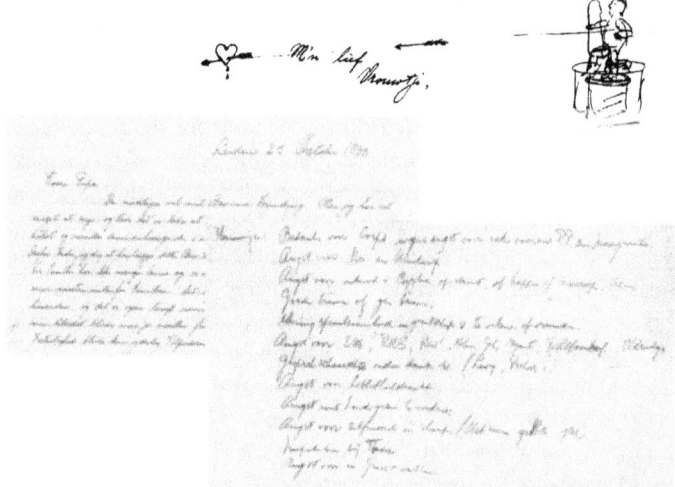

Figure 7.1 Examples of documents which drew the author's attention to the existential
dimension of Jerne's life. Left: In a long letter to his father of October 25, 1933,
Jerne tries to explain why he failed in his chemistry studies at the University of
Leiden. Upper right: illustration in letter to his onetime girl-friend, in which Jerne
tries to explain to her why he wants to study instead of getting engaged. Right:
undated note (around 1932) in which Jerne lists all the things he is afraid of in life,
for example of committing suicide in his sleep, of falling into a canal, of being
impotent, and of not having anything to say when he was with people whom he did
not know.

Source: Royal Library, Copenhagen.

material: few scientists have had the impulse to search their souls like Jerne did,
even fewer have written their observations down, and fewer still have agreed to let
such private notes and letters go into a public archive together with the scientific
documents (at least not without clauses on their future use). In this sense the Jerne
archive is exceptional. A lesson to be drawn from this case is that the historian of
recent science should make strong efforts to secure documentary material that allows
for fine-grained reconstructions of the interaction between life and science. These
efforts become especially worthwhile if we realize that writing biography has other
aims than using individuals as means for understanding science in its cultural,
political, and social context.

3.   *Scientific biography and the popular understanding of science*

Biographies of recent scientists also can contribute to the public understanding of
science. Few historians of science and science studies scholars, however, seem to
appreciate this function of the genre. Jane Gregory and Steve Miller's *Science in Public*
(1998) makes the point: they include chapters on science in museums, in popular
books and magazines, and in newspapers; they discuss science on television and radio,

and devote many pages to the popular science lecture culture—for example, in the Royal Institution and the meetings of the British Association for the Advancement of Science. But there is not a single reference to biography in this book, either in the text or in the index. It is as if biographies of scientists have nothing to do with public science, or with "communication, culture, and credibility," as they subtitle the book.

The omission is noteworthy. To popularize science has perhaps been the most common use for the genre of scientific biography for more than a century. Even before the Life-and-Letters dinosaurs began to die out in the early twentieth century, publishers like Longmans-Green, John Murray, and Macmillan poured out popular biographies about scientists, often in series format with series titles like "Immortals of Science," "Famous People, Famous Lives," and "British Men of Science." Some of the most impressive series were published in Germany: these included "Grosse Männer" (subtitled "Studien zur Biologie des Genies") and the fifty volumes in the "Große Naturforscher" series published by Wissenschaftliche Verlagsgesellschaft in Stuttgart. The East Germans after the Second World War were even more ambitious with respect to using biography as a medium for popularizing science. Teubner's in Leipzig published several popular biographical series—"Biographien hervorragender Physiker," "Biographien hervorragender Naturwissenschaftler, Techniker und Mediziner," etc.—more than a hundred titles altogether, each book roughly a hundred pages in length.

Not all of these were about recent scientists, of course. Popular biographies of living scientists nevertheless have been part of the popular science literature for almost a century. A classical example of a journalist-cum-biographer was R. L. Mégroz, authorized in the late 1920s to do the full biography of the malaria researcher and Nobel Prize winner Ronald Ross (who was so keen on getting a posthumous life that he paid for some of the publication costs out of his own pocket).[44] His pecuniary efforts were not in vain: Mégroz hailed not only Ross's scientific achievements but also his literary genius. In *Ronald Ross: Discoverer and Creator* (1931), he concluded that his subject's fame as a scientist had in fact "overshadowed his reputation as a creative artist in literature."[45]

Some biographers working in this vein have been extraordinarily productive. For example, Ronald William Clark published at least fifteen major biographies of scientists between 1960 and 1985, compiling around 8,000 pages of high-quality popular science writing, often based on considerable original research, primarily about then recent people such as J. B. S. Haldane, Julian Huxley, Ernst Chain, Bertrand Russell, and Sigmund Freud. About the same time, John Rowland wrote some fifteen biographies about recent or near-recent scientists with titles like *The Mosquito Man, The Radar Man, The Penicillin Man, The Insulin Man, The Jet Man, The Microscope Man*, and *The Polio Man*. They were not based at all on original research (they were not very good either), but since they continued to be published for two decades, between 1955 and 1975, they evidently sold well. And they certainly informed the public about the latest advances in recent science of the time.

Today, biographies of recent scientists seem to occupy a middle ground between science journalism, didactics, and history of science. For instance, William Lanouette— a political scientist and journalist who has been Washington correspondent for the *Bulletin of the Atomic Scientists* and energy policy advisor in the Clinton administration—has provided an indepth account of the political aspects of Leo Szilard's life.[46] Similarly, science journalist George Johnson, a frequent contributor to the *New York Times*, has written a successful biography of the physicist Murray Gell-Mann.[47] After a captivating *in media res* opening chapter, Johnson delivers Gell-Mann's chronological story, using published documents and interviews with Gell-Mann (who was still alive when the book came out). Former Pulitzer Prize winner and *The Sciences* editor Jonathan Weiner has provided a biography of molecular biologist and recent Nobel Prize winner Seymour Benzer, built on interviews with some 150 scientists (like Johnson, he uses no archival sources).[48] Finally, Shane Crotty, at the time an MIT student in biology and writing, faithfully renders the scientific work of David Baltimore, the Nobel Prize winner and former president of the Rockefeller University who became embroiled (and later exonerated) in a major late twentieth-century investigation of scientific fraud.[49] He simultaneously makes recent molecular biology and science politics accessible to a broader readership.[50]

In recent decades, children and adolescents too have gotten their own series of so-called juvenile biographies. Blurbs describing "Makers of Modern Science" proclaim this as "a new biography series for young adults ... [E]ach volume depicts the human drama of scientific work, the excitement and frustration of research, and the exhilaration and rewards of discovery." To these should be added the Paul Strathern-phenomenon: snapshot-like portraits including *Newton & Gravity* as well as portraits of recent scientists such as *Crick, Watson & DNA* and *Oppenheimer & the Bomb*. We ought also to consider the many short volumes written by John and Mary Gribbin in the Constable "90 minutes" series—which librarians in the Science Museum Library in London claim are mostly borrowed by students who want a "short cut to the science."

I am certainly not arguing that all these popular and juvenile biographies involve serious scholarship. In my humble opinion, some are absolutely awful, such as *Listen to the Music: The Life of Hilary Koprowski* (2000) by Robert Vaughan who (according to the blurb) otherwise "writes for print, television, and film." His book is entirely based on interviews with no references to the scientific literature, perhaps the worst kind of celebrities-in-science writing imaginable.[51] But some are much better, and they remind us of the fact that the popular understanding of science is a strong aim among publishers of biographies. Many of these books do not always live up to the scholarly standards that historians of science impose on themselves, but they nevertheless provide intellectual food for members of the general public. Indeed, the public today gets their understanding of the recent history of science from journalists rather than from historians.[52]

I'm not sure if I ever had any particular wish to inform the general reader. But as it turned out, this was a feature of my Jerne biography that some of the reviewers nonetheless noticed. Even if we do not aim to contribute to the popular understanding of science by writing a biography, we are probably doing it anyway—by default.

4.   Scientific biography as belles-lettres

One can be easily turned into a novelist by default as well. But discussions about the use of scientific biography rarely bring up its status as belles-lettres—that the text is an end in itself, with no other functions than the aesthetic. Even if few biographies are read solely for their aesthetic value, life-writing is nevertheless a genre where form traditionally plays a major role. Historians may get away with bad writing if they present previously unknown archival material about an important historical event, or if they construct new and interesting interpretations and explanations. But biographers can hardly eschew their care for the literary qualities of their text. It is difficult to imagine a successful biography which is a bad read. In this sense biography bears a interesting resemblance to the art of portraiture in painting. It is also a common knowledge in today's publishing world that readers tend to choose biographies instead of novels—not only because of an alleged nostalgic yearn for the narrative format of the classical nineteenth-century novel, but probably also because readers value the skill of authors using their fictional tool-set on a world restricted by the available historical sources.

Even if biographies of scientists rarely match the highest literary standards of the genre as a whole, there are good exceptions. Janet Browne's recent Darwin biography, for example, is not only an excellent piece of historical scholarship—it rightly received the History of Science Society's Pfizer Prize in 2004—but it has also been been awarded two prestigious literary prizes: The National Book Critics Circle Award in 2003 and the James Tait Black Award in 2004.[53] Yet historians of science tend to value such literary qualities as just an extra bonus on top of the allegedly more important historical functions of the genre.

Book reviewers are partly to blame for this ignorance of the literary aspects of the genre. Indeed, most reviews of scientific biographies in history of science journals seem to follow the same standard format. In the first paragraph, the reader is reminded that biography is making a comeback in history of science. Then comes a long descriptive summary of the narrative, followed by a ritual complaint about the lack of (or, vice versa, praise for the successful) historical contextualization. Finally, the review ends with one or two sentences about how good (or bad) a read it is. But rarely do reviewers of scientific biographies expand on the composition, style, or other textual, literary, or aesthetic qualities of the book; they rarely become objects of literary criticism. It is as if scientific biographies by default are classified under the rubric of "history of science" and are supposed to be read for their informative and contextualizing value only.

Crafting the life of a recent scientist poses special problems for the aesthetically inclined writer. Biographers of Einstein, Newton, and Edison can build on dozens of earlier lives, on updated critical editions and well-ordered archival collections, and finally on a huge historical corpus dealing with the times of their subjects. Many biographies of securely dead scientists do not really have much new factual material to add. These biographical works therefore mainly involve new, often interesting interpretations of the portraits produced by previous scholarship. As a consequence, biographers of well-researched figures of the past can focus more freely on the form, composition, and aesthetic qualities of the text.

The biographer of a recent scientist, on the other hand, often has to create the archive from scratch. Usually we have no earlier biographies or even critical editions to lean on. Nor do we have much additional historical work to rely on for contextualization, for the simple reason that historians of science have not yet dealt with this period, or field, or area. Thus the biographer of recent scientists has other, usually more pressing, problems to deal with than aesthetic qualities of the text. Furthermore, if pressed by deans or publishers, the author may have to sacrifice certain aesthetic ambitions in order to move more rapidly toward publication (the archival work for a first-time critical biography is forbiddingly time-consuming, which is why PhD candidates are regularly discouraged from trying their hands on the genre). I was indeed lucky to be employed in a regional university where a low publication output had only marginal effects on one's scholarly status and therefore felt free to spend whatever time I felt necessary on the composition and style of the book. While I am pleased that several reviewers have noted these qualities, I am also grateful for not having had a Research Assessment Exercise or tenure review hanging over my head to curb my literary ambitions. Other biographers of recent scientists may not be so lucky.

The latent literary ambitions of biographers of recent scientists are probably also restrained by the fact that scientists themselves form a large proportion of the readership. I am not saying that scientists do not care about aesthetics. But as the "science wars" of the 1990s indicated, any tendency to shift the balance from accuracy to style in writing about science is usually met by disapproval by a noticeable segment of the readers. By contrast, biographies of seventeenth- to early twentieth-century scientists are primarily read by an educated and literary sophisticated public which holds composition, style, and the elegant use of literary tropes in higher regard than the precise rendering of mathematical formulas, experimental details, and conceptual precision (as writer Dava Sobel's success indicates).[54] In addition, recent scientists have not yet entered the general cultural canon. This means that a biography of James D. Watson (irrespective of its literary qualities) is bound to remain on the bookstore's science shelves together with textbooks on flow cytometry and gene expression, whereas a Darwin or Wallace biography will almost automatically find its way into sections for customers with an interest in Victorian studies.

Readers also should not underestimate the specific compositional problems in writing the biography of a recent scientist. Not because it is almost always involves a first-time rendering of the person, or because there is always too little historical schol-arship or too few archival documents available to allow a rich contextualization (both historical and personal contexts are excellent means for introducing literary qualities into a biographical text). Rather, it is primarily because the readership's alleged lack of familiarity with recent science forces the author to spend undue amounts of textual space on explicating the technical content—pages which are difficult to imbue with aesthetic qualities and even more difficult to weave seamlessly into the less-technical parts.[55] This is perhaps the greatest challenge for a biographer who has chosen to write about a recent scientist but who also wants to express some of the belles-lettres ambi-tions which most scholars have. Who does not want to write well?

5.   *Scientific biography as public commemoration*

Nor can any discussion about the use of biography of recent scientists avoid the issue of commemoration. Professional historians of science, often rightly, tend to distance themselves from commemorative practices.[56] This is probably because such practices have been, and still are, common among scientist-biographers and scientist-historians of science. To pay one's respect with good words (*eulogoi*) to the deceased, to erect a gravestone in speech and paper, is not only the oldest use of the biographical genre as a whole, it was also the function of the first *vitae* of natural philosophers and astronomers that appeared in the seventeenth century. It has been a strong aspect of the genre of scientific biography ever since. This commemorative use of biography has always had strong institutional and political overtones. Scientific lives were often written as part and parcel of national, professional, or organizational interests. To take one single example: Georges Guillian's *J.-M. Charcot 1825-1893: sa vie, son œuvre* (1955) was propelled by a strong nationalistic sentiment: "My mission in writing these pages is to show that J.-M. Charcot should not be forgotten... The glory that Charcot brought to French medicine in the nineteenth century should survive; our country should derive from it a genuine pride."[57] Hundreds of statements with similar intentions can be found in the biographical literature.

The use of biography for public commemoration has been kept in low regard by professional historians of science in recent decades. Many regard the eulogy as a phenomenon of the past or, perhaps, as something produced at the margins by contemporary amateurs: scientists writing about their heroes and publishing their results in obscure provincial journals. Surely these do exist, and they are not uncommon. But this does not mean that the eulogistic component is absent from mainstream scientific biography. For example, progressive women scientists are perfectly acceptable subjects for praise, as shown by the laudatory reviews for Linda Lear's hagiographical account of Rachel Carson.[58] Georgina Ferry's recent biography of Dorothy Hodgkin provides another example of an unashamedly eulogistic biography.[59] Put simply, nationalistic and professional eulogies have given way to commemorative biographies written for gender or ethnic identity political reasons.

I cannot entirely brush off the possibility that I did not write a eulogy of Niels Jerne because of fear I would be tabooed by my academic clan for praising a dead, white, arrogant, elitist, male scientist. But I had a better reason—I had no interest whatsoever in using Jerne for any kind of commemorative purpose. Being an ethnic Swede living in Denmark, I felt no special nationalistic reasons for hailing him. I had once taken a course about antibodies but had left a scientific career to become a historian of science and thus had no further stakes in the success of immunology. I was breathtaken by the progress of science but did not see it as a *sine qua non* for our civilization and thus felt no urge to glorify him for being a scientist. Likewise natural circumstances disinclined me from indulging in life-writing for gender or ethnic identity politics. The bottom line was simply that I was fascinated by Jerne as the begetter of a stream of documents and autobiographical stories, allowing me to reconstruct a life in narrative detail.

Lack of interest on my part in engaging in public commemorative practices did not prevent me, however, from taking advantage of the prevalence of such practices in the surrounding culture. I shamelessly emphasized Jerne's Nobel status in my application to the Swedish Humanities Research Council, knowing that the agency was eager to promote this most Swedish of all scientific prizes. And although Jerne had no clear national identity—in one of his autobiographical notes he described himself as a citizen of the North Sea, writing that "[f]our of its languages Danish, English, Dutch & German are my mother tongues that I speak with equal ease & delight"[60]—I solicited the manuscript to my future Danish publisher with the argument that he was one of the most famous "Danish" scientists of all times. (Likewise the *Oxford Dictionary of National Biography* accepted an entry on Jerne, not only because he was a British subject by birth, but also on the anecdotal evidence that he once wanted to acquire a pied-a-terre in London!) I also took advantage of the fact that the Nordic immunologists wished to claim him for their cause, and therefore accepted the invitation to give a keynote about Jerne to the International Immunological Congress in Stockholm 2001, simply because this was a brilliant opportunity to promote the book.[61] Hence, even though as a professional modernist biographer of recent scientists, I was supposed to have the anti-eulogistic debunking virtue steadily in mind, the spectre of public commemoration always tends to water down the purest of hearts.

6.   Scientific biography as a labor of love

Commemoration is not just a manifestation of identity politics, however. It also can be a private impulse, without any political connotations. The classical nineteenth-century Lives-and-Letters were often written for nationalistic and professional purposes, but they were also written by family members—sons, daughters, or widows-who wished to think back on the life of a recently deceased family member, or by scientific colleagues who wanted to celebrate the memory of a close friend, dear teacher, or mentor. Evelyn Sharp, a friend of the physicist Hertha Ayrton (1854–1923), held that "since friendship boasts of being stronger than death, there can surely be no more appropriate task for the friend who survives than to reconstruct, however, imperfectly, the life-story of the friend who has passed on in advance."[62] Fifty years later, Anne Sayre similarly delivered an energetic, passionate, and wonderfully ironic defense of her old friend Rosalind Franklin (1920–1958) against Jim Watson's not very flattering picture.[63]

Scientific biography as a labor of love is more common than most historians of science may wish to acknowledge. It is a practice largely specific to the writing of the lives of recent scientists, whether a Pasteur biography written in the late nineteenth century or an engineer's biography written in the 1990s. These personal memorials are often wonderful reads. They should not be read as public commemorations or historical documents, or for deeper insights into the science, but for exactly what they are. Some are forbiddingly naïve, but some express a high degree of reflexive awareness such as journalist Vanda Sendzimir's portrait of her father, the steel engineer. One day, at the age of 92, Tad Sendzimir asked her if she would like to write his biography. As she later reflected:

> The challenge encompassed a search—revealing, painful nourishing—through my past and my relationship with him. As a journalist, and as a feminist, I had to discover and show not only the great man, but the great man's faults, the burdens on his loved ones, the failures. And I said to him, straight out, "You're no saint, you know." And he said, "Oh, I'm well aware of that. The book would be pretty boring if I was. Write what you like."[64]

She interviewed him for weeks, found his memory "sharp for stories from his childhood, and crystalline for the details of machinery, but blurred and superficial as to personal relationships." The task was both fascinating and frustrating. "Would a more patient, sympathetic, and above all removed person be better able to elicit honest and heartfelt responses?" Sendzimir wrote. "But perhaps an outsider would go home satisfied with Tad's glib and often contradictory answers. Perhaps our tension was necessary, at least for this version of the story." Gradually her understanding of her father's character and of their relationship changed. She began to accept him "as he was: a man, a man of some greatness and some weakness" who just happened to be her father: "I began to accept our relationship, and accept the difficult dynamic between two hard-headed individuals who can argue but can still love each other."[65] This is a labor of love—but it is not a eulogy.

In my case, I was neither a student of Jerne nor member of the family. I hardly knew about him or his scientific status before the project started; it began with a fascination about his autobiographical stories and his archive as means for rethinking the narrative rules of scientific biography—a fascination in turn fueled by my reading of literary theorists and sociologists of science who advocated discarding traditional biographical realism in favor of a more reflexive and deconstructive metabiographical stance. My instrumental view of Jerne was further reinforced by many interviews with his friends and earlier colleagues that supported the picture of a culturally constructed and decentered self. Everyone told me a different story about him. Soon there were as many Jernes as there were interviewees. There seemed to be no true self to find, no stable relation between life and work to decipher. He became a kind of human chameleon, like Zelig in Woody Allen's mock documentary film from 1983, and I therefore imagined I could depict him as a narrative self, constructed from the multiple accounts he had written and from my interviews with him.

Jerne rapidly accepted this interpretation of his life. Once, after I had let one of his old friends in on my plans, Jerne told me that the friend had just called him up saying that:

> You don't exist at all . . . [Jerne laughs], the only thing that exists, is what others have said or say about you and also what you say about yourself, but that is on the same level as what others say about you. But . . . that you yourself have a privileged position in that you actually exist, that's something one averts one's gaze from in this post-modern biographical philosophy.

Jerne added: "It's a construction, yes."[66]
This incident (although it conveyed a fairly distorted version of postmodernism and poststructuralism) contributed to my abandoning the thought of writing about Jerne as a decentered and evasive subject. His evident delight with the idea corresponded all too easily with his own nihilistic self-image and worldview. In time I concluded that the idea of culturally constructing a subject makes sense only as a commentary on a naïve realism. If reality comprehends itself in terms of the linguistic turn, then the constructivist approach becomes, in Richard Rorty's words, "madness in the most literal and terrible sense."[67]

Our many interviews, our lunches, and our late afternoon chats undoubtedly contributed to my changing attitude about Jerne. Initially, after brief interviews with him (and his colleagues), he did indeed appear as a human chameleon. But after some 100 hours of interaction within him at his home, amid his everyday settings, the idea of a decentered subject became difficult to sustain. His self became more embodied. Increasingly I began to familiarize myself with his daily habits, his body language, how the tone of his voice reflected his mood. All these attributes are largely constant throughout life and contribute to what we call a "character." I thus began to see him as a much more stable self, a man with anxieties and dreams, hopes and fears, a man who spent a lifetime struggling with the same basic existential themes. (My own reading of Kierkegaard probably also contributed to this interpretation of him as "existential man.")[68] In sum, Jerne ceased to be just an instrument for my biographical constructive ambitions. Indeed, I concluded that, unless we are writing about people with multiple-personality disorders, the hermeneutics of suspicion at some point must be suspended. Every individual has a stable core-self. On occasion even this stable core can yield a number of narratives, each referring to a different social persona in different social contexts, but a coherent biographical subject nevertheless remains.

Both historians and biographers have rightly warned against too much personal closeness between biographer and subject.[69] The longer the association, the more the outcome risks being affected by strong emotional ties. Again, the disturbing specter of hagiography comes to mind. But the problem of attachment is more subtle. First, my major problem was not one of reverence. True, I admired Jerne's intellectual capacity and abilities as a life scientist, but I rapidly became much more critical of him as a human being. In addition, I was increasingly uneasy about how to handle the less virtuous parts of his character.[70] A short session in April 1994 with Chicago therapist George Moraitis, who specializes in working with biographers, led me to release some of these emotional ties. It was a breakthrough: I allowed myself to not like Jerne.[71] Only after his death did I begin to see him with the attitude that Iris Murdoch calls "attention"—a more detached view which allows one to see both the brighter and darker sides of the other with detachment.[72] Put another way, he changed from being nothing but a means for my experiments in biographical construction to being as well a "Thou" in Martin Buber's sense.[73]

I doubt that any biographer of a living scientist can altogether avoid being involved in some kind of labor of love. The trick is not to try to repress the demons but transform them. One can hardly set out to write a biography without being in some way emotionally involved with the central figure. But one has to work hard on establishing a more distant yet attentive stance in the process of writing. The final result should emerge as a happy divorce: the book should be a certification that the writer has freed herself or himself from the central figure.

7.   *Scientific biography as research ethics*

The six good reasons for writing biographies of recent scientists outlined above, all competed for my attention while I was interviewing Jerne and working in his papers at Castillon-du-Gard. But as a plowed myself through his documents, I began to consider yet another and, as I later understood, much older idea of what biography is good for. This seventh reason partly grew out of my slowly growing awareness that Jerne was not just a means for my own intellectual endeavor, but, as I mentioned above, also a "Thou."

Above all, however, this perception stemmed from my discovery of documents that testified to his character and personality—most significantly letters and notes which disclosed his views on the personal dimension of scientific work. For instance, in a letter to his first wife in July 1943, after he had gained employment in the Danish State Serum Institute, Jerne asked himself if it was sound, "humanly speaking . . . to invest such great energy and powers of concentration in something so specialized, something that doesn't even slightly impinge on your personal sense of life." Was it sensible, he wondered,

> to employ your time in the demanding assignment of familiarizing yourself with a thought-structure that others have already built up to completion; to develop this part of your life as a dilettante in peripheral abstractions, while the pulsing purple-red blood in your veins and the feelings in your heart have to take care of themselves until "later."[74]

Such personal reflections led me to consider a new set of biographical research questions. Biographers of scientists usually focus on the intellectual or material products of their subjects. If they pay any attention at all to the other aspects of their subjects' lives, these are usually considered secondary to the work; the scientific work and its outcome is judged as the rationale for doing a biography of a scientist. Yet—stimulated by the documents already mentioned—I increasingly found myself turning the usual biographical priorities around. Indeed, I made his entire life situation, including his personal life, the center of attention. I began to ask new questions: what choices did Jerne make during his life? What brought him to pursue science instead of a career in business, or becoming a physician, or a writer—or making a life of caring for family and children? And how did he bring together (or separate), his life inside versus outside science?

In other words, Jerne's life *as a whole* became the primary aspect of the enterprise, his scientific activities secondary. Instead of seeing a successful scientist who also happened to be a troubled man, I saw a troubled man who also happened to be a successful scientist. I found support for inverting the traditional priorities in British biographer Robert Skidelsky's declaration that "with the life, rather than the deeds, the achievement" we enter "a new biographical territory, still largely unexplored."[75] Philosopher Søren Kierkegaard likewise gave a foundation for this move to decenter (but certainly not devalue!) the scientific aspects of the biographical narrative: "The scientist and scholar has his personal life in categories quite different from those of his professional life," he wrote, adding that "it is precisely the first [categories] which are the most important."[76]

Skidelsky overstates the newness of this biographical territory. Plutarch claimed this ground almost 2000 years ago in his *bioi paralleloi* (*Parallel Lives*). But today this territory is largely forgotten, especially when it comes to biographies of scientists. I prefer to call it (continuing to draw on Kierkegaard) an "existential" approach to biographies of scientists: the aim here is not psychobiography (which aims to explain the work) but biography that focuses on the life as an achievement in itself. From this perspective any life (including a life mainly spent in science) is a deed—a deed which incorporates the scientific achievements as a special case.

This way of decentering science in a biographical narrative in favor of the personal life is analogous, I believe, to the way that some cultural historians have decentered science in favor of the cultural context. For instance, both Adrian Desmond in *The Politics of Revolution* (1989) and Jim Secord in *Victorian Sensation* (2000) decenter Darwin and Darwinism by focusing their narratives on the cultural and political contexts of evolutionary thinking.[77] Darwin's evolutionary theory is not devalued, but instead flows out of narratives which center on larger contexts. Skidelsky's "new biographical territory" similarly can be seen as a narrative about the personal context (what Kierkegaard regarded as the most important context) from which flows the rest—including the work and work results.

Such a reversal of the priorities between life and work has other interesting consequences for the way we judge the use of lives of recent scientists. Life stories also can contribute to an ethical discourse, rather than simply aid a historical or didactical discourse, explore aesthetics, or serve as eulogies or labors of love. This suggestion reflects the revival of a third position in contemporary moral philosophy. Philosophers have conventionally based their argument on two different metaethical positions. The first is a deontological position resting on questions about what is fundamentally the "right" or "wrong" thing to do: for example, is it right or wrong to do research on fetal stem cells? The second is a consequentialist position: what kind of medical treatment will be best for a majority of patients? In the 1990s, however, a growing number of philosophers have revived a third metaethical position, held by thinkers from Socrates to Foucault. This is a *virtue ethical theory*, where moral reasoning also involves reflection about the way one lives, carves out a life-course, builds a personality and character, and cultivates or wastes one's talents.[78]

Virtue ethical reasoning further supports my argument above for decentering the work and outcome of the work in favor of reviewing the whole life of the scientist. From the perspective of virtue ethics, biographies of scientists can thus be written (and read) to answer a crucial question: how to live a life in science in a good way? The question of "the good life" is a complex one, which I have written about elsewhere.[79] Different answers emerge. Some are hedonistic (let's have fun!) while other more serious answers emphasize cultivating virtues as a prerequisite of the good life. This view goes back to Socrates and was widely shared by Hellenistic philosophers and later philosophers inspired by the Hellenistic tradition.

Some modern moral philosophers have suggested that the way we cultivate the virtues necessary for the good life is by emulating other good persons around us. Others have suggested that we nurture them by reading novels about the vexations of life. I believe it is possible to combine these two answers: we can cultivate the virtues (and thereby live good lives) if we read biographies. Specifically, academics can cultivate the good life in scholarship and science (among other things) if they themselves read biographies. The format of the biography—in print or in film—does not matter here. For example, John Bailey's portrait of the aging and Alzheimer-stricken philosopher Iris Murdoch cannot but invite the reader to reflect upon the role of intellectual work at the brink of mental break-down. Kate Winslet's portrayal of young Iris and Judy Dench's of the old Iris Murdoch (in Richard Eyre's film *Iris*, 2000) demonstrate that biographical movies can do the same job as text versions. Similarly, Ed Harris's dramatization of American artist Jackson Pollock (*Pollock*, 2000), lays bare the painful contradictions between living a life in art and living a life with a family. (I do not include Russell Crowe in the role of John Nash in

A Beautiful Mind (2001) here because I think it is a shallow and sentimental portrait that does not serve the function of cultivating the viewer's virtues well.)[80]

Such a use for biography is not new: Plutarch advocated this in the first century, and it remained one of the major purposes of biographical writing until the late nineteenth century. Biographies were thought to be edifying. Biographies of scientists were no exception, and the eighteenth and nineteenth centuries abound with potentially edifying scientific lives. Victorian biographers especially valued them. For instance, the Scottish author Samuel Smiles saw industrious and thrifty scientists and engineers as role models for the rest of society.[81] But around the turn of the last century, this kind of biography began to die out. One rarely sees it today. Other uses, especially the historical ones, took over.

But is the existence of this long tradition for edifying biography an argument for revitalizing it? Might this risk an uncritical resurrection of Victorian moralism and uncritical hero-tales? I see no reason why the fear of ghosts of the past should scare biographers from writing lives for ethical purposes today. After all, today's biography authors live in a culture with a much more sophisticated and relaxed view on moral issues, and with quite different criteria for what constitutes an edifying story. In fact, I believe scientific biographers have no choice but to reinvent and modernize the classical edifying tradition for life-writing. No age (not even this one) can avoid asking the perennial questions asked by philosophers from Socrates to Foucault about how one should live and craft one's life.

Hence the question is not *whether* to ask these kinds of questions about how to live or not, but rather *how* to ask them. We should not exclude the idea of role models altogether from biography, but reinvent it for twenty-first-century sensibilities. A twenty-first-century scientific role model does not have to be uniformly good and positive, provide simple answers, or even offer a "model" in the positive sense of the word. A twenty-first-century edifying life-story will probably instead refer to complex human characters and fates. Stories of troubled lives will have many dark zones, narratives that present the reader with genuine moral dilemmas. A "model" today is a life-story that other people can relate to for better or for worse, a story that provokes the readers to think about the way they are living. In this particular case, I began to pose questions to myself like: how do we make sense of the loss of love in our lives? How do we handle feelings of guilt? How do we handle the longing for the sublime and the scorn for the mob?

Scientific biography and the care of self

Since eventually I embraced this seventh use of life-writing as the locomotive driving the narrative in my Jerne biography (even as I more-or-less deliberately incorporated aspects of the six previous uses), let me add a possible philosophical underpinning of the virtue ethical function of biography by mentioning an important reinterpretation of ancient philosophy made by the French classical philologist, Pierre Hadot—one that underscores a deeper significance for biography.[82]

Hadot argues that already in classical antiquity there was a pronounced difference between doing philosophy in the sense of systems, concepts, and theoretical discourses on the one hand, and philosophizing as a mode of life, on the other: a practice based on the classical maxim *gnothi seauton* (know thyself) and the Socratic recommendation in Plato's *Apology* that the unexamined life is not worth living. Hadot traces this distinction through the history of philosophy, from Plato, via Petrarch, Montaigne, and Descartes, to Kant, Nietzsche, Wittgenstein, and Foucault. They all agree that it is one thing to think about what the world is like—or what justice and goodness may be—or what characterizes true knowledge: questions at the center of academic philosophy today. But it is another and very different thing to *live* and *practice* justice, goodness, truth, and other virtues. Reviewing this history, Hadot suggests that modern academic philosophy has largely gone astray in its attempt to objectify (externalize) its object of study, instead of being more concerned about how philosophical practice influences its practitioners. It is worth noting that Hadot's discussion of the two kinds of philosophy had a seminal influence on the thinking of the late Michel Foucault, especially his notion of "souci de soi" (care of self) in the third volume of *L'histoire de la sexualité* (indeed, the book is actually called *The Care of Self* in the translated American edition).[83]

Hadot restricts his analysis to philosophy. But there is no reason why his argument cannot be used to review and challenge a number of scholarly activities today, in the sciences as well as in the humanities. Following Hadot's reasoning we could then say that it is a good and admirable thing to do science or medicine in order to understand the physical world and the human body, but it is another, and equally good and venerable thing, to be a scientist as a mode of life. And with respect to the practice of historiography of recent science, we could say that it is a good thing to do it in order to understand the science of the recent past, but it is another, and equally good thing, to do it as a way of practicing "souci de soi."

In the same way, Hadot's conceptual scheme can also be applied to the genre of biography, including the practice of writing lives of recent scientists. Thus it is a good thing to write biographies of recent scientists in order to understand their work and their lives (as my analysis has shown, one can discern several good reasons for doing so). But in the light of Hadot's and Foucault's notion of philosophizing as a mode of life, it is an equally good thing to write biographies of recent scientists as a way of practicing the care of one's scholarly self. Both writing the history of recent science and writing biographies of recent scientists is a way for historians, biographers, and scientists alike to explore the perennial questions of how to craft a life-course out of talent and circumstances. In this respect, historians, biographers, and scientists have a common ground for self-reflection that transcends the "science wars" of the 1990s.

Conclusion

The genre of scientific biography certainly has many different uses today. So different, in fact, that we should perhaps speak about several genres, rather than one genre with different purposes. What makes it attractive to think in terms of a broad, multipurpose genre, however, is that single biographies often display several of these aims simultaneously.

In my own case, it turned out that in addition to my original aims—to experiment with the juxtaposition of biographical and autobiographical voices, to understand the biographical origin of my subject's work, to contribute to a contextual understanding of the history of recent immunology, and to write an aesthetically satisfying book—the archival material provided other good reasons as well. Likely, I am not the only biographer to have so many aims running in parallel. But even so, I am surprised that there is so little written about the problems of writing lives of recent scientists. It is probably a reflection of the fact that scientific biography generally is a under-analyzed genre of science past, compared to the prolific commentary that now exists on the historiography of science. I hope this chapter can help remedy this situation.[84]

Acknowledgments

A shorter version of this chapter was presented on a lecture tour in September-October 2004, including the keynote lecture at the NIH History Day, National Institutes of Health, Bethesda, MD; the Historical Seminar on Contemporary Science and Technology, National Air and Space Museum, Washington DC; Department of History and Sociology of Science, University of Pennsylvania; Department of Medical History and Bioethics, University of Wisconsin; Department of History of Medicine and Science, Yale University; and the Boston Colloquium in the History and Philosophy Science, Boston University. I am grateful to the organizers and to the participants in these meetings for very lively discussions and a wealth of critical and constructive remarks. Also my warmest thanks to Ron Doel for very helpful comments on the manuscript, and above all his patience.

Notes

1 Throughout this chapter I use the term "historiography" in the literal meaning as "writing about history" to emphasise the parallel to "biography" for "writing individual lives." Furthermore, in spite of my earlier misgivings about the term "scientific biography" (Thomas Söderqvist, "Existential Projects and Existential Choice in Science: Science Biography as an Edifying Genre," in *Telling Lives: Studies of Scientific Biography*, ed. R. Yeo and M. Shortland (Cambridge, MA: Cambridge University Press, 1996), 45–84), I use it here as a synonym for biographies of scientists in a broad sense, analogous to "political biography," "literary biography," etc.

2 Rachel Laudan, "Histories of Sciences and their Uses: A Review to 1913," *History of Science* 31 (1993): 1–34.

3 For example, Pierre Gassendi's lives of Peiresc (1641) was followed by the lives of Tycho Brahe, Copernicus, Puerbach and Regiomontanus in 1654. The first Descartes-biography came half a century later (Adrien Baillet, *La vie de monsieur Des-cartes*, 2 vol. (Paris: Daniel Horthemels, 1691)).

4 This estimate is based on Edouard-Marie Oettinger, *Bibliographie biographique universelle: dictionnaire des ouvrages relatifs á l'historie de la vie publique et privée des personnages célèbres de tous les temps et de toutes les nations, depuis le commencement du monde* jusqu'á *nos jours* (Bruxelles: J. J. Stienon, 1854); Leslie Howsam, *Scientists since 1660: A Bibliography of Biographies* (Aldershot, England: Ashgate, 1997); Leslie T. Morton and Robert J. Moore, *A Bibliography of Medical and Biomedical Biography*, 2nd ed. (Aldershot, England: Scholar Press, 1994); *Isis Cumulative Bibliography* 1913–1965, 1966–1975, 1976–1985, and 1986–1995; and a survey of the collections and catalogues of the Science Museum Library, The Wellcome Library, and the library of the Royal Society, all in London. Biographies published in

several smaller European languages (e.g. in Finnish, Czech, Greek, and Portuguese) have not been included in this estimate.

5 See for instance the special issue on "Le Biografie Scientifiche," in *Intersezioni: rivista di storia delle idee*, vol. 15, ed. Antonello La Vergata (1995); Ramesh S. Krishnamurty, ed., *The Pauling Symposium: A Discourse on the Art of Biography* (Corvallis, OR: Oregon State University Libraries, 1996); Michael Shortland and Richard Yeo, eds, *Telling Lives: Essays on Scientific Biography* (Cambridge, MA: Cambridge University Press, 1996).

6 The following review of Jerne's life is based on Thomas Söderqvist, *Science as Autobiography: The Troubled Life of Niels Jerne* (New Haven, CT: Yale University Press, 2003). For an authorized short version, see the entry on Jerne in *Oxford Dictionary of National Biography* (Oxford: Oxford University Press, 2004).

7 Nobelprize.org, April 5, 2004, http://nobelprize.org/medicine/laureates/1984/presentation-speech.html, accessed on April 6, 2004.

8 Niels K. Jerne, "The Natural-selection Theory of Antibody Formation," *Proceedings of the National Academy of Sciences* 41 (1955): 849–857.

9 F. Macfarlane Burnet, "A Modification of Jerne's Theory of Antibody Production Using the Concept of Clonal Selection," *Australian Journal of Science* 20 (1957): 67–68. The theory was elaborated in F. Macfarlane Burnet, *The Clonal Selection Theory of Acquired Immunity* (Cambridge, MA: Cambridge University Press, 1959).

10 Niels K. Jerne, "Towards a Network Theory of the Immune System," *Annales d'Immunologie* 125C (1974): 373–389.

11 Thomas Söderqvist, "A Nobel Prize is no Guarantee Against the Anxiety of Being Forgotten: Niels K. Jerne, the 'king of Theorists' in Modern Immunology," in *Neighbouring Nobel: The History of Thirteen Danish Nobel Prizes*, ed. H. Nielsen and K. Nielsen (Aarhus, Denmark: Aarhus University Press, 2001), 523–552.

12 Niels K. Jerne, "Antibody Formation and Immunological Memory," in *Macromolecules and Behavior*, ed. J. Gaito (Amsterdam: North-Holland, 1966), 151–157.

13 Carl Pletsch, "On the Autobiographical Life of Nietzsche," in *Psychoanalytic Studies of Biography*, ed. G. Moraitis and G. H. Pollock (Madison, CT: International Universities Press, 1987), 415.

14 Now, fifteen years later, the situation is beginning to change. Linus Pauling's papers, which are currently being put on-line by the Special Collections division of the Valley Library of Oregon State University, is perhaps the most complete and best-curated collection of a twentieth-century scientist so far (see http://osulibrary.orst.edu/specialcollections/ accessed on May 6, 2005). It too reflects a neurotic bent to save everything.

15 For example, Arthur M. Silverstein, *History of Immunology* (San Diego, CA: Academic Press, 1989); Alfred I. Tauber, *The Immune Self: Theory or Metaphor?* (New York: Cambridge University Press, 1994); Scott H. Podolsky and Alfred I. Tauber, *The Generation of Diversity: Clonal Selection Theory and the Rise of Molecular Immunology* (Cambridge, MA: Harvard University Press, 1997); Alberto Cambrosio and Peter Keating, *Exquisite Specificity: The Monoclonal Antibody Revolution* (New York: Oxford University Press, 1995); and others.

16 Warwick Anderson, Myles Jackson, and Barbara Gutman Rosenkrantz, "Toward an Unnatural History of Immunology," *Journal of the History of Immunology* 27 (1994): 575–594.

17 Dale F. Eickelman, *Knowledge and Power in Morocco: The Education of a Twentieth-century Notable* (Princeton, NJ: Princeton University Press, 1985).

18 To found a biography of a recent scientist on a private collection supplemented with autobiographical interviews presupposes, of course, that the subject is willing to participate. This lack of compliance is one of the reasons why Victor K. McElheny's biography of Jim Watson (*Watson and DNA Revolution: Making a Scientific Revolution* (Chichester, England: Wiley, 2003)) in my view is a failure: Watson neither put his collection of papers nor his persona at the author's disposal. Jerne eventually did both—but only, as described here, after several years of negotiations.

19 The interview notes, tapes, and transcriptions are presently deposited in the archives of the Medical Museion, University of Copenhagen.

20 In one conversation, I confronted him with the possibility that his motives for a specific event might be a rationalization after-the-fact, and he answered, without consulting his sixty year old notes and letters: "Oh yes, surely, that's what I am saying now. If you had asked me on that occasion, I don't really know what I would have answered." Jerne, interview by author, October 17, 1988.

21 For instance, he once said: "What I'm saying on this tape recorder … it's not necessary true, because I'm also playing games with you, with myself and so … I'm not sitting here like witness for the prosecution or something, I'm just saying what suddenly occurs to me as appropriate at this moment." Jerne, interview by author, February 11, 1988.

22 Thomas Söderqvist, *Hvilken kamp for at undslippe* (Copenhagen: Borgen, 1998). The English edition is cited in Note 6.

23 See Note 6. [*Note added in proof*: A Japanese edition is forthcoming (Igaku Shoin, 2006)].

24 Genres and subgenres are not fundamental book essences as the ancients and the classicists in the seventeenth and eighteenth centuries thought, and some genre theorists still believe. Most literary theorists today emphasize that genres are constructed in on-going, largely tacit, negotiations between authors, publishers, reviewers, librarians, and readers about how to classify and label a specific book. So one can expect boundary cases and genre transgressions, not only between biography on the one hand, and history and novels on the other hand, but also between the subgenres discussed here. For further discussion see David Duff, ed., *Modern Genre Theory* (Harlow, England: Pearson, 2000).

25 Elizabeth Garber, ed., *Beyond History of Science: Essays in Honor of Robert E. Schofield* (Bethlehem, PA: Lehigh University Press, 1990), 9.

26 See Thomas Söderqvist, "Wissenschaftsgeschichte á la Plutarch: Biographie über Wissenschaftler als tugendethische Gattung," in *Biographie schreiben*, ed. H. E. Bödeker (Göttingen: Wallstein Verlag, 2003), 287–325.

27 Thomas L. Hankins, "In Defence of Biography: The Use of Biography in the History of Science," *History of Science* 17 (1979): 1–16.

28 Hankins, "Defence," 5.

29 See Jill Lepore, "Historians who Love too much: Reflections on Microhistory and Biography," *Journal of American History* 88 (2001): 129–144, for a discussion of similarities and differences between biography and microhistory.

30 Garber, *Beyond History of Science*, 9.

31 Michael Shortland, "Bonneted Mechanic and Narrative Hero: The Self-modelling of Hugh Miller," in *Hugh Miller and the Controversies of Victorian Science* (Oxford: Clarendon Press, 1996), 17.

32 Adrian Desmond, *Evolution's High Priest* (London: Michael Joseph, 1997), 235–236.

33 Arnold Thackray, "Preface," *Osiris* 7 (1992), viii-ix.

34 This was one of George Sarton's major arguments for biography: "Full and honest biographies," he wrote in 1936, "should be encouraged by all means" because they help us to know "our fellow men and ourselves" better (found in "The Study of the History of Mathematics," reprinted in George Sarton, *The Study of the History of Mathematics and the Study of the History of Science* (New York: Dover, 1957), 23). This view of scientific biography as a humanizing genre was repeated in the first issue of the journal *Annals of Science*, founded by Douglas McKie in 1936. Science was indeed "'High and dry light'," but it also had "emotional associations," said McKie and his co-editors: "To recall the story—nay, the romance, of love for natural knowledge and devotion to its pursuit cannot fail to appeal to the higher emotions and help to evoke a widened sympathy and interest for the subject. … Indeed, the biographies of students of Nature furnish just as many examples of kindly feelings and loyal attachments as do those of any other type of man." D. McKie, Harcourt Brown, and H. W. Robinson, "Editorial," *Annals of Science* 1 (1936): 1–3, 3.

35 L. Pearce Williams, "The Life of Science and Scientific Lives," *Physis* 28 (1991): 199–213, 207.

36 There are of course several ways of understanding the construction of scientific knowledge on the market, not least sociological ones, like the SSK-school (the "social construction of scientific knowledge"), which have had a strong impact on the history of science community. Some historians of science would even say that the social and cultural construction of knowledge is the only kind of knowledge construction worth considering. I do not agree: I see no reason why a sociological or cultural approaches to scientific knowledge should rule out the power of individualistic approaches, using motivations, ambitions, ideas, feelings, personal experiences, and the like. In other words, the social construction of scientific knowledge doesn't have any epistemological priority over individual or personal construction. The only reason for such an epistemological priority is the Zeitgeist—that is, studies of social and cultural construction have been more popular in the last twenty years, that is all.

37 See for instance, Ann Jefferson, "Saint-Beuve: Biography, Criticism, and the Literary," *Mapping Lives: The Uses of Biography*, ed. Peter France and William St Clair (Oxford: Oxford University Press, 2002), 133–155.

38 Frederic L. Holmes, *Hans Krebs. Vol.1: The Formation of a Scientific Life. 1900–1933* (New York: Oxford University Press, 1991); idem. *Hans Krebs. Vol.2: Architect of Intermediary Metabolism, 1933–1937* (New York: Oxford University Press, 1993).

39 Eugene Garfield, "The Articles most Cited in 1961–1982. II: Another 100 Citation Classics Highlight the Technology of Science," in *Essays of an Information Scientist: The Awards of Science and Other Essays* v. 7 (Philadelphia, PA: ISI Press, 1985), 218–227.

40 On the role of memory and history see Hoddeson, this volume.

41 Söderqvist, *Science as Autobiography*, ch. 13–14.

42 For an evaluation, see Thomas Söderqvist, "The Architecture of a Biographical Pathway," *Historical Studies in the Physical and Biological Sciences* 25 (1994): 165–175.

43 See Söderqvist, *Science as Autobiography*, Ch. Parabasis.

44 Edwin R. Nye and Mary E. Gibson, *Ronald Ross, Malariologist and Polymath: A Biography* (Basingstoke, England: Macmillan, 1997), cf. 216–217.

45 R. L. Mégroz, *Ronald Ross: Discoverer and Creator* (London: Allen and Unwin, 1931), 12.

46 William Lanouette, *Genius in the Shadows: A Biography of Leo Szilard: The Man Behind the Bomb* (Chicago, IL: University of Chicago Press, 1992).

47 George Johnson, *Strange Beauty: Murray Gell-Mann and the Revolution in Twentieth-century Physics* (New York: Random House, 2000).

48 Jonathan Weiner, *Time, Love, and Memory: A Great Biologist and his Quest for the Origins of Behavior* (New York: Knopf, 1999).

49 Shane Crotty, *Ahead of the Curve: David Baltimore's Life in Science* (Berkeley, CA: University of California Press, 2001).

50 On science journalists and the history of recent science see Davidson, this volume.

51 For a discussion of celebrity in science, see Janet Browne, "Charles Darwin as a Celebrity," *Science in Context* 16 (2003): 175–194.

52 I am grateful to Ron Doel for pointing this out to me.

53 The awards were given for the second volume, Janet Browne, *Charles Darwin: The Power of Place* (London: Jonathan Cape, 2002).

54 For example, Dava Sobel, *Galileo's Daughter: A Drama of Science, Faith, and Love* (London: Fourth Estate, 1999).

55 Abraham Pais's much celebrated biography of Einstein is a good example of a biography of a (then near-recent) scientist that fails in integrating the technical exposition of the work with the life. Abraham Pais, *"Subtle is the Lord...": The Science and the Life of Albert Einstein* (Oxford: Clarendon Press, 1982).

56 See, for example, Pnina Abir-Am, *La mise en mémoire de la science: Pour une ethnographie historique des rites commémoratifs. Responsable scientifique* (Amsterdam: Éditions des Archives Contemporaines, 1998).

57 Georges Guillain, *J.-M. Charcot 1825–1893: sa vie, son œuvre* (Paris: Masson, 1955). Quoted from the English translation: *J.-M. Charcot, 1825–1893: His Life, His Work* (London: Pitman, 1959), xvi.

58 Linda Lear, *Rachel Carson: Witness for Nature* (New York: Henry Holt, 1997).

59 Georgina Ferry, *Dorothy Hodgkin: A Life* (London: Granta, 1998).

60 Undated note (probably late 1980s). Jerne Collection, Royal Library, Copenhagen.

61 I partly spoiled the promotion campaign when, in my lecture, I dismissed using Jerne for professional and nationalistic purpose and suggested that the main reason to remember him was that he illustrated what a life in science can be like (for more, see Thomas Söderqvist, "The Life and Work of Niels Kaj Jerne as a Source of Ethical Reflection," *Scandinavian Journal of Immunology* 55 (2002): 539–545.

62 Evelyn Sharp, *Hertha Ayrton 1854–1923: A Memoir* (London: Edward Arnold, 1926), viii.

63 Anne Sayre, *Rosalind Franklin and DNA* (New York: Norton, 1975).

64 Vanda Sendzimir, *Steel Will: The Life of Tad Sendzimir* (New York: Hippocrene, 1994), 18.

65 Sendzimir, *Steel Will*, 18–20.

66 Author's interview with Jerne, August 28, 1989.

67 Richard Rorty, *Philosophy and the Mirror of Nature* (Princeton, NJ: Princeton University Press, 1980), 366.

68 See Söderqvist, "Existential Projects."

69 For example, Leon Edel, *Writing Lives: Principia Biographica* (New York: Norton, 1984); Barbara W. Tuchman, "Biography as a Prism of History," in *Biography as High Adventure: Life-Writers Speak on Their Art*, ed. S. B. Oates (Amherst, MA: University of Massachusetts Press, 1986), 93–103; and Vassiliki B. Smocovitis, "Living with your Biographical Subject: Special Problems of Distance, Privacy and Trust in the biography of G. Ledyard Stebbins Jr," *Journal of the History of Biology* 32 (1999): 421–438.

70 I dedicated the English version of the book to his elder son, Ivar, hoping that he would read it in order to work through his memories of his father.

71 George Moraitis, "The Psychoanalyst's Role in the Biographer's Quest for Self-awareness," in *Introspection in Biography: The Biographer's Quest for Self-Awareness*, ed. Samuel H. Baron and Carl Pletsch (Hillsdale, NJ: Analytic Press, 1985), 319–354.

72 Iris Murdoch, *The Sovereignty of Good* (London: Routledge & Kegan Paul, 1970).

73 Martin Buber, *I and Thou* (New York: Scribner, 1958).

74 Jerne to Tjek Jerne, July 12, 1943. Jerne collection. Royal Library. Copenhagen.

75 Robert Skidelsky, "Only Connect: Biography and Truth," in *The Troubled Face of Biography*, ed. E. Homberger and J. Charmley (London: Macmillan, 1988), 1–16.

76 Søren Kierkegaard, *Søren Kierkegaard's Journals and Papers*, vol. 1. (Bloomington, IN: Indiana University Press, 1967), 408–409; document #928.

77 I am grateful to Betty Smocovitis for drawing my attention to this point.

78 For a critical review, see for example, Daniel Statman, ed., *Virtue Ethics* (Edinburgh: Edinburgh University Press, 1997).

79 This argument is developed further in Thomas Söderqvist, "Immunology à la Plutarch: Biographies of immunologists as an ethical genre," in *Singular Selves: Historical Issues and Contemporary Debates in Immunology*, ed. A. M. Moulin and A. Cambrosio (Paris: Elsevier, 2001), 287–301.

80 For additional discussion on this point see Davidson, this volume.

81 Adrian Jarvis, *Samuel Smiles and the Construction of Victorian Values* (Phoenix Mill, England: Sutton, 1997).

82 Pierre Hadot, *Philosophy as a Way of Life: Spiritual Exercises from Socrates to Foucault* (Oxford: Blackwell, 1995).

83 Michel Foucault, *The History of Sexuality, vol. 3: The Care of Self* (New York: Vintage Books, 1988).

84 See also Thomas Söderqvist, ed., *The History and Poetics of Biography in Science, Technology, and Medicine* (Aldershot, England: Ashgate, 2006).

8   Scholarship as self-knowledge

A case study

Alfred I. Tauber

Introduction

Do historians inscribe their own lives in their narratives? Do the threads of their life stories reveal themselves in their work? To varying degrees, I believe they do—both in intellectual and existential ways. Many scholars have already commented on this issue.[1] Rather than delineate the contours of this position, or explore its implications, I will review my own scholarship in the history of immunology—two books since 1994 and many articles—to add another case example to this literature.

I do so at a time when I seem to have reached a plateau in my own writing on immunology, for the basic idea I wished to explore has for me, at least for the time being, been exhausted. The project began in 1987, when I took a few months off from my laboratory activities to review the genesis of my particular area of expertise, phagocyte biology. I also re-educated myself in evolutionary and developmental biology, which seemed to have undergone significant changes since I had last studied these subjects twenty years previously. Little did I know, at least consciously, that this sabbatical would initiate a career shift from active basic laboratory research to philosophy and history of science.

In this chapter, my remarks are directed less as to *how* I wrote my histories than to *why* they took the conceptual form they did. Indeed, before describing my history of immunology and offering an "explanation" of its underlying *telos*, I should offer my readers a short orientation.

First, let me comment on my views of immunology from a theoretical point of view. While I was a practicing scientist, duly elected to the societies devoted to biochemistry, cell biology, and immunology, my specialty interests were remote to the pivotal debates about immunology's theory. As an *immunologist*, I was an informed "outsider" to the central action of the discipline; I interpreted its conceptual development with no particular partiality toward one theory or another based on my own research. But now I can acknowledge a certain "bias." For, as a *biologist*, I maintained an organismic orientation—by which I mean that my conception of biology made the organism the orienting site of study. So while my research in free radical chemistry and cellular activation mechanisms was firmly committed to a reductionist research program, my broader concerns were how to integrate these elemental functions back into a holistic construct. This was a position in contrast to those who embraced what

I considered to be a radical (exclusive) molecular or genetic approach and (on the other hand) those who were committed to an ecological perspective. In my view, the organism ought be placed between these two grand styles of study. In the twentieth century, however, my view was relatively neglected as reductionists and ecologists pursued their own agenda.[2]

My second general comment pertains specifically to the heterodox historiography I have pursued. It seems evident that not only does science bestow a worldview, it in turn probes nature under the subtle guidance of underlying metaphysical and cultural currents. For me, the more interesting half of this dialectical relationship is the dependence of science on its complex supporting intellectual infrastructure. I therefore began to write a history of a science that sought to link research programs to deeper conceptual issues: in this instance, those that pertained to the general philosophy of biology followed by leading immunologists. Thus my narrative was built on a scaffold with three sections: first, the fate of a countervailing holistic attitude relative to a dominant reductionist scheme; second, the theoretical significance of the movement of immunology's theory from a Darwinian orientation to a cognitive one; and, finaly, the fate of the immune self-concept that had guided the discipline implicitly from its inception in the late nineteenth century (explicitly since the 1950s). Each of these issues intersected the others. Indeed, this thematic construction was itself divided into three historical phases: origins, which examined the emergence of twentieth-century biology from its Romantic roots and emphasized the philosophical influence of Darwinism on the new discipline; a middle period, in which the self-model became the defining theory for a science that sought to discern self from non-self; and a third phase, where the self-metaphor collapsed under the weight of conflicting laboratory and clinical findings, leaving immunologists groping for a theory to explain the organization of the immune system. Far from what some immunologists claimed as the "end" of the science, I envisioned horizons yet to be approached.

While not claiming my interpretation was comprehensive, I maintained that contemporary immunology was a product of all the thematic issues I had identified. Thus my historical account reflected an interpretation of both the development of the science and its current standing from a somewhat unorthodox orientation. Relative to most other immunologists, I was (and am) more critical of the self-metaphor as immunology's basic scientific model or theory; I was more skeptical about the promise of molecular biology to answer questions critical for understanding immune organization; finally, I was more enthused about the cognitive paradigm and its associated theories of self-organization and nonlinear dynamics upon which new models of immune function could be built. Thus my critical historical insights cannot be separated from those opinions arising from my scientific judgment. In this instance, I hoped to make a case for a historiography of science where science and philosophy synergistically met to create a hybrid in the history of ideas that is fecund for the discourse of both disciplines. Consequently, I regard myself as practicing a particular historiographical style, which in this pluralistic era joins ranks with those other varieties that have made history of science such a dynamic and interesting enterprise.

Nineteenth-century origins of immunology

My entry into writing history of science assumed a certain autobiographical posture. Seeking the roots of my own laboratory research based on phagocytes, my work in the history of immunology began with an intellectual-scientific biography of Elie Metchnikoff (1845–1916). Metchnikoff, a prominent Russian embryologist, had discovered the phagocyte's function in immunity by a circuitous route: he first identified its role in the development of diverse animals, from the way that sponges consume food to the tadpole's phagocytosis of its own tail during its metamorphosis. This early work reflected debates about Darwinism, and thus he was poised to join in the momentous discoveries of infectious diseases. Unlike those who saw Darwinian struggle between adult animals as paradigmatic of the evolutionary process, Metchnikoff, as a developmental biologist, sought genealogical relationships by tracing germ lines and their functions in diverse animals. In this developmental context, he perceived how normal physiological maintenance might be turned to defensive functions. Although he shared the Nobel Prize for Physiology or Medicine (1908) with a leading German competitor, Paul Ehrlich, Metchnikoff's phagocytosis theory never attained the central importance he sought for it.

In reconstructing the genesis and structure of Metchnikoff's ideas, my first book on this issue, co-authored with the Russian émigré philosopher Leon Chernyak, explored the conceptual contrasts between Metchnikoff and the German microbiologists and later immunochemists who came to the nascent field of immunology from divergent theoretical and methodological perspectives.[3] He was intrigued with the problem of how divergent cell lineages were integrated into a coherent, functioning organism, and thus he was preoccupied with understanding development as process. He regarded these investigations as inspired by Darwin: since cell lineages were inherently in conflict to establish their own hegemony, he hypothesized that a police system was required to impose order, or what he called "harmony," on the disharmonious elements of the animal.[4] He found such a system in the phagocyte, which retained its ancient phylogenetic eating function: to devour effete, dead, or injured cells that violated the phagocyte's sense of identity. Thus the phagocyte was initially viewed as a purveyor of identity.

When Metchnikoff became engaged in the nascent field of infectious diseases at the beginning of the 1880s, he was poised to champion his phagocytes in the role of protecting the organism from pathogens (i.e. maintaining integrity).[5] He presented a grand scheme in a series of public lectures in Paris in the spring of 1891. Later published as *Lectures in the Comparative Pathology of Inflammation*, Metchnikoff argued here that the phagocyte had preserved its most ancient physiological functions. In simple animals, this was to serve as the nutritive organ (eating resident microbes): in animals with a gut, to continue to eat, but now for defense. He was thus the first to identify the defensive functions of the phagocyte.[6] More generally, he identified a primordial physiological process adapted to a new function—that is, a first line of defense against invading microbial pathogens. Thus, in Metchnikoff's theory, immunity was a particular case of physiological inflammation—a normal process of animal economy. But there was a subtler message. First, immunity was an active process with the phagocyte's response seemingly mounted with a sense of independent arbitration. Second, organismal identity was a problem bequeathed from a Darwinian perspective that placed all life in an evolutionary context. This Metchnikoff extended to the individual animal. The agency quality of his argument, and the radical sense of self-definition, reflected major Nietzschean themes. Intrigued, I attempted to make this parallel explicit.[7]

Let me put modesty aside; this historical interpretation was novel. Metchnikoff had been brushed aside by his German detractors as a hopeless Romantic, with outdated teleological precepts. They caricatured his phagocytes as possessing volition and intention and thus vitalist independence. Metchnikoff's polemics with the Germans was complicated by both political and personal issues, but the conceptual differences dominated our history.[8] We were empathetic to his stance against the strong reductionist program of his contemporary immunochemists. Later historians, however, had generally followed the initial German assessment and discounted Metchnikoff's role in the development of the science. For instance, Paul Baumgarten, a leading microbiologist and pathologist in the late nineteenth century, rejected the phagocytosis theory.[9] In addition, the new positivist science of the late nineteenth century rejected teleology as explanatory of biological function, seeking instead to ground phenomena in a materialistic schema, reducing organic functions to physics and chemistry.

In response (and defense), I documented Metchnikoff's problematic status in the scientific community by examining the Nobel archives and public testimony, as well as by contrasting his views with other scientists who were involved in similar research.[10] I argued that his scientific posture employed emergent and dynamical thinking appropriate to an organismic orientation of a biologist who is keenly aware of the problem of identity in a post-Darwinian age.[11]

Perhaps with unique insight, Metchnikoff deeply comprehended how the Darwinian revolution applied to the organism itself. He maintained that throughout the life experience of the plant or animal, there are changing environments, new insults, encounters with novel challenges. He understood that adaptability and versatility determine overall success. This is a key lesson of evolutionary biology— and it is a radically different conception of the organism from that of the pre-Darwinian era. Prior to *On the Origin of Species*, the organism was a "given." Naturalists viewed it as essentially unchanging and stable. By the late twentieth century, we now appreciate a more dynamic picture, where the organism is in a dialectical relationship with its world. In an ever-changing set of relationships, at many different levels of engagement, the organism lives both in *response* to its environment and in turn *alters* its environment, both passively and actively. Reacting and adjusting to external stimuli and conditions, vital processes are characterized by continuous dialectical exchange, both of nutrition and of information. For Metchnikoff, the phagocyte captured both functions—by eating pathogens and more generally through its ability to perceive, process information, and react to the environment. These primary cognitive functions are fundamental to even the most primitive animals, and for him, the phagocyte embodied this "intelligent" behavior and thus emerged as the agent responsible for organismal integrity.

But *identity* in this dialectical world becomes a problem. If the organism is in constant exchange—if it continuously adapts to its environment and changes accordingly—how can it maintain any notion of essence? Indeed, what is its core identity? There must be "boundaries," but in a post-Darwinian construct, where everything is in evolutionary flux, how are those limits drawn? If one is a radical genetic reductionist, the answer is simple: the genes "program" development. But given the flexibility of gene expression and the extra-genetic factors in development and the even more dominant epigenetic effects endured during adult life, such a "solution" is hardly satisfactory. So irrespective of the genetic revolution, recognizing how identity becomes a *problem* in nineteenth-century biology set the foundation for all of my subsequent study.[12]

Wary of an anachronistic interpretation, Chernyak and I nevertheless regarded Metchnikoff's case as prescient for a science yet to come: that is, the foundation of *current* immune theory and thus highly relevant. We made the metaphysical argument explicit.[13] But the broader philosophical implications were largely ignored and only elliptically noted by a single review, which refrained from assigning any labels.[14]

The problem of the self

Perhaps I was seeking a contemporary voice, or at least a resonance to Metchnikoff's formulation, when I stumbled upon a book called *Theoretical Immunology*, published by the Santa Fe Institute.[15] The Institute had been founded in 1984 to specifically examine complex systems from a multidisciplinary perspective, attracting to it the likes of Murray Gell-Mann, Geoffrey West, and George Cowan, and it quickly became a dominant voice for the application of computer modeling to nonlinear dynamical analyses. This immunology text happened to be the first systematic analysis in the Institute's series of publications (which continues to give immunology a high billing). I found only a few papers in the 1988 volumes interesting. The one written by Francisco Varela, Antonio Coutinho, and their colleagues immediately struck a responsive cord.[16] They evocatively espoused a self-determinism closely related to Metchnikoff's dialectical vision of the organism, one I found so conducive to my own thinking. As they wrote:

> The self is not just a static border in the shape space, delineating friend from foe. Moreover, the self is not a genetic constant. It bears the genetic make-up of the individual and of its past history, while shaping itself along an unforeseen path.[17]

Varela was a cognitive theorist—co-author with Huberto Maturana of *Autopoiesis and Cognition* (1980)—and Coutinho was a disciple of Niels Jerne, who had won the Nobel Prize in 1984 for various theoretical and practical contributions to immunology (discussed later).[18] Their respective intellectual heritages were clues that might have alerted me to the intellectual "baggage" of their unorthodox orientation. But at the time I read this paper with an innocence that allowed me to hear echoes of my own views, clearly articulated in a framework that was novel in my reading of contemporary immunology.[19] As Thomas Söderqvist and Craig Stillwell later, and appropriately, noted,

> [Tauber and Chernyak] have been challenged by the recent so-called autonomous network approach [*sic*. Varela/Coutinho] to the immune system, an epistemologically radical extension of Niels Jerne's well-known theory of idiotypic networks proposed in 1973. Accordingly, notions of anti-reductionism, dialectics, and self-determining systems (albeit not autopoiesis) reverberate throughout this [book]. The authors' main claim is that the real novelty in Metchnikoff's phagocyte theory of inflammation and immunity was his reformulation of the notion of organismic integrity.[20]

Although we were accused of following a Whiggish tradition, they simultaneously absolved us of historiographical sin: "But there is nothing intrinsically wrong with using historical and biographical material for present purposes. (What other legitimation has history of science in the long run?[!])"[21] We had thus received critical credos, but I also smarted: my intellectual foil had been pierced. At the same time, I was pleased to emerge out of my initial explorations—out of my ideological mists—relatively unscathed. To be sure, there was already a fair amount of historiographic posturing going on amongst the fledgling immunological historical community, and I determined not to let these criticisms bother me.[22] And soon I explicitly claimed a particular niche.

The foundations of my historiography were thus well in place when, in June 1992, I attended a meeting organized to encourage dialogue between historians and a generation of immunologists whose key research was conducted in the late 1950s and 1960s. Sponsored by the Stazione Zoologica Anton Dohrn and entitled "From Immunity to Cellular and Molecular Immunology: History of Immunological Thought and of Discoveries in Immunology," this was an impressive assembly. Gus Nossal, the emerging doyen of international immunology and the meeting's organizer, had invited a distinguished group of scientists and virtually all the various science studies students of immunology. Held on the Bay of Naples, resplendent with superb food, the conference proved to be an extraordinary (if not notorious) example of two communities speaking past each other. The historians regarded the scientists as posturing themselves for the Pantheon. Conversely, the immunologists felt that the historians "simply did not get the message." Others have commented on that symposium with a more generous (if not sterile) appraisal and less jaundiced eyes.[23] But it is noteworthy that the *Proceedings* of that meeting were exclusively written by the scientists. Their critics were dismissed without further ado.[24]

My Naples paper on Metchnikoff, a summary of the thesis that I described earlier, provoked little controversy. Indeed, it was largely ignored.[25] One cogent comment I recall was from Noel Rose and Rolf Zinkernagel (later a Nobel laureate), both of whom opined that if Metchnikoff did not deal with immune specificity and memory (this was true), he could hardly even be considered an immunologist at all! This was a position I had fought both on a strictly scientific basis and more extensively in the historical context.[26] Although my conceptual concerns did not enter the debate, the forum proved to be most useful for my own purposes: I saw that the question of immune identity, in the guise of the immune self, had not been critically explored by either camp. Selfhood was assumed as a governing construct, yet there was no systematic analysis of its genesis nor its current meaning. My sequel to *Metchnikoff* now seemed apparent. It was to afford me the vehicle to bring up-to-date the nascent ideas I found implied a century before.

Immunology during the first half of the twentieth century was preoccupied with the chemical questions of immune specificity. As a result, the biological questions concerning immune identity were set aside and never formally articulated.[27] But after the Second World War, transplantation and autoimmunity became increasingly relevant both to basic immunologists and clinicians. It was at this juncture that the Australian immunologist, later 1960 Nobel Prize winner, Macfarlane Burnet introduced the "self" into the immunological lexicon, and upon that metaphor erected a theory of immunological tolerance that was to dominate the field to this date.[28] Indeed, the triumph of Burnet's theory defined immunology as the science of self/non-self discrimination. Thus Burnet's clonal selection theory, by which selfhood is understood "with only slight modification," in the words of two recent practitioners, "has passed from the status of theory to that of paradigm."[29] Even though certain historians may feel uncomfortable with such sweeping notions as "paradigms," there is a general consensus, as another recent textbook author asserted, that clonal selection "is no longer a theory but a fact."[30] The Immune Self has indeed become dogma. The "self" versus "other" axis has assumed the role of an operative thought style that organizes the entire discipline.[31]

Burnet, originally trained as a virologist, came to immunology from a biological perspective quite different from the immunochemists then dominating the field. He was ambitious to integrate developmental biology, genetics, and immunology into a cohesive theoretical whole, and he did so by explicitly drawing both upon Metchnikoff and later ecological theory to devise a view of the immune system as the purveyor of organismal identity.[32] I read Burnet's personal history through a prism similar to the one I had used to dissect and reconstruct the Metchnikovian saga, namely with an appreciation for the dynamical and hierarchal properties of biological systems. But Burnet was not the best champion to carry the Russian's mantle. The "self" was a complex construction, and I argued that immunologists had different visions of selfhood as borrowed from various philosophical and psychological formulations. In my 1994 *The Immune Self, Theory or Metaphor?*, I argued that the "self" concept was developed along a continuum, stretching from "punctual" (i.e. defined, demarcated) to "elusive" views of identity.[33] The dominant view among immunologists was that there is a "self" and that it has borders defined by a genetic signature. The immune system is designed to react against the "foreign" and not against the host. When the immune system was in fact directed against the body, this was generally regarded as pathological *autoimmunity*, a condition Paul Ehrlich called "dysteleological in the highest degree" and which generations of immunologists believed to be true.[34] Not surprisingly, Metchnikoff thought that autoimmunity was expected because the immune system was always sensing the inner environment of the animal, seeking abnormal cells to destroy, whether originating from the host or invading pathogens.

Burnet, assuming the Ehrlich precept, sought a mechanism that demarcated "self" from "other" and thereby establish a "punctual" definition of the immune self. His formulation of immune identity has been the guiding principle of immunology, an orientation that I have critiqued extensively and in summary form.[35]

In brief, Burnet's theory held that during neonatal development, the animal exercised a purging function of self-reactive lymphocytes. All antigens (substances which evoke an immune response) encountered during this period would be ignored by the immune system and thus the "self" was defined negatively (i.e. tolerated). First presented in 1949 with Frank Fenner, this hypothesis was later developed into the clonal selection theory ("clonal" refers to all those cells that develop from a single cell—or a cluster of stimulated cells—so that a population of lymphocytes are developed that react with specific antigens).[36] According to this theory, lymphocytes with reactivity against host constituents are destroyed during development. Only those lymphocytes that are nonreactive would be left to engage the antigens of the foreign universe.[37] These potentially deleterious substances would select lymphocytes with high affinity for them, and through clonal amplification this population of lymphocytes would differentiate and expand to combat the offending agents, either directly through cytotoxic mechanism or through the production of antibodies.

By the 1970s most immunologists assumed that the theory was proven, but I regarded this vision of immune identity with considerable skepticism. Bountiful evidence in recent years had suggested that autoimmunity was a normal finding. Thus I was delighted to find Coutinho and Varela championing a novel orientation that accounted for the bi-directionality of immune reactivity. As they wrote in 1988:

> Clearly, one can define "self" from a biochemical or genetic or even *a priori* basis. But from our vantage point, the only valid sense of immunological self is the one defined by the dynamics of the network itself. What does not enter into its cognitive domain is ignored (i.e. it is non-sense). This is in clear contrast to the traditional notion that the IS [immune system] sets a boundary between self in contradistinction to a supposed non-self. From our perspective, there is only self and its slight variations.[38]

They repositioned the immune self, leaving it intact: but the critical turning point—their main contribution—was appreciating that the immune system in fact recognizes selfness as natural autoimmunity. Such host-directed reactivity is thus physiologically normal. In my developing view, this was the conceptual step that would lead to the immune self's ultimate deconstruction.

The significance of regarding autoimmunity as normal (as opposed to "dysteleological") has taken some time to sink into the collective consciousness of the discipline. Many practitioners still do not appreciate its wider ramifications. As Coutinho and Michel Kazatchkine later wrote:

> During this century, the evolution of concepts on autoimmunity could be summarized by "never, sometimes, always." Thus from the early "horror autotoxicus" [Ehrlich] to the 1960s, immune autoreactivity was simply not

considered.... With the first identification of autoreactive antibodies in patients and the subsequent conceptual association with autoaggressive immune behaviors, the "sometimes" phase was entered, necessarily equated with disease.[39]

Their position concurred with my own scientific judgment. I appreciated its contrast with the "one-way" definition of selfhood, where there is a genetic self whose constitutive agents see the foreign (immune reactivity arises from this polarized view with attack directed only against non-self). In his own formulation, Varela drew upon a definition of immune selfhood as analogous to the mind, which has no firm genetic boundaries, but rather takes form from experience and self-creative encounter.[40] Not surprisingly, since they emphasized the cognitive nature of immune function, Coutinho and his colleagues argued that the global properties of the immune system cannot be understood from analysis of component parts alone. Their conceptions encompassed "emergent properties," "global co-operativity," nonlinear network or complex systems, and other terms borrowed from the neurosciences, underscoring their affinity to methods already adopted for describing other complex cognitive systems.

Perceptive readers will note that Coutinho and his colleagues were still committed to the notion of selfhood, and I was becoming increasingly concerned with such a designation ruling immune function. There were then at least half a dozen different conceptions of what constituted the immune self, arrayed along a continuum ranging from a severe genetic reductionism to a complex construct employing different principles of organization.[41] With so much dispute surrounding the definition of self, I began to believe the "self" might be better regarded as only a metaphor for the immune system's silence, that is, its *non-reactivity*.[42] The theory that was built upon "the self" now appeared to have many *ad hoc* caveats and paradoxes. Perhaps the evolution of the original metaphor into theory was now yielding to another, different metaphorical construction.

Not knowing where that road might lead, I followed my intuition, based on a deep philosophical resonance with Coutinho. I engaged with a small group of immunologists who believed that they were part of, as Coutinho and one of his colleagues later wrote, "a major shift in the central paradigm of immunology."[43] My empathy for their re-configuration of immune theory stemmed from three prominent sources. First, I was repelled by the severe genetic reductionism heralded by the "molecular biology revolution," both within the Human Genome Project itself and the broader influence it was exercising as the Holy Grail of Biology on the practice of immunology.[44] Second, I was attracted to new concepts concerning hierarchical, self-organizing systems; geneticism was looking in the opposite direction. And third—perhaps directing the other tributaries of my thought—I embraced a philosophical bias that "the self" was a moral category, not epistemological, believing it highly probable that the immune self would deconstruct much as postmodern notions of self-hood have dissolved. I was particularly influenced by Nietzsche's multi-perspectivism, William James's *Principles of Psychology*, and Edmund Husserl's phenomenology. Whatever separated these philosophers, at least they held in common that Kant's cohesive self was a fiction. So what could the Immune Self be?

Figure 8.1 Niels Jerne sitting on a bench, reading, somewhere in Europe in the late 1960's. (Photographer unknown.)

Source: Medical Museion, University of Copenhagen.

With these rather ill-formed notions and intuitive leanings, I turned to Niels Jerne, the father of the "paradigm" shift—if in fact there was one.

Niels Jerne and the deconstruction of the self

I had interviewed Jerne in 1978, and of course knew his theory from reading one of its early expositions.[45] But I never had taken it seriously before this project (Figure 8.1). As an experimentalist, I found the theory well beyond my narrow research interests. I had not previously assessed its standing as a challenge to the prevailing model of immune function. Jerne had gone well beyond the then current notion of the immune network composed of lymphocyte subsets secreting immuno-stimulatory and inhibitory substances (essentially a simple mechanical model with interlaced, first-order feedback loops) to propose a novel conception of immune regulation. His network theory was born in the hope of modeling the immune system as analogous to the nervous system. The agenda behind Jerne's theory, from its very inception, was a complex amalgam: fitting the pieces of the regulation puzzle in place, along with an overriding desire to understand the immune system as a cognitive enterprise.

Organized along principles that exhibited some deep similarity with the brain, and manifesting behaviors that might be modeled as analogous to the mind, Jerne's hypothesis was in many respects a meta-theory. For beyond its efforts to elucidate the self-organizational principles that regulated the immune response, there lurked the larger concern: defining the immune system as a cognitive entity or process.

Jerne had embarked on his theoretical odyssey as early as 1960.[46] He embraced the cybernetic enthusiasm of the period, writing of the antibody-producing system as being "analogous to an electronic translation machine."[47] By the mid-1960s, he was dealing explicitly with the metaphorical meaning of immunological "memory" and "learning." Before his immune network hypothesis was formally proposed, he noted how immunologists used metaphors such as "recognition" that were obviously derived from psychology.[48] He drew even more explicit comparisons and contrasts with the nervous system.[49] Jerne saw that each system has a history of encounters with the world that remain present in two ways: in the form of irreversible changes and as memories that always affect the next response. Thus both the immune and nervous systems change with, and learn from, experience. Over the next decade, Jerne continued to draw explicit parallels and eventually used language to illustrate his own understanding of immune recognition.[50]

Jerne's idiotypic network theory hypothesis, extensively presented in 1974 and proposed in outline earlier, regarded antibodies as forming a highly complex interwoven system, where the various specificities "referred" to each other.[51] Under the general rubric of "cognition," Jerne conceived of the immune system as self-regulating, where antibody not only recognizes foreign antigen but is capable of recognizing self constituents as antigens. In Jerne's view, there was no essential difference between the "recognized" and the "recognizer," since he thought that any given antibody might serve either function, or both. Put another way, immune regulation was based on the reactivity of antibody (and later lymphocytes) with its own repertoire. Accordingly, the immune system formed a set of self-reactive, self-reflective, self-defining immune activities. Thus the key structure of Jerne's vision of the immune system consisted of interlocking recognizing units, where the foreign was perceived as a *perturbation* of that system. "Foreignness" was recognized by an already developed "vocabulary," whose immunogenic image was already represented in the library of lymphocytes (and their antibodies) and thus could be recognized.[52] From this formulation, Jerne's metaphors not only implied approximate parallels with the functions of the mind, but they also served as powerful directives for his theoretical conceptions—specifically, functional parallels with the brain and later with language.[53] (This is a matter I have discussed elsewhere.)[54]

Jerne's theory presented a radically altered view of immune selfhood. If the biological world were so easily divided between "host" and "foreign" constituents, then anything an antibody (or lymphocyte) encountered would be suitable for destruction according to Burnet's clonal selection theory. In that simplified world of self–non-self discrimination, the immune system learned these distinctions, generated an army of reactive antibodies and lymphocytes, and acted aggressively when an "antigen" was encountered. But Jerne coupled the simple antibody–antigen interactions to the far more complex and nondiscriminatory functions of the immune system that built upon *self*-recognition. Thus "autoimmunity" became the organizational rule to explain immune function. The Jernian network was fundamentally "dynamic" and "self-centered." It generated antibodies to its own antibodies, which he thought constituted the overwhelming majority of antigen present in the body. Strikingly, there is no explicit mechanism for self–non-self discrimination—and this apparent lacuna served as the nexus of critiques.[55] However, for Jerne, the need to define the self as distinct from the other seemed to recede from his primary theoretical concerns. This posture was to have important repercussions.

Jerne regarded the immune system as essentially self-reactive and interconnected; the "meaning" of immunogenicity, that is, *reactivity*, then must be sought in some larger framework. Antigenicity is only a question of degree, where "self" evokes one kind of response, and the "foreign" another. "Foreign" is based not on its intrinsic foreignness, but rather because the immune system sees that foreign antigen in the context of invasion, damage, or degeneracy. There is no foreignness per se, because if a substance were truly foreign, it would not be recognized. There would be no image by which the immune system might engage it. So the "foreign" becomes a perturbation of the system. As observers, we record the ensuing reaction. Only as third parties do we designate "self" and "non-self." From the immune system's perspective, it only knows itself. Thus reaction to the foreign is a by-product of this central self-defining function.

This hypothesis served as an exciting, fecund nexus for reformulating the entire question of how the immune system is organized. If there is a "self" in Jerne's theory, it is the entire immune system as it "senses" itself. Jerne's theory thus appeared radically different from the dominant theories of interlocking inhibitory–stimulatory activities that described immune function built from Burnet's original self–non-self dichotomy.

For me, Jerne's theory commanded a critical reaction to the entire conception of self–non-self. His idea resonated strongly with my own intuitions. Philosophically, I had become sensitive to the elusive character of personal identity: the self "existed" only when self-consciously invoked. We live in the world essentially unaware of ourselves as agents of action. We simply *are* and we simply *do*, and in the attempt to define the self as an entity we must self-reflect to produce a construct. Certainly, constructs are useful, but the lessons of analytic psychology reenforced my notions that such stories are essentially autobiographical narratives. Taking various forms, such stories characterize something called "the self" as a projection of what seems self-justifying to ourselves and explanatory to others. These stories are crucial for grounding behavior, establishing goals, conferring responsibility, bestowing coherence, etc., but I very much doubted their consistency, comprehensiveness, and veracity. In short, we are most certainly persons, individuals, and moral agents, but what is gained by claiming to be "a self?" The persona of selfhood is a mask— sometimes assigned, sometimes projected—and the idea had served as the map of my social and psychological world since I had first articulated it at age of 14 in my school's literary magazine. I saw Jerne struggling with the same imbroglio.

No wonder I was excited by Jerne's theory once I appreciated its implications. The immune system made no claim of defining "the self." For him, the inability to differentiate self from non-self was only to forfeit a false conceit. The system could only know itself, and the self beyond that system, an entity as it were, vanished. That his challenge was yet to be fully realized by the broader research community did not deter my enthusiasm. His writings propelled me to seek whether another theoretical basis for immunology might better accommodate contemporary findings than Burnet's older theory. In the letters he wrote me concerning certain historical details, Jerne appeared rather jubilant about my project and my alliance with his conceptual progeny.[56] Although I was enlisted in support of Jerne's general orientation, I had no intention of defending the network hypothesis as he formulated it; not because it did not have experimental standing, but rather because I believed its significance as an explanation of immune regulation remained problematic. Avoiding an alignment with Jerne's specific theory, I sought instead later developments that recommended the key conceptual turn proposed by his dismissal of self–non-self discrimination as the critical fulcrum of immune responses.

Jerne's legacy

Jerne had suggested analogies of the immune system with the mind and with language more specifically, and in his wake others followed these leads, specifically in semiotic terms[57] and more generally as a complement to the nervous system.[58] Indeed, contemporary theorists now represent immune function using models similar to those proposed for understanding neural cognition. In their view, to engage its targets, the immune system must first perceive them and then, in a sense, *decide* whether to react. This is a cognitive model, where the immune and nervous systems are regarded analogously. The immune and nervous systems each have perceptive properties; each of them have capabilities to discern both internal and external universes; each processes information so that their respective perceptive properties are linked to effector systems (muscles or lymphocytes, the basic process was the same).[59] Models based on neural networks, complete with analogous computer program simulations, suggested new research directions by the late 1980s.[60] Yet thus far, these efforts have generated little interest, either in devising experimental strategies or success in predicting research outcomes. In short, their utility remains to be shown.

Even if these models have not been particularly influential, investigators with a cognitive perspective have pursued exciting laboratory findings and extrapolated them to support this new orientation. Indeed, the cognitive formulation has become an explicit mode for organizing theoretical discussion among a small group of immunologists. Irun Cohen, probably the first to explicitly declare that a new "cognitive paradigm" had eclipsed the clonal selection theory, organized the first conference dedicated to exploring this theoretical shift.[61] Held at the Weizmann Institute of Science in Rehovot, Israel in April 1994, the "Symposium on Immunology as a Cognitive Science" attracted a diverse group of immunologists (both experimentalists and theorists), cognitive scientists, psychologists, historians, and philosophers of science.

Listening to these presentations, I quickly realized that there was little agreement about the application of "cognition" to immune theory. I profited enormously from the meeting nevertheless. I obtained a well-directed update in cognitive psychology (Benny Shanon), self-organizational dynamics (Henri Atlan), computer simulations (Alan Perelson and John Stewart), and post-Jernian theorizing (Cohen and Coutinho).

But perhaps most importantly, I realized that few, if any, were willing to dispense with immune selfhood, and from that point onward my own contribution began to emerge in clear form.

By then I had written the first draft of *The Immune Self*. My revisions were based on my experience in Rehovot, which had convinced me that "self-hood" still framed discussions of the immune system's organization. Yet the immune self was no better characterized than the self of our everyday experience. The term refracted many meanings and perhaps thereby had lost its original function. So my book, which had begun as a narrow historical account of the "self" concept in immunology, became a study of scientific thinking. Specifically it became an explication of how a metaphor was constructed, and why. Later I extended my history to a more expansive philosophical interpretation.[62]

At the Rehovot meeting, Jerne's idiotypic theory had been mentioned only in passing. Yet it was apparent to me that his ideas had filtered into the immunological community in diverse ways. I saw this as the problem of "meaning": that is, how does an antigen become antigenic and evoke a response?[63] For me, contextual meaning seemed to hold together modern immunology. Jerne's ideas of the network being "perturbed" suggested this, as did the dominant model concerning lymphocyte activation. This latter view held that specific recognition of an antigen by a lymphocyte receptor is not sufficient for activation; additional signals determine whether a cellular response or cell inactivation follows. In short, an antigen is neither self nor non-self, except as it attains its "meaning" within a broader construct.[64]

In recent years, researchers have debated what constitutes the milieu of "meaning" of antigenicity, and these discussions have spawned provocative and potentially important models of immune regulation. I believe that these more recent developments in immune theorizing inspired by Jerne's formulation herald a shift in the very foundations of how immune regulation might be understood. It is here that I detected the exposition of a postmodern ethos. The entire contra-Burnetian perspective rests on recognizing the "relativity" of perspectives. For the context of the immune encounter is paramount in conferring meaning on any antigen. Once the self–non-self structure is weakened, more radical perspectives may be entertained.[65] Indeed, the *New York Times* reported that the Burnetian paradigm was being threatened. Reporting on three different experimental scenarios that had appeared in a single issue of *Science*, the *Times* informed its readers about the apparent failure of previously well-accepted self–non-self discriminatory boundaries.[66]

Echoes from this major challenge to the self–non-self paradigm quickly spread well beyond the esoteric musings of a few investigators. Some immunologists embraced the news with the enthusiasm of a palace revolt. "We're challenging 50 years of immunological thought," exclaimed Paul Lehrmann. Others equivocated: "In a way, the new studies undermine what has been taken as a pillar of the self-nonself model," mused Albert Bendalac. "That doesn't mean the model is necessarily wrong. But the reports undermine its foundations." Others simply denied it: "This is being blown so far out of proportion...I don't think the studies fundamentally challenge the self-nonself theory," Alfred Singer argued. Charles Janeway, Jr took the cautious middle ground: "I think the work is an extension of the theory rather than a direct contradiction."[67] In the early twentieth-first century, this issue remains unresolved. Whether the self–non-self paradigm falls or stands, these new findings at a minimum have highlighted paradoxes that demand explanation.[68] Irrespective of the final verdict, the challenge to the self model has become big news and a vindication of sorts for my project. Some commentators were more generous and thought me prescient.

Why did I embrace this particular interpretation of immunology's history? Why did I champion Jerne's cognitive orientation and its later advocates? I can hardly offer a comprehensive explanation, but here is an outline of a rationale: by assuming a cognitive stance, they offered an organismic orientation to immunology with which I sympathized. I saw traces of the dialectical biologist, one whose anti-reductionist thinking reflected my own thought style. Their arguments required more complex and subtle models than simple mechanical systems extrapolated from nineteenth-century biology. Resisting this earlier reductionist approach, I sought to champion a history of ideas that had lapsed but was now invigorated by respected—but clearly unorthodox—laboratory-based immunologists. For me, they were heralds of a new biology: a contemporary science that tapped into a neglected tradition, one I traced back to Metchnikoff. This brought me to their camp, but their science convinced me of the fecundity of their ideas for my own purposes. In the end, they became, in some cases inadvertently, allies to my own argument against the centrality of the self, which had originated from a philosophical analysis of immunology's theory, but which new findings supported.

I would summarize my critique of late twentieth-century immunology as arising from two tributaries: one an assessment of autoimmunity and the inadequate theory to explain it, and the other closely linked problem of understanding the failure of the key model/metaphor of immunology—the immune self—to sustain the discipline's dominant theory. The immune self could only survive by adopting *ad hoc* modifications to its standing definition, and therefore it was supported neither by its own scientific criteria, nor by—as I analyzed the theory—its metaphorical baggage. Historical analysis evolved into philosophical studies and was finally sustained by a reinterpretation of the laboratory findings.

Reflections on my writing: an existentialist note

As readers will recall, this project began through my collaboration with Leon Chernyak, whose own commitment to philosophy revealed itself in attempts to define a "metaphysics" for Metchnikoff's scientific orientation.[69] Metaphysics is used loosely here. I simply wish to indicate that we sought to situate Metchnikoff in that elusive domain where deeply embedded philosophical assumptions and orientations mediate, if not determine, our thought. Chernyak's conviction that we cannot escape our metaphysics sharpened a view I too had held. In believing that we might delineate the underlying philosophical basis of Metchnikoff's science, we were preoccupied with exposing the deepest intellectual sources of his experimental program. We recognized Metchnikoff's own holistic view of the organism, including his integrated and comprehensive approach to his biology as theory-driven and his neo-Romantic outlook, which permeated his larger views of biological thought and humankind. In short, we endeavored to place him within the broader intellectual currents of his period. It was from this perspective that we discerned the question of identity at the heart of his theory—and more specifically, the notion of disharmony. This idea of struggle turned inward was most characteristic of Metchnikoff's phagocytosis theory. Much of our study was devoted to tracing his theory's multiple sources. Although we discussed Metchnikoff's personality and how it might have figured in his rivalries and professional growth, ours was hardly a psycho-biography (our primary concern here was the original reading he offered of Darwin).[70] The history of Metchnikoff's idea of *identity* framed the discussion. On this foundation, I expanded my own history of immunology.

As a physician, I was sensitive to a holistic orientation for biology—having learned from personal experience the dangers of compartmentalizing the patient into disordered functions, each approached piecemeal. Deeply concerned with identifying how contemporary medicine developed a reductionist view of the ill, I regarded this issue as a key problem of my profession. Soon after we initiated the Metchnikoff project, these wider implications for medicine were charted and extended.[71] Thus the problem of understanding organismal identity was always on my agenda. As I continued to write my history of immunology, concern with tracking the origins of our contemporary biological reductionist attitudes remained a high priority for me.

In my 1994 *The Immune Self*, one of my goals was an attempt to show the limits of a reductionist approach to what was essentially a problem of systems regulation. To achieve that goal, I again employed the strategy of placing this question in its widest intellectual context. My discussions were framed by the dynamical, dialectical, and indeterminate nature of biological identity. I ended this book with a rather circumspect view of defining selfhood in the immediate scientific context and, even more broadly, as an *epistemological* agent. Here the resonance with postmodernism became transparent. Further, and more importantly, I began to recognize identity as a moral category. I wrote several essays on the ethical construction of the self.[72] My concern with delineating medicine's "loss of the patient" had finally presented itself as an ethical issue. It is this problem that has most intrigued me lately, resulting in another published set of "confessions."[73]

Looking back, I can see that my philosophical historiography was driven by twin concerns: a self-evident epistemological exercise and, in a more latent form, an exploration in moral philosophy (a dimension that I became only aware of in retrospect). By "moral," I refer to the value-ladenness of what we see and understand, whether we engage in scientific or historical discourses. Not only is knowledge itself valued— that is, used for particular ends—but knowledge is constituted by an ordering, a prioritization of interest, which confers a particular character on observations, facts, and theory. If we admit that epistemology is constituted, at least in part, by a value-driven perspective or *telos*, then we must measure this factor when we reflect on an epistemological project. This close connection between epistemology and moral agency has also served as the dominant theme of my more general writings on science in culture.[74] In particular, I am concerned about the need to recognize the incipient ethical orientation that may easily influence the interpretation of scientific data used for social ends. For instance, debates about the biological determinism of complex social behaviors such as alcoholism, homosexuality, or violence have found ideologues using scientific data for their own purposes, but whose rationales cannot be finally decided by such appeals to "scientific objectivity."[75]

Many scientists seek an "objective" history by standards they believe applicable from their experience as researchers.[76] Yet the science studies community widely accepts that we can offer only interpretation. The "only" is ironic, for it *is* interpretation that we seek. The most fecund narratives will expand our understanding of the science and its cultural matrix. It is from this vantage that we begin interpretative history; one that breaks from the intellectual confines of any particular science narrowly construed to allow a more expansive and contextualized reading. I regard my own historiography as "layered" with various kinds of interpretation. In tracing Metchnikoff's genesis of the phagocytosis theory, or Burnet's development-of-the-self notion, or Jerne's network theory, I endeavored to show how laboratory data were organized by deeper theory or metaphor. These in turn may reflect "extra-curricular" intellectual or cultural influences. As I affirmed at the beginning of this essay, such analysis must begin with some foundation, an orienting perspective. By recognizing the limits of an intellectual or ideological orientation, and revealing identifiable cultural and metaphysical tethers, the historian can claim some ability to fathom the vague limits of his or her efforts, while at the same time acknowledging that no matter the degree of self-consciousness, one is left with interpretation.

This self-reflective consciousness was clearly enunciated by the Dutch historian Peter Geyl, whose observation concerning the historian's craft was published precisely at the same time as Burnet proposed his theory of the self:

> [As] soon as there is a question of explanation, of interpretation, of appreciation, though the special method of the historian remains valuable, the personal element cannot be ruled out, that point of view which is determined by the circumstances of his time and by his own preconceptions. Every historical narrative is dependent upon explanation, interpretation, appreciation. In other words we cannot see the past in a single, communicable picture except from a point of view, which implies a choice, a personal perspective.[77]

These caveats seem to me to be characteristic of a postmodern sensibility and are assessments that also apply to me. I appreciate, perhaps in an ever recursive spiral, how self-reflection itself influences the history I have attempted to capture,[78] and from a less personal vantage I would concur with Edward Hallett Carr's circumspect assessment of the historian's art:

> Study the historian before you begin to study the facts.... The facts are really not at all like fish on a fishmonger's slab. They are like fish swimming about in a vast and sometimes inaccessible ocean; and what the historian catches will depend partly on chance, but mainly on what the historian catches will depend partly on chance, but mainly on what part of the ocean he chooses to fish in and what tackle he chooses to use – these two factors being, of course, determined by the kind of fish he wants to catch. By and large the historian will get the kind of facts he wants. History means interpretation.[79]

Acknowledgment

This essay is dedicated to the memory of Professor Lily E. Kay, a colleague of like mind and a friend in arms.

Notes

1 For instance, I have in mind "European Ego-Histories: Historiography and the Self, 1970–2000," *Historein* 3 (2001).

2 Alfred I. Tauber, "Introduction: Speculations Concerning the Origins of Self," in *Organism and the Origins of Self*, ed. Alfred I. Tauber (Dordrecht: Kluwer Academic Publishers, 1991), 1–39. John Tyler Bonner, *Lives of a Biologist* (Cambridge: Harvard University Press, 2002).

3 Alfred I. Tauber and Leon Chernyak, *Metchnikoff and the Origins of Immunology* (New York and Oxford: Oxford University Press, 1991).

4 Ibid.; Alfred I. Tauber, "The Immunological Self: A Centenary Perspective," *Perspectives in Biology and Medicine* 35 (1991): 74–86; Tauber, "Speculations." The modern resurrection of competing cell lineages vying for hegemony as an evolutionary problem has been explored by Leo W. Buss, whose views Chernyak and I viewed with mixed reaction (Alfred I. Tauber, "Metchnikoff, the Modern Immunologist," *Journal of Leukocyte Biology* 47 (1990): 560–566; Tauber and Chernyak, *Metchnikoff*. Noteworthy for our own project, Buss's *The Evolution of Individuality* (Princeton, NJ: Princeton University Press, 1987) was far closer to our thematic concerns than any other scholarship in immunology or in science studies for that matter.

5 Alfred I. Tauber, *The Immune Self: Theory or Metaphor?* (New York and Cambridge: Cambridge University Press, 1994), 20: 62–63. Metchnikoff's evolutionary biology papers are now available in English: Helena Gourko, Don Williamson, and Alfred I. Tauber, eds, *The Evolutionary Biology Papers of Elie Metchnikoff* (Dordrecht: Kluwer Academic Publishers, 2000).

6 Metchnikoff's most explicit statement describing the protean roles of the phagocyte can be found in his short paper, "The struggle for existence between parts of the animal organism," published in 1892 shortly after he delivered his famous Paris lectures on comparative inflammation. While his later work emphasized the role of the phagocyte in combating pathogens and repair of injury in adult animals, his 1892 paper gave a broad overview of phagocyte function in normal development and body economy. By drawing explicit parallels between phagocytes devouring the tadpole's tail—"eaten" at the appropriate time of metamorphosis—and wound repair or bacterial killing, Metchnikoff clearly regarded the phagocyte's role in the evolutionary drama as essentially unchanged in these various settings or by the species in which they were observed. See E. Metchnikoff, "The struggle for existence between parts of the animal organism," in Gourko, *Biology Papers*, 207–216; see also E. Metchnikoff, *Lectures on the Comparative Pathology of Inflammation*, trans. F. A. Starling and E. H. Starling ([1892] 1893 reprinted 1968 by New York: Dover).

7 Tauber, *Immune Self*; Alfred I. Tauber, "The Organismal Self: Its Philosophical Context," in *Selves, People, and Persons*, ed. Leroy Rouner (South Bend, IN: Notre Dame University Press, 1992), 149–167; Alfred I. Tauber, "A Typology of Nietzsche's Biology," *Biology and Philosophy* 9 (1994): 24–44.

8 Alfred I. Tauber, "A Case of Defense: Metchnikoff at the Pasteur Institute," in *L'Immunologie: l'Heritage de Pasteur*, ed. P. A. Cazenave and G. P. Talwar (New Delhi: Wiley Eastern Limited, 1991), 21–36.

9 P. Baumgarten, "Referte," *Berl. Klin. Woch.* 21 (1884): 802–804 and 818.

10 Alfred I. Tauber, "The Birth of Immunology: III. The Fate of the Phagocytosis Theory," *Cellular Immunology* 139 (1992): 505–530; Tauber, *Immune Self*.

11 Tauber, "Immunological Self"; Tauber, "Speculations."

12 Heavily indebted to Richard Levins and Richard Lemontin's *The Dialectical Biologist* (Cambridge: Harvard University Press, 1985), I proceeded to explore the false security that the genome offered to establish organismal identity: Tauber, *Immune Self*; Tauber, "Speculations"; Alfred I. Tauber, "The Molecularization of Immunology," in *The Philosophy and History of Molecular Biology: New Perspectives*, ed. Sahotra Sarkar (Dordrecht: Kluwer Academic Publishers, 1996), 125–169; Alfred I. Tauber and Sahotra Sarkar, "The Human Genome Project: Has Blind Reductionism Gone Too Far?" *Perspectives in Biology and*

Medicine 35, no. 2 (1992): 220–235. Alfred I. Tauber and Sahotra Sarkar, "The Ideology of the Human Genome Project," *Journal of the Royal Society of Medicine* 86, no. 9 (1993): 537–540. The limits of the genetic program are exposed by Lily E. Kay, *Who Wrote the Book of Life? A History of the Genetic Code* (Palo Alto: Stanford University Press, 2000).

13 Tauber and Chernyak, *Metchnikoff*; Tauber, "Modern Immunologist."

14 Tauber and Chernyak, *Metchnikoff*; Thomas Söderqvist and Craig Stillwell, "Immunological Reformulations," *Science* 256, no. 5059 (1992): 1050–1052.

15 Alan S. Perelson, ed., *Theoretical Immunology: The Proceedings of the Theoretical Immunology Workshop held June, 1987 in Santa Fe, New Mexico*, 2 vols (Redwood City, CA: Addison-Wesley Publishing Co., 1988).

16 F. J. Varela, A. Coutinho, B. Dupire, and N. M. Vaz, "Cognitive Networks: Immune, Neural, and Otherwise," in *Theoretical Immunology*, vol. 2, ed. Alan S. Perelson (Redwood City, CA: Addison-Wesley Publishing Co., 1988), 359–375.

17 Ibid., 363.

18 Huberto R. Maturana and Francisco J. Varela, *Autopoiesis and Cognition. The Realization of the Living* (Dordrecht and Boston, MA: D. Reidel Publishing Co., 1980).

19 The entire autopoietic argument that oriented Varela's approach to the immune system was criticized in Tauber and Chernyak, *Metchnikoff*, but the radical "decenteredness" and "innerdirectedness" of the Paris School notion of selfhood were conducive to my own thinking, a position Coutinho had developed before his collaboration with Varela. See Antonio Coutinho, L. Forni, D. Holmberg, F. Ivars, and N. Vaz, "From an Antigen-Centered Clonal Perspective of Immune Response to an Organism-Centered, Network Perspective of Autonomous Activity in a Self-Referential Immune System," *Immunological Reviews* 79 (1984): 151–168.

20 Söderqvist and Stillwell, "Immunological Reformulations," 1050.

21 Ibid., 1051.

22 Thomas Söderqvist, "How to Write the Recent History of Immunology—Is the Time Really Ripe for a Narrative Synthesis?" *Immunology Today* 14, no. 11 (1993): 565–568; Myles Jackson, Warwick Anderson, and Barbara Rosenkrantz, "Toward an Unnatural History of Immunology," *Journal of the History of Science* 27, no. 3 (1994): 575–594.

23 Horace F. Judson and Ian R. Mackay, "History in the Bay of Naples," *Immunology Today* 13, no. 11 (1992): 459–461. For a severe criticism of their evaluation, see Söderqvist, "How to Write."

24 *Immunology: The Making of a Modern Science*, ed. Richard B. Gallagher, Jean Gilder, Gustav J. V. Nossal, and Gaetano Salvatore (London and San Diego, CA: Academic Press, 1995).

25 I was better received at other scientific forums. For instance, at the National Institutes of Health I delivered a well-received lecture (March 31, 1993) to a large and sophisticated scientific audience. The experience encouraged me to publish the general thesis in a scientific journal: Alfred I. Tauber, "The Immune Self: From Theory to Metaphor," *Immunology Today* 15 (1994): 134–136.

26 Tauber and Chernyak, *Metchnikoff*; Tauber, "Defense"; Tauber, "Speculations"; Tauber, "Modern Immunologist."

27 Arthur M. Silverstein, *A History of Immunology* (San Diego, CA: Academic Press, 1989).

28 Tauber, *Immune Self*; Tauber, "Molecularization."

29 Edward S. Golub and Douglas R. Green, *Immunology: A Synthesis*, 2nd ed. (Sunderland, MA: Sinauer, 1991).

30 Jan Klein, *Immunology* (Boston, MA and Oxford: Blackwell Scientific Publications, 1990).

31 The first critical appraisal of how the self concept structures immunology was offered by Ilana Lowy, "The Immunological Construction of the Self," in *Organism and the Origins of Self*, ed. Alfred I. Tauber (Dordrecht: Kluwer Academic Publishers, 1991). It is further developed with a different strategy in Tauber, *Immune Self*; a recent expansion of the heuristic basis of the self concept appears in Eileen Crist and Alfred I. Tauber, "The Phagocyte, the Antibody and Agency in Immunity: Contending Turn-of-the-Century

Approaches," in *Singular Selves: Historical Issues and Contemporary Debates in Immunology*, ed. A. M. Moulin and A. Cambrosio (Paris: Elsevier, 2001), 115–139.

32 Tauber, *Immune Self*; Scott H. Podolsky and Alfred I. Tauber, *Generation of Diversity: Clonal Selection Theory and the Rise of Molecular Immunology* (Cambridge: Harvard University Press, 1997). See also the biography by Christopher Sexton, *The Seeds of Time: The Life of Sir Macfarlane Burnet* (Oxford: Oxford University Press, 1991).

33 Tauber, *Immune Self*.

34 Ibid.

35 Ibid. I have further developed my original critical stance in Podolsky and Tauber, *Diversity*, chapter 9, and Alfred I. Tauber, "Conceptual Shifts in Immunology: Comments on the 'Two-Way Paradigm'," *in Paradigm Changes in Organ Transplantation*, ed. Kenneth F. Schaffner and Thomas E. Starzl, *Theoretical Medicine and Bioethics* 19 (1998): 457–473. See also Alfred I. Tauber, "The Elusive Immune Self: A Case of Category Errors," *Perspectives in Biology and Medicine* 42 (1999): 459–474; Tauber, "Molecularization."

36 Frank Macfarlane Burnet and Frank Fenner, *The Production of Antibodies*, 2nd ed. (Melbourne: Macmillian and Co., 1949).

37 Frank Macfarlane Burnet, *The Clonal Selection Theory of Acquired Immunity* (Nashville, TN: Vanderbilt University Press, 1959).

38 Coutinho, Forni, Holmberg, Ivars, and Vaz, "Cognitive Networks."

39 Antonio Coutinho and Michel Kazatchkine, "Autoimmunity Today," in *Autoimmunity: Physiology and Disease*, ed. Antonio Coutinho and Michel Kazatchkine (New York: Wiley-Liss, 1993).

40 F. J. Varela and A. Coutinho, "Second Generation Immune Networks," *Immunology Today* 12 (1991): 159–166.

41 Polly Matzinger, "Tolerance, Danger, and the Extended Family," *Annual Review of Immunology* 12 (1994): 991–1045.

42 This position is also problematic, since this silence might be actively attained through tolerance mechanisms, as witnessed by chimeric transplants. See Thomas E. Starzl and Anthony J. Demetris, "Transplantation Milestones Viewed with One- and Two-Way Paradigms of Tolerance," *Journal of the American Medical Association* 273, no. 11 (1995): 876–879.

43 Coutinho and Kazatchkine, "Autoimmunity Today."

44 Tauber and Sarkar, "Blind Reductionism"; Tauber and Sarkar, "Ideology." My critique of the reductionist approach was based on the concern of its hegemony over other modes of investigation. Certain questions require elemental analysis; other kinds of problems refer to organization, where a reductionist approach is generally unhelpful. See Alfred I. Tauber, "The Ethical Imperative of Holism in Medicine," in *Limits of Reductionism*, ed. Marc H. V. Van Regenmortel and David L. Hull (New York: Jossey-Bass, 2001). Thus a key contribution, and triumph, of molecular biology was the successful explanation of how antibody diversification is achieved, as documented in Podolsky and Tauber, *Diversity*. But to understand the "meaning" of an antigen, that is, how it is to be processed requires a different level of analysis, one that looks at the system as a whole.

45 Niels K. Jerne, "The Immune System," *Scientific American* 229, no. 1 (July 1973): 52–60.

46 For a comprehensive reconstruction of Jerne's scientific life and work, see Thomas Söderqvist, *Science as Autobiography: The Troubled Life of Niels Jerne* (New Haven, CT: Yale University Press, 2003).

47 Niels K. Jerne, "Immunological Speculations," *Annual Review of Microbiology* 14 (1960): 341–358.

48 Ibid.

49 Niels K. Jerne, "Antibodies and Learning: Selection Versus Instruction," in *The Neurosciences*, ed. Gardner C. Quarton, Theodore Melnechuk, and Francis O. Schmitt (New York: The Rockefeller University Press, 1967), 200–205.

50 Niels K. Jerne, "The Generative Grammar of the Immune System," *EMBO Journal* 4, no. 4 (1985): 847–852; Niels K. Jerne, "The Generative Grammar of the Immune System," in *Nobel Lectures, Physiology or Medicine 1981–1990*, ed. Jan Lindsten (Singapore: World

Scientific Publishing Co., 1993), 211–225; Niels K. Jerne, "Idiotypic Networks and Other Preconceived Ideas," *Immunological Reviews* 79 (1984): 5–24.

51 Jerne, "Immune System"; Niels K. Jerne, "What Precedes Clonal Selection?" in *Ontogeny of Acquired Immunity* (Amsterdam: Elsevier Science, 1971): 1–15.

52 Jerne, "Immune System."

53 Jerne, "Generative Grammar," *EMBO*; Jerne, "Generative Grammar," *Nobel*; Niels K. Jerne, "Towards a Network Theory of the Immune System," *Annals of the Institute Pasteur/Immunology* 125, no. C (1974): 373–389.

54 Alfred I. Tauber, "Historical and Philosophical Perspectives on Immune Cognition," *Journal of the History of Biology* 30 (1997): 419–440; Podolsky and Tauber, *Diversity*.

55 Jerne, "Generative Grammar."

56 Niels K. Jerne, September 21, 1992 (privately held by author).

57 *The Semiotics of Cellular Communication in the Immune System*, vol. H23, *NATO ASI Series*, ed. E. E. Sercarz, F. Celada, N. A. Mitchison, and T. Tada (Berlin: Springer-Verlag, 1988).

58 David L. Felten, Robert Ader, and Nicholas Cohen, ed., *Psychoneuroimmunology*, 2nd ed. (San Diego, CA: Academic Press, 1991).

59 Ibid.

60 Irun R. Cohen and Henri Atlan, "Network Regulation of Autoimmunity: An Automaton Model," *Journal of Autoimmunity* 2, no. 5 (1989): 613–625; Francisco J. Varela, Antonio Coutinho, and John Stewart, "What is the Immune Network for?" in *Thinking About Biology: An Invitation to Current Theoretical Biology*, ed. W. D. Stein and F. J. Varela (Reading, MA: Addison Wesley, 1993); John Stewart, Francisco J. Varela, and Antonio Coutinho, "The Relationship between Connectivity and Tolerance as Revealed by Computer Simulation of the Immune Network: Some Lessons for an Understanding of Autoimmunity," *Journal of Autoimmunity* 2 (Suppl.) (1989): 15–23; Perelson, *Theoretical Immunology*; Franco Celada and Philip E. Seiden, "A Computer Model of Cellular Interactions in the Immune System," *Immunology Today* 13, no. 2 (1992): 56–62.

61 Irun R. Cohen, "The Cognitive Paradigm Challenges Clonal Selection," *Immunology Today* 13, no. 11 (1992): 441–444.

62 Tauber, "Immune Cognition"; Podolsky and Tauber, *Diversity*.

63 The "representational" sense of an antigen, that is, that it carried its meaning with an intrinsic property, was dismissed and replaced with the notion that meaning was derived from the antigen's *context*. Although we begin with a representational foundation, each word has a spectrum of definitions, both the particular choice and dimension, or latitude of meaning is conferred by the surrounding context in which the word is used. See Benny Shanon, *The Representational and the Presentational: An Essay on Cognition and the Study of Mind* (London: Harvester-Wheatsheat, 1993). When I say "Let's go to the bank," do I mean to go to a building and get some money, or to the river and go fishing? Multiple contextual elements confer specificity to words, and interpretations are delimited by those supporting structures.

64 Tauber, "Immune Cognition"; Podolsky and Tauber, *Diversity*.

65 This credo is developed most explicitly by Polly Matzinger who, in abandoning the "structural" model of selfhood, arrives at a purely functional formulation where the immune system decides what is insultive to the organism, that is, what causes distress, destruction, or non-programmatic death, and through signals of such aberrancy, immune reactions are initiated. Selfhood, per se, recedes as the basis of immune definition; immunity becomes organismally driven (i.e. functional), immunocytes become dependent on extra-immune factors and context. Matzinger, "Tolerance." For a fuller discussion see Tauber, "Elusive"; Tauber, "Immune Cognition"; Podolsky and Tauber, *Diversity*.

66 Podolsky and Tauber, *Diversity*; for the *New York Times* quotes, see George Johnson, "Findings Pose Challenge to Immunology's Central Tenet," *New York Times*, March 26, 1996, sec. C, p. 1.

67 Johnson, "Findings."

68 Elizabeth Pennisi, "Teetering on the Brink of Danger," *Science* 271, no. 5267 (1996): 1405–1408.
69 Leon Chernyak and Alfred I. Tauber, "The Idea of Immunity: Metchnikoff's Metaphysics and Science," in *Journal of the History of Biology* 23 (1990): 187–249. Discussed in more detail in Tauber, "Ethical Imperatives."
70 See Gourko, *Biology Papers* for Metchnikoff's scientific reaction to Darwin. But beyond his scientific attitudes, Metchnikoff was existentially almost preoccupied with self-realization (or self-actualization), a Darwinian mode characteristic of his time.
71 Alfred I. Tauber and Leon Chernyak, "Metchnikoff and a Theory of Medicine," *Journal of the Royal Society of Medicine* 82 (1989): 699–701; Alfred I. Tauber, "Darwinian Aftershocks: Repercussions in Late Twentieth Century Medicine," *Journal of the Royal Society of Medicine* 87 (1994): 27–31; Tauber, "Immunological Self"; Alfred I. Tauber, "Tales of the Neglected (Orphaned?) Historiographies," in *Singular Selves: Historical Issues and Contemporary Debates in Immunology*, ed. A. M. Moulin and A. Cambrosio (Paris: Elsevier, 2001), 247–258. The ethical implications were developed in Alfred I. Tauber, *Confessions of a Medicine Man: An Essay in Popular Philosophy* (Cambridge, MA: The MIT Press, 1999).
72 See, for example, Alfred I. Tauber, "From the Self to the Other: Building a Philosophy of Medicine," in *Meta Medical Ethics, the Philosophical Foundations of Bioethics*, ed. Michael A. Grodin (Dordrecht: Kluwer Academic Publishers, 1995): 149–195. All of these themes concerning the self are developed in Alfred I. Tauber, *Henry David Thoreau and the Moral Agency of Knowing* (Berkeley, CA and Los Angeles: The University of California Press, 2001).
73 Tauber, *Confessions*.
74 Alfred I. Tauber, "Ecology and the Claims for a Science-Based Ethics," in *Philosophies of Nature: The Human Perspective, in Celebration of Erazim Kohak*, ed. Robert S. Cohen and Alfred I. Tauber (Dordrecht: Kluwer Academic Publishers, 1998), 185–206; Tauber, *Thoreau*; Tauber, "Immune Cognition."
75 Tauber and Sarkar, "Ideology"; Tauber, "Ecology"; Alfred I. Tauber, "Is Biology a Political Science?" *Bioscience* 49 (1999): 479–486.
76 In contrast, the history offered in Gallagher is a good example of the "internalist" or "linear" genre that makes little allowance for interpretative latitude: Richard B. Gallagher, *Immunology: The Making of a Modern Science* (London: Academic Press, 1995).
77 Peter Geyl, *Napoleon, For and Against* (New Haven, CT: Yale University Press, 1949). My reflections on this matter are presented in Alfred I. Tauber, "Book Review Essay of *The Historiography of Contemporary Science and Technology*, ed. Thomas Söderqvist (Amsterdam: Harwood Academic Publishers, 1997)," *Science, Technology, and Human Values* 24 (1999): 384–401; see also Tauber, *Thoreau*.
78 I have attempted when writing on immunology in the postmodern context to differentiate its history writing from the science itself. The former may easily be understood in that genre, for example: Alberto Cambrosio and Peter Keating, *Exquisite Specificity: The Monoclonal Antibody Revolution* (Oxford: Oxford University Press, 1995) and Alfred I. Tauber, "Postmodernism and the Immune Self," *Science in Context* 8 (1995): 579–607. But this is not to say that the "science" is postmodern, that is, governed (or even inspired) by newer notions of causality, for example: Stuart A. Kaufman, *The Origins of Order: Self-Organization and Selection in Evolution* (New York and Oxford: Oxford University Press, 1993) and organizational design, for example: Henri Atlan, *L'organization biologique et la theorie de l'information* (Paris: Hermann, 1992).
79. E. H. Carr, *What is History?* (New York: Vintage Books, 1961): 26.

Part IV

Secrecy, politics, and science

Probing the meaning of the Cold War

9 The politics of phosphorus-32

A Cold War fable based on fact

John Krige

Prologue

Two—and only two—of the characters in my story, Giuseppe Di Tomaso and Prof. Dr Angelo Rincaldi, are fictional. Yet they are historically plausible people. Di Tomaso is an Italian crane driver, injured in an accident at the docks. It is 1949, and he is being treated in Rincaldi's sanatorium in Trieste, Italy. Though their words and actions are fabricated, the people with whom they interact, and who are constitutive of the networks in which they are embedded, are not (as the notes make clear).

Most studies of the relations between recent science and US foreign policy follow, often in great detail, the evolution of the policy as the negotiated outcome of debates between the elites who were involved in framing it. They are usually limited to the views of American scientists (with inputs from their senior colleagues abroad) and of American administrators: they provide a view of "the world according to Washington."[1] The central section of this paper follows this traditional pattern.

While essential, such approaches tell us little about how those general policies, once implemented, were interpreted and given meaning in specific local circumstances, and how they were used to promote local aspirations and agendas that sometimes overlapped, and sometimes did not, with the aims of the American scientists and administrators who originally hammered out their general framework.[2]

This chapter aims to overcome this limitation.[3] Science, American foreign policy, and the cultural politics of the Marshall Plan materially intersect with each other in the thoughts and lives of our two heroes, Giuseppe and Angelo. The primary node drawing these disparate strands together and giving them meaning in their lives was the need to cure a potentially lethal disease. That anchoring, however, also engaged their attitudes towards the United States, their love for the then-occupied city where they lived—Trieste—and their desire to see it free from external domination. The radioactive isotope phosphorus-32 (^{32}P) was a polyvalent, plastic substance invested with a multitude of meanings by diverse actors depending on the local context in which they encountered the element, be that Washington or Rome, Oak Ridge or Trieste . . . or one of the many other places in the world where, thanks to the nuclear piles built during the Manhattan project, it was cheaply produced in large quantities and made available for research or medicinal purposes. This story is thus no less a contribution to the social history of the Cold War.

Introduction

Giuseppe Di Tomaso tossed and turned in his hospital bed. It was June 1949. The ward was suffocatingly hot. He longed to feel the gentle caress of the breeze off the Adriatic Sea that kept Trieste livable in the summer. He longed to be back at work in the shipyards, to linger at the harbor when his brother brought in the day's catch, to then go home and enjoy a huge bowl of *brodeto a la triestina* which his wife had expertly prepared from fresh fish and vegetables.

Instead he was trapped in a stuffy room with several other men in the city's Provincial Neurological Sanatorium. What was more, Giuseppe Di Tomaso was dying.

The headaches had triggered the alarm. Well, not just the headaches. Giuseppe had had a totally unexpected seizure several weeks earlier. He had lost control of the crane he had operated flawlessly for years: the boom had swung violently, and its load had crashed to the ground, almost killing one of his fellow workers. He had immediately been sent to the sanatorium for observation and diagnosis. The doctors were not encouraging. Giuseppe, they discovered, had a particularly malignant kind of primary brain tumor, a glioblastoma. The doctor explained that this form of tumor originated in the brain itself, and then spread rapidly to other parts of the brain tissue. The glioblastoma could not easily be removed by surgery. A different kind of therapy would be needed to save Giuseppe's life.

As Giuseppe lay prone, mentally wandering through the streets of the city, his ward suddenly came to life. Doctors and nurses, all in a great hurry, gathered around his bed. He was moved to an isolated room. Heavy shielding was placed over his eyes and upper body, leaving his cranium exposed. The doctors and technicians gathered in a protected area, from where they could watch him through a thick-glassed window. A cylindrical source, containing 30 millicuries of radioactive ^{32}P, was lowered onto the top of his head. His therapy was underway.[4]

The treatment over, Giuseppe was wheeled back to the ward. As he flopped back onto his bed, he noticed a flag fluttering from a window in a nearby building. It was the Stars and Stripes, the flag of the occupier. He was struck by its imposing size, and then he remembered that it was the 4th of July. Normally he would have cursed. What concerned him now, though, was not politics, but what had been done to him, and what his chances were of survival. His doctor had explained that he had been treated by radiation from a disintegrating atom that would burn away his spreading glioblastoma, and perhaps cure him. The material had just arrived from America. This was his only hope.

Indeed, Giuseppe Di Tomaso was a benefactor of the decision taken in August 1947 by the United States Atomic Energy Commission (AEC) to make small quantities of radioisotopes available to foreign clients for research and medicinal purposes. Trieste's ^{32}P had been produced at the AEC's Clinton Laboratories at Oak Ridge, Tennessee.[5] The sample was shipped on June 28, 1949.[6] As its half-life was only 14.3 days, treatment began as soon as possible after delivery.

America! The atom! Giuseppe pondered the irony that his life depended on an American atom. It was not that he was against modern science and technology. Increasingly he was aware of their wonders. But how could he place his faith in the atom and in radiation, both of which were bearers of death, not life?[7]

Nor was he anti-American. On the contrary, his wife's brother lived in New York. Just the year before he had written to tell them how wonderful it was, and implored them not to vote for the Communists. But how could he trust the Americans? Had the United States not intervened heavily in the Italian elections in April 1948? Had the people not been promised by the Americans and the British, who governed Trieste, that their troops would leave his soil and that his beloved city would be reintegrated into Italy if the Christian Democrats were elected? Had they kept their promises? No! Perhaps the Communists were right. The United States and its British lackey were imperialist powers intent on colonizing Western Europe.[8] This was the true intent of the Marshall Plan and was behind the integration of Italy into the North Atlantic Treaty, which had been signed just a few days before Giuseppe had had his accident.

Giuseppe groaned audibly. The science wasn't to be trusted. The donor wasn't to be trusted. He had thought he was here to be cured. But the "cure" would surely kill him.

The Italian elections of April 1948 and the status of Trieste

The Italian elections in the spring of 1948 were among the most bitterly fought in the history of the country, and laid the basis for half a century of Christian Democratic, centrist domination that succeeded them. They occurred against a backdrop of the implementation of the Marshall Plan, announced by President Truman's Secretary of State in June 1947. Put simply, the Plan aimed to create an "economic United States of Europe" which, through a single market, mass production and mass consumption, would bring increasing prosperity to all, undercut the lure of communism and dissolve class tensions.[9] While everyone welcomed the financial support, neither government, nor industry, nor the unions were convinced that the kind of economic restructuring it called for was appropriate in Italy.[10] What was more, it was violently opposed by the powerful Italian Communist Party. Fears that the Communists would win the elections and install a regime hostile to western interests were heightened by the Prague coup. The Communists took control of the Czech government legally in February 1948 when twelve ministers resigned, whereupon they rapidly and brutally crushed their democratic opposition. The United States was determined not to let the scenario repeat itself in Italy.

Every imaginable form of persuasion and pressure was brought to bear on the electorate to ensure that it did not vote Communist. Senior policymakers in Washington secretly shipped military supplies to the Italian government, and warned that Marshall aid would not be forthcoming if enough Communist delegates were elected to require their presence in a coalition government. As an eminent diplomatic historian later noted, the Truman administration allocated as much as $10–20 million to pay for local election campaigns, anti-Communist propaganda, and bribes.[11] The Justice Department announced that anyone who voted Communist would be denied entry into the United States. Officials encouraged Italian-Americans to write home begging their friends and relatives not to vote for the Communists; 10 million letters and cables inundated the country from abroad, helped by "Freedom Flights" cosponsored by the US Post Office and the airline company TWA.[12] Frank Sinatra and Gary Cooper recorded special radio programs and a huge propaganda effort was launched (by both sides) using film.[13] In his Christmas message of 1947, Pope Pius XII warned, "he who gives his support, his services and his talents to those parties and forces that deny God is a deserter and a traitor."[14] Bishops and cardinals told priests in their dioceses not to administer sacraments to anyone who voted for the pro-Communist slate. The issues before the electorate were reduced, in the words of one historian, to "a series of simple choices: democracy or totalitarianism, Christianity or atheism, America or the Soviet Union, abundance or starvation."[15] When the dust had settled, the Christian Democrats, whose symbol on the ballot-paper was a shield with one word on it, *libertas*, had 48.5% of the vote (while the Communists and Socialists together had 31%), and the center-right had secured over half the seats in the Chamber of Deputies.

In the midst of the election campaign the United States—along with Britain and France—pledged to return the disputed area around Trieste to Italy if the Communists were defeated. The existing Trieste agreement, adopted the year before, left significant parts of the region under Yugoslav control. It had been "Italy's biggest blow" in the post war settlements. "Areas that were indisputably Italian were torn away. It left bitterness and rancor inside the country, for Trieste was the one foreign-policy issue really felt by a large number of Italians, even those in small villages and remote rural areas."[16] Inevitably, therefore, it emerged as a key site of political struggle in the run-up to the April 1948 elections.

The occupation of Trieste had begun at the end of the Second World War. In May 1945, Tito's Communist partisans had overrun this historic port on the Adriatic, occupying the city and most of the surrounding region.[17] Yielding somewhat reluctantly to British and American pressure, and following Stalin's advice, Tito had agreed to withdraw temporarily to a provisional boundary, the so-called Morgan line. This had left the city under allied control and located a provisional Yugoslav border just south of Trieste. However, it also left most of the beautiful Istrian Peninsula under Yugoslav control, an arrangement that Italy found quite unacceptable. Subsequent negotiations deadlocked. A committee of experts from the four powers was established. When it reported in April 1946, each proposed a different boundary. The British and the American lines pleased Rome, as they integrated Trieste and the Istrian Peninsula unambiguously into Italy. The French line roughly followed the US and UK lines in Venezia Giulia to the north but differed from them in Istria, where it conceded far more to Tito than did to the British and American lines. Under pressure from Stalin, the Anglo-Saxons agreed to retreat to the French line, incensing the Italians. Even this did not please the Yugoslav leader, who was being asked to give up a strip of coast running for about 20 miles south of Trieste to satisfy the French boundary. The compromise reached in 1947 was to designate the area around Trieste as a Free Territory, having two zones. Zone A, under an Anglo-American military government, included the city. It was a strip of shoreline between 2 and 5 miles wide and about 20 miles long, contiguous with Italy at its northern end. Zone B, occupied by Yugoslavia, begun just below Trieste. This strip was about as long, but broader than Zone A, and almost entirely contiguous with Yugoslavia (Figure 9.1). The UN Security Council was supposed to quickly appoint a neutral governor for the entire Free Territory of Trieste.

Figure 9.1 "Troubled Trieste." From *Time*, August 3, 1953, p. 28. © 1953 Time Inc.
Source: Time Inc; reproduced by permission.

Cold War tensions made this impossible, and the occupying powers remained in place in their respective zones until an agreement was reached in 1954.[18]

The victory of the Christian Democrats in April raised hopes in Rome that the United States, along with Britain and France, would soon reopen negotiations with Stalin and Tito on the status of Trieste, as they had promised during the election campaign. But the Allies never delivered. Two months after the elections, in June 1948, simmering tensions between the two Communist leaders came to a head. The directorate of the Yugoslav Communist Party was officially accused of following policies that were hostile to the Soviet Union. Wanting to exploit this rift in the Soviet camp, Britain and America let the Trieste issue die. Their priority now was to draw the rebellious Yugoslav leader into their orbit; they hoped other Soviet satellite states would follow his example. They were not prepared to risk a confrontation with him over an issue which had been essentially a propaganda ploy, and that had now been overtaken by considerations of far greater geopolitical import.

The issue may have been trivial to the Allies in the light of their anti-Communist struggle that was being played out in Europe at the time. To Giuseppe Di Tomaso, however, it was just the opposite. It confirmed him in his view that Italy was simply a pawn in big-power politics, and this hurt his national pride. What is more, an American-inspired pledge that the occupying powers would leave Trieste, and fuse the two zones into Italy had been broken. Giuseppe was a man who kept his word. The United States had presented itself as a friend of Trieste and its aspirations. If his life were not in danger, he said to himself, he would refuse to be treated with these American rays. Who could trust a "friend" who broke his word so readily to suit his own interests?

Behind the scenes in Trieste

Prof. Dr Angelo Rincaldi felt extremely pleased.[19] When the Di Tomaso file had arrived on the Director's desk, he had seen in this patient a wonderful opportunity to put his Provincial Neurological Sanatorium on the map—and simultaneously to make Rome sit up and take notice of what Trieste had to offer. The city had suffered from economic decline, political neglect, and cultural stagnation since the 1920s. The British and the Americans had been fickle, certainly, but so too had the Italian government, which should have opposed the 1947 deal with Tito more aggressively. No one cared about Trieste.

But now Rincaldi would do something about this. He was a man of some importance in this city of 300,000 people, a member of an elite that "sought to project Trieste internationally in order to overcome the city's chronic recession and isolation."[20] To date, only one other Italian institution, the internationally renowned zoological station in Naples, had taken delivery of an American-produced radioisotope, and that was for research purposes.[21] His Sanatorium would be the first in the country to use ^{32}P imported from Oak Ridge for medicinal purposes.

To get hold of the phosphorus, Rincaldi had exploited his network of relationships with the University, and through them, with the British and American occupation authorities.[22] The new Physics Institute, set up in 1945, was of course enrolled in the affair. He discussed procedures with Luzzato Fegiz, the university statistician who had participated in the peace talks on the future of Trieste in New York in November 1946, who was well respected in both diplomatic and financial circles. His friend Diego de Castro, another statistician and university professor, was also an influential diplomat. He served as a counselor to the Allies and provided a link between them and Rome. This had smoothed over the bureaucratic procedures required to secure the isotopes. What was more, one of de Castro's collaborators, the young lawyer Guido Gerin, was an assistant to the rector of the University, himself a militant supporter of Italian nationalism in Trieste.[23] Without access to this network, it would have been impossible for Rincaldi to procure the isotope.

For the procedures were complicated. One did not simply write to Oak Ridge and ask for the material, one man of science to another. First, his government had to nominate its representative in the United States—typically a member of its diplomatic corps—and authorize him to "maintain liaison with the US Atomic Energy

Commission in all matters pertaining to allocation, purchase orders, payments for isotopes and containers, transportation arrangements and charges."[24] This official had to deposit a formal note with the State Department, stating his credentials. He had to acknowledge that the detailed technical information provided to the USAEC by the scientific user was accurate, recognize that the use of radioisotopes involved health and safety hazards, and exonerate the US government and its instruments of any responsibility for injury or damage caused by the radioactive substance. This diplomatic *demarche* was to be followed by a direct request to the AEC. This would specify the person and the institution that would use the material, the name of the isotope, the amount required, the purpose for which it would be used, and the frequency of delivery—short-lived isotopes could be mailed by air at weekly intervals if required.

The client also had to make a number of promises. First, that the material would only be used in the manner specified in the request, and none other. Second, that the user would provide the AEC with three copies of a report every six months on the progress of the work and, if publishable, would announce his findings in the appropriate open scientific or technical literature. Finally, the user had to agree "that qualified scientists irrespective of nationality will be permitted to visit the institutions where the material will be used and to obtain information freely with respect to the purposes, methods and results of such use, in accordance with well-established scientific tradition."[25]

Even after Rincaldi and his friends had navigated successfully through these diplomatic and administrative channels, they had to find a way of getting their phosphorus to Trieste. The city was poorly served by road or rail. Its "airport" was a strip of tarmac on a small military base at Ronchi, nearly 20 miles away. To secure authority for its use, which was imperative if the isotope was not to be worthless by the time it arrived, intricate arrangements had to be made with the occupying authorities and the military.

On the morning of July 4, 1949, Angelo Rincaldi, together with Diego de Castro, the Rector of the University, and the British governor of Trieste, had driven out to Ronchi to collect their precious shipment. It arrived in a small wooden box roughly 9" × 9" × 12", the size of a child's suitcase.[26] They carefully loaded it in the car and rushed back to the offices of the occupying authority. There they peeped inside. They could see little. The radioactive phosphorus was in the small aluminum can in which it had been irradiated in the nuclear pile. Completely surrounding it was a cylindrical shield of lead and steel. The Director of the sanatorium continued on alone to the hospital, and put emergency procedures in place. A few hours after the phosphorus had arrived at Ronchi, Rincaldi, along with his senior doctors and nurses, watched from behind the protective glass as Giuseppe Di Tomaso's glioblastoma was treated for the first time. Rincardi smiled proudly. It was a supreme moment.

* * *

Reflecting that evening over a glass of the best local *grappa* Angelo Rincaldi could not but be satisfied. He had given a dying man a ray of hope. He had worked closely with his colleagues and friends in the university, he had consolidated his network of local relationships, and he had built important links with the occupying authorities and the military. He had raised the visibility and status of his sanatorium and of its Director. The bureaucratic procedures were onerous, but they had obliged Rome to engage itself on Trieste's part, and that was always a good thing. What is more they had demonstrated that this supposedly backward city was embracing modernity and demanded to be taken into consideration both nationally and internationally. This would surely carry weight when the future of Trieste was settled, as soon it must be.

Just one thing left a bad taste in the mouth. Why did he have to make a written engagement to the AEC to allow "qualified scientists irrespective of nationality" to visit his institution and to "obtain information freely with respect to the purposes, methods and results of [...] use" of the material shipped to him? His request had stated clearly what he wanted to do with his radioactive phosphorus: "treatment of an infiltrated glioblastoma." His six-monthly reports would describe his results. If visits to one another's research sites were "in accordance with well-established scientific tradition," as the AEC claimed, why did the Director have to agree formally to an on-the-spot investigation of what he was doing? That morning on the way to the airport he had learnt from the British administrator that his government was also making radioisotopes available to foreign researchers for peaceful purposes. As far as disclosure was concerned, however, all that the British asked was that the recipient be willing to "facilitate [...] exchange between qualified scientists of information and visits in connection with the work carried out with the radioisotopes."[27] The Americans wanted the right to know first hand the "purposes, methods, and results" of his work.[28]

Uffa! Obviously the Americans didn't trust him. Did they think he would use ^{32}P for military purposes? Were they afraid that he would turn the material over to Tito's Communists? Were the American authorities scientifically and politically illiterate, he wondered? Did they have the slightest understanding about the aspirations of the people of Trieste, and the political and economic agenda of its bourgeois elite? Who had yielded to Tito's demands in 1947 that Zone B be kept under Yugoslav control? Who had been soft on communism, and had failed to deliver on promises made in 1948? Not the people of the Trieste, but the American and British negotiators, with the collusion of a supine Rome! Angelo Rincaldi sighed cynically. One day, he said to himself, we will be freed from dependence on this overbearing, untrustworthy, and untrusting, paranoid power. One day we will shape our own destiny.

Behind the scenes in Washington

While Giuseppe Di Tomaso lay dying in his bed in Trieste, little did he realize that the fate of the phosphorus used in his treatment was being decided in a drama being played out at the time in Washington. On June 13, 1949 the Congressional Joint Committee on Atomic Energy, chaired by the conservative Midwestern Senator Bourke B. Hickenlooper, was investigating a charge of "incredible mismanagement" against the AEC, notably the loss of large quantities of uranium-235 from the Argonne National Laboratory. Under scrutiny too was the policy of shipping radioisotopes abroad, a policy that from its inception had been criticized as posing a threat to national security and which was now, once again, under attack. Fortunately for our afflicted patient, the foreign isotope program had emerged from the hearings stronger than it was before.

The distribution of radioisotopes to foreign users for research or therapeutic purposes had been first discussed by the AEC in August 1947. It was an extension, on a more restricted basis, of an extremely popular domestic program that had got under way in mid-1946. Laboratories with nuclear piles could produce far more radioactive material far more quickly and cheaply than the cyclotrons that had been used before the war. The AEC estimated that the Oak Ridge pile, for example, could produce 200 millicuries of carbon-14 in a few weeks for about $10,000; it would take one thousand cyclotrons, and operating costs of well over $1 million, to do the same.[29] The program was so popular that within a year researchers and medical centers in the US and Hawaii had received more than 1,000 shipments of 90 regularly available radioisotopes.[30] By June 1947 the AEC had also received 96 inquiries from 28 foreign countries, 75% of them for radioisotopes for medical research and therapy.[31] It suggested making twenty of the most popular of these radioactive elements available to doctors and scientists abroad, carefully excluding any that might be used to develop atomic energy for military or industrial purposes.

The Commissioners who favored the program gave a variety of reasons for going ahead with it.[32] They noted that many American scientists supported the scheme, being embarrassed when resentful colleagues abroad found it difficult to understand why the United States did not release radioisotopes to them. It tended to confirm them in the view that American science was being financed and controlled by the military—indeed, as another scientist put it, led some to go "so far as to class us in the same light as Russia on scientific and political matters."[33] Thus as J. Robert Oppenheimer, the immensely prestigious chairman of the General Advisory Committee, wrote on June 1, 1947, the program would probably "have a great effect in restoring the confidence of scientists, and educated men generally, in foreign countries, in their colleagues in the United States."[34] This was also one way in which science could contribute to the aims of the Marshall Plan, "the number one key to the policy of the United States" at the time, and "designed to strengthen what can be friendly and peaceful countries." In any case, Canada and Britain would soon be making isotopes available for foreign users; countries like France would shortly be able to do so too. If the US ever wanted isotopes from them, it could hardly refuse not to reciprocate. More to the point, it was preferable for the United States to have a major presence in this market if it were to retain its position of world leadership in this line of research.

AEC Commissioners backing this internationalist view had to admit that "some shipment abroad might fall into the hands of capable persons who wish to develop atomic weapons." However, they felt that this risk was extremely slight and that the "recognizable advantages outweigh the possible dangers." They pointed out that the materials they proposed to distribute could be made by foreign nations in cyclotrons anyway, that the quantities were small, that the isotopes were dispersed between many different research and medical groups, that many decayed rapidly, and that not one had been significantly useful in the development of atomic weapons in the United States. In any event, procedures had been put in place to monitor the use to which the foreign shipments were put, including having US scientists "visit the institution where the materials are used and observe the methods and results of such use."

Proponents of the scheme also had to admit that the materials might *indirectly* strengthen the military capability of another nation. A lot of people would be trained in the skills of handling radioisotopes, and these would be useful for the subsequent development of atomic energy. Significant discoveries might be made abroad and this would strengthen, in the long run, the prosperity (and thus military capability) of other nations, especially if these discoveries were themselves of direct military relevance. They felt, however, that the rules of disclosure and the rights of inspection ensured that American scientists would quickly learn of any such developments. "With its superior technological potential," they pointed out, "the United States can expect to profit more quickly and more fully than any other nation from the exploitation of published findings." In short, the Commissioners backing the foreign isotope program argued that it would reinforce, rather than undermine, "the common defense and security of the United States." It would dispel the image at home (and abroad) that the US atomic energy program was under tight military control, so ensuring "the wholehearted support of United States scientists and medical doctors for our national program for atomic energy." It would show that the United States was far more open than the Soviet Union in this domain, and it would garner goodwill and support among an influential elite for US policies in Europe, notably the Marshall Plan. Finally, through reporting and inspection procedures, it would ensure that the US had access to any major discoveries, so enhancing "our national security, which depends upon continued progress in this field."

Four of the five Commissioners favored the scheme. Lewis Strauss did not. A lawyer rather than a scientist, convinced that the Communist threat was a grave menace to the US, Strauss rejected the argument that America would win scientific "hearts and minds" with the program. It overlooked the point that many scientists "worked and will work just as zealously for dictatorships of the Right and Left as they did for democracy." He dismissed the assertion that foreign governments and scientists would eventually get the substances anyway: this was not a reason for breaching US security by giving it to them. Even Strauss's opponents admitted that, whether or not the isotopes could be used by other countries to produce atomic weapons, "they would be useful as tools in biological research, metallurgical research, petroleum chemistry, and other areas which are part of the war-making potential of nations." This was unacceptable. As far as Strauss was concerned, "until the satisfactory international control of atomic energy is assured," the United States could not "help scientists who may work for a putative enemy one jot or tittle without displaying naiveté and imperiling our own security."[35]

Strauss was overruled. His associates were not persuaded that the risks were not worth taking, and felt that the procedural safeguards put in place would ensure that the program was not abused. Leading scientists at home and in foreign countries were in favor of it for the research opportunities it provided and for the positive image of the country it projected abroad. The State Department accepted the majority view that the program did not prejudice national security and was gratified that the atom was being promoted as an instrument of peace and goodwill. As Richard Lovett,

Acting Secretary of State, noted, "these valuable products of United States atomic energy plants will now be available in the services of mankind and that, to this extent at least, we are able to advance toward the beneficient [*sic*] use of this new force. This initiative," Lovell went on, "should promote harmony and good feeling among nations."[36] President Truman officially announced the program on September 3, 1947 at the opening of the 4th International Cancer Research Congress in St Louis.[37]

The putative dangers to national security, and Strauss's hostility, hung like a shadow over the program until 1950. The desire to promote it within the framework of the Marshall Plan meant that the isotope program was formally open to the USSR and what were called "Russian-dominated" countries, identified as Albania, Bulgaria, Czechoslovakia, Finland, Hungary, Poland, Rumania, and Yugoslavia.[38] In practice, even before Truman had made his announcement, the Soviet Union had decided not to participate in the Plan and forbade its satellites to accept American aid. However, some countries on this list, like Finland—which were relatively loosely coupled to Moscow, although forced to refuse Marshall aid—did ask for American radioisotopes. The Finnish Legation in Washington deposited its request on August 31, 1948, asking for 300 millicuries of ^{32}P for research at the University of Helsinki.[39] This caused another bruising struggle with the State Department, and this time Strauss was not alone among the Commissioners in having serious doubts about satisfying the request.

The State Department's official policy on Finland was that the "native political forces of Finland can be counted upon to reject, to the maximum possible extent, the efforts of the USSR to maintain or increase its tutelage." The "central objective" of US policy was to "ensure the maintenance of effective independence for Finland as a sovereign state." This required prudence and sensitivity, since "While Finland may be expected to continue its resistance to the Communist Party domestically and to act with considerable independence internationally, it will, in situations in which the Soviet Union regards as crucial, most probably have to follow Soviet directives"—as it had done regarding participation in the Marshall Plan.[40] The State Department, in short, sought to encourage Finnish relationships with the US but only in relatively non-controversial areas that did not provoke the Soviet Union to intervene in Finnish affairs. The delivery of 300 millicuries of ^{32}P was, in its view, a suitable instrument for furthering this policy.

When the Commissioners had met on December 15, 1948 Strauss asserted at once that he strongly opposed the inclusion of Finland in the program. Soviet influence over the country was so great, he said, that it could "order transfer of the isotopes from Finland to Russia should it so desire."[41] Though he was willing to concede that the export list contained isotopes that likely had no direct military application, he emphasized again that they "strengthen through research the war potential" of unfriendly countries. Strauss also embarrassed the proponents of a liberal policy by pointing out that the much-vaunted safeguards to impede abuse or diversion were not being implemented systematically and that, in practice, the AEC actually had no means at its disposal for policing its agreements with foreign researchers. Strauss was overruled again, but by the time the decision was conveyed back to Finland, the Helsinki university researcher had become impatient and had turned to another supplier, the British, for his radioactive phosphorus.[42] Suspicious minds noticed, however, that whereas the Finnish researcher had asked the US for six shipments of 50 millicuries each (an amount which many Commissioners already found unduly excessive), he had requested only 1 millicurie per month for an indefinite period from the British—"presumably," US officials noted, "because of the limited production facilities in the U.K."[43]

Within months the simmering tensions surrounding the foreign isotope campaign had been catapulted into the public domain via the June 1949 Congressional hearings into charges that the AEC was responsible for "incredible mismanagement." As far as the isotope program was concerned, the proximate cause was the request by Norway (which had actually just signed the North Atlantic Treaty) for 1 millicurie of Iron-59 for metallurgical research. This time it was suggested that national security had been breached because material had been provided to a military client, the Norwegian Royal Defense Research Establishment.[44] In addition, Lewis Strauss had discovered that a member of the Norwegian research team "could be described as a Communist."[45]

The 1949 hearings were noteworthy for Oppenheimer's testimony. For him the most important reason for distributing radioisotopes internationally was, as he put it, that

> [Isotopes] were discovered in Europe; they were applied in Europe; they are available in Europe, and the positive arguments for making them available [...] lie in fostering science; in making cordial effective relations with the scientists and technical people in western Europe; in assisting the recovery of western Europe; in doing the decent thing.[46]

"Decency" and "fostering science" had to be weighed against national security, however. Time and again, members of Congress pressed Oppenheimer whether or not the isotopes being sent abroad might not fall into the hands of the Soviets and accelerate the development of nuclear weapons behind the iron curtain. He dismissed the suggestion, provoking guffaws in the room:

> No one can force me to say that you cannot use these isotopes for atomic energy. You can use a shovel for atomic energy. In fact you do. You can use a bottle of beer for atomic energy. In fact you do. But to get some perspective, the fact is that during the war and after the war these materials have played no significant part and in my knowledge no part at all.

He was equally contemptuous of concerns that, while not of direct military use, the radioisotopes could contribute to the war-making potential of a foreign power:

> My own rating of the importance of isotopes in this broad sense is that they are far less important than electronic devices, but far more important than, let us say, vitamins, somewhere in between (laughter).[47]

This "spectacular provocation" soured the already hostile relationship between Oppenheimer and Strauss.[48] However, it did not harm the foreign isotope distribution program. Subsequent amendments to the Atomic Energy Act, rather than tightening up the provisions, clarified the authority of the AEC to distribute radioisotopes to foreign clients.[49] The program was also extended to include all isotopes available for domestic users, and industrial applications were included within its scope, always excluding the few that could evidently pose a national security threat (notably tritium).[50]

There were a number of reasons for this relaxation. Demand was increasing. To meet it, in February 1950 the AEC created "a sort of atomic pharmacy" at Oak Ridge that "puts radioisotope processing, packaging and shipping on an assembly-line basis, eliminating for the most part the time-consuming method of handling radio-isotope [*sic*] shipments manually."[51] In addition, by then the United States had lost its monopoly on the atomic bomb. After the first Soviet test, announced by President Truman on September 23, 1949, Communist states could acquire war-related isotopes directly from within their bloc if they wished to. And the AEC had lost its monopoly on the market. The British and the Canadians were making a far wider range of isotopes available than was the US, including for industrial applications.

The geopolitical physiognomy of the foreign isotope program was not shaped by "decency" and the desire to "foster science," but by the changing balance between the demands of national security and the scientific and propagandistic advantages to be gained by making a valuable resource available to "friendly" nations. In its early years, the desire to maintain control over the foreign users and the use of the isotopes dominated all other considerations. National security was defined as strengthening Western Europe through the Marshall Plan, combating communism and retaining military superiority. No "Russian-dominated" country ever received any of the material. (As noted: only one quasi-satellite state, Finland, forbidden by Moscow to receive Marshall Aid, dared ask for material, but Washington dithered and the University of Helsinki turned to Britain instead.) Even a shipment to a NATO ally like Norway was deemed highly undesirable by some because the request came from a military establishment.

By 1950, the geopolitical context in which the program had been launched had changed completely. The Soviet Union had the atomic bomb and two of America's allies, Britain and Canada, were far less obsessive than were Strauss and Hickenlooper about the risks of sharing their isotopes with others. Maintaining control and leadership in this competitive market meant expanding the program dramatically, ensuring that US business could capitalize on industrial demand and limiting the constraints imposed by national security concerns to a minimum. Announcing the expanded program in May 1951, a government press release stressed the benefits of radioactive isotopes to advancements in "basic science, medicine, agricultural and industry." It deemed them the "single most important contribution of atomic energy to peacetime welfare."[52]

* * *

It does not appear that the request in summer 1949 for 30 millicuries of ^{32}P by the Provincial Neurological Sanatorium in Trieste raised eyebrows in Washington. The use was purely medicinal. The situation in Trieste was not front-page news in the anti-Communist struggle being waged in Europe at the time. Even people who had heard of it probably did not grasp the exotic political situation that prevailed in the Free Territory, nor the reasons for its existence. Which was all for the best for the patient afflicted with an infiltrated glioblastoma and for the director who wanted to take advantage of the American offer and use radiotherapy in his institution to treat the tumor.

Epilogue

More than five years after he was first hospitalized, Giuseppe Di Tomaso got out of bed with a sprightly step and walked into the kitchen to make his first *espresso* of the day. He was excited. For one thing, miraculously, his treatment had been successful. A second batch of 40 millicuries of ^{32}P had been shipped from Oak Ridge in January 1950 to reinforce the improvement he had shown over the previous six months.[53] By the end of that year the staff at the clinic was persuaded that his tumor had been dealt with. He had felt fine ever since.

Today was also a special day: October 5, 1954. At noon a Memorandum of Understanding on Trieste would be signed by senior diplomats in London. Some 6,500 Allied troops would soon be withdrawn, and the city would at last be reunited with Italy.[54] The streets were festooned with red, white and green Italian flags. Giuseppe was determined to make the best of this historic moment.

As his coffee percolated noisily, Giuseppe began to be assailed by doubt. Some of his friends were saying that this was not such a wonderful day after all. The occupying authorities had provided jobs, and the fear that absorption into Italy would destroy the already-weak economy of the city had led to a flight of capital (as much as 25 million lire a day was mentioned).[55] At the docks, the Marshall Plan inspired productivity drive had backfired due to worker opposition, low wages, and a lack of cargo to handle; Giuseppe had a specialized job but even that was not secure. Things at home were not what they used to be either. More often than not his wife bought manufactured pasta instead of making it herself, and the wonderful fresh tomatoes she could buy in the market were being stealthily replaced by canned produce. She insisted on going to see Hollywood movies in the local cinema with her sister-in-law every week (she loved being immersed momentarily in the "American dream," she said) and always came home enthusing about the latest kitchen gadget she had seen on screen.[56] Giuseppe sighed. This was the price of freedom, he supposed—and what freedom! The next day, which had also been declared a holiday, he and his brother planned to drive off in his new Fiat *cinquecento* to the rolling vineyards of Venezia Giulia. There they would sample a range of last season's superb Pinot Grigio at the producers' roadside stalls and buy a few bottles for themselves. He began feeling better again.

Giuseppe and his wife were in the Piazza dell'Unita at noon when a huge Italian flag was hoisted to cheers and the singing of Italian patriotic songs. They moved with the throng to San Giusto, whose bells pealed joyfully.[57] He noticed Prof. Dr Angelo Rincaldi across the nave and smiled. The two men embraced tearfully after the service, before Giuseppe's wife dragged him over to a side altar to light a votive candle—"to thank God for the freedom of Trieste," as she put it. Giuseppe was not sure that God should be given credit for this. He was not even sure that God existed. He lit a candle all the same. He remembered that an American radioisotope had saved his life. He also remembered that it was a personal letter from President Eisenhower to Marshall Tito that had broken the deadlock in the recent negotiations that had finally settled the status of the region.[58] As the flame fluttered into life and pierced the gloom Giuseppe Di Tomaso said softly, *Grazie, America.*

Several months earlier, in Washington, J. Robert Oppenheimer had been stripped of his security clearance—the victim of a campaign against him launched with unrelenting determination by Lewis Strauss and the physicist Edward Teller.[59] Strauss's deep personal animosity to Oppenheimer had been fueled by the scientist's contemptuous dismissal of his and Hickenlooper's security concerns during the Congressional hearings in June 1949. Seventeen year later, AEC Commissioner David Lilienthal would still recall Strauss's tight-lipped reaction to Oppenheimer's testimony in 1949. "There was a look of hatred there," he said, "that you don't see very often in a man's face."[60] Oppenheimer's spirited support for the foreign isotope program had helped save Giuseppe Di Tomaso's life and made him "pro-American," it had enabled Angelo Rincaldi to develop his links with the local elite and to promote his sanatorium as a site exemplifying Trieste as an "international city," worth 'fighting for against Tito's claims. But it had also cost him his security clearance.[61]

Acknowledgments

I would like to thank Alexis de Greiff, Ron Doel, Howard Kushner, and Bruno Strasser for valuable comments on an earlier version of this paper.

Notes

1 Sally Marks, "The World According to Washington," *Diplomatic History* 11 (summer 1987): 265–282.
2 My approach is informed by the "new diplomatic history" promoted above all by John Lewis Gaddis, Geir Lundestad, and Charles Maier. Its core claim is that the US established a hegemonic regime in Western Europe after the Second World War, but a hegemony that was "consensual" and that built an "empire by invitation." It is central to this approach that an American hegemonic regime was acceptable to European elites who managed to secure and to consolidate mass political support for their actions by adopting and adapting the American agenda for the region in consultation with US policymakers. See, for example, John Lewis Gaddis, *We Now Know: Rethinking Cold War History* (Oxford: Oxford University Press, 1997); Geir Lundestad, *"Empire" by Integration: The United States and European Integration, 1945–1997* (Oxford: Oxford University Press, 1998); and Charles S. Maier, "Alliance and Autonomy: European Identity and U.S. Foreign Policy Objectives in the Truman Years," in *The Truman Presidency,* ed. Michael J. Lacey (Cambridge: Woodrow Wilson International Center for Scholars, and Cambridge University Press, 1989), 273–298. While diplomatic historians in general have focused on a broad range of social and political factors, those concerned with the role of science and technology thus far remain

most interested in elite policy issues; see Joseph Manzione, "'Amusing and Amazing and Practical and Military': The Legacy of Scientific Internationalism in American Foreign Policy, 1945–1963," *Diplomatic History* 24, no. 1 (2000): 21–56; and Kenneth A. Osgood, "Form before Substance: Eisenhower's Commitment to Psychological Warfare and Negotiations with the Enemy," *Diplomatic History* 24, no. 3 (2000): 405–434. See also John Krige, *American Hegemony and the Postwar Reconstruction of Science in Europe* (Cambridge, MA: MIT Press, 2006). Pathbreaking work in this field has of course been done by one of the editors of this volume; cf. Ronald E. Doel, "Scientists as Policymakers, Advisors and Intelligence Agents: Linking Contemporary Diplomatic History with the History of Contemporary Science," in *The Historiography of Contemporary Science and Technology*, ed. Thomas Söderqvist (Amsterdam: Harwood Academic Publishers, 1997), 215–244.

3 The original inspiration for this paper came from Angela N. H. Creager, "Tracing the Politics of Changing Postwar Research Practices: The Export of 'American' Radioisotopes to European Biologists," *Studies in the History and Philosophy of Biological and Biomedical Sciences* 33 (2002): 367–388. See also Angela N. H. Creager, "The Industrialization of Radioisotopes by the U.S. Atomic Energy Commission," in *The Science-Industry Nexus: History, Policy, Implications: Nobel Symposium 123*, ed. Karl Grandin, Nina Wormbs, Anders Lundgren and Sven Widmalm (New York: Science History Publications/USA, 2004). I selected Trieste because of its unusual political situation in the late 1940s, and thanks to Alexis de Greiff, "The International Centre for Theoretical Physics, 1960–1979: Ideology and Practice in a United Nations Institution for Scientific Co-Operation and Third World Development," (PhD dissertation, University of London, Imperial College of Science, Technology and Medicine, 2001). See also Alexis de Greiff, "The Tale of Two Peripheries: The Creation of the International Centre for Theoretical Physics in Trieste," *Historical Studies in the Physical and Biological Sciences* 33, no.1 (2002): 33–60.

4 Bruno Strasser of the University of Geneva has pointed out (personal communication) that after the war one of the most innovative and distinctive uses of radioisotopes in therapy was to have them injected or ingested into the body. I am assuming (perhaps incorrectly, but we have no way of knowing) that the ^{32}P was irradiated externally, as was the case for X-ray radiotherapy before the war.

5 On Clinton Laboratory's biomedical activities, and those of the AEC in general, see Peter J. Westwick, *The National Laboratories: Science in an American System, 1947–1974* (Cambridge: Harvard University Press, 2003), chapter 7.

6 National Archives and Records Administration (NARA hereafter), AEC Records, RG 326, E67A, Box 46, Folder 3, "Reports of Foreign Shipments of Isotopes," Shipment 572, 6-28-49, Provincial Neurological Sanatorium, Trieste, Treatment of an infiltrated glioblastoma.

7 Spencer Weart, *Nuclear Fear: A History of Images* (Cambridge: Harvard University Press, 1988).

8 On the propaganda of the Italian Communist party at this time, see Marc Lazar, "The Cold War Culture of the French and Italian Communist Parties," in *The Cultural Cold War in Western Europe, 1945–1960*, ed. Giles Scott-Smith and Hans Krabbendam (London: Frank Cass, 2003), 213–224.

9 For an introduction to the Plan see David W. Ellwood, *Rebuilding Europe: Western Europe, America and Postwar Reconstruction* (London: Longmans, 1992).

10 David W. Ellwood, "The Propaganda of the Marshall Plan in Italy in a Cold War Context," in *Cultural Cold War*, ed. Scott-Smith and Krabbendam, 225–236; David W. Ellwood, "Italian Modernisation and the Propaganda of the Marshall Plan," in *The Art of Persuasion: Political Communication in Italy from 1945 to the 1990s*, ed. Luciano Cheles and Lucio Sponza (Manchester: Manchester University Press, 2001, 23–48). For the unions see Federico Romero, *The United States and the European Trade Union Movement, 1941–1951* (Chapel Hill, NC: University of North Carolina Press, 1992).

11 Melvyn P. Leffler, *A Preponderance of Power: National Security, the Truman Administration and the Cold War* (Stanford, CA: Stanford University Press, 1992), 197.

12 Ellwood, *Rebuilding Europe*, 116.

13 David W. Ellwood, "The 1948 Elections in Italy: A Cold War Propaganda Battle," *Historical Journal of Film, Radio and Television* 13, no. 1 (1993): 19–33. See also David W. Ellwood, "From 'Re-Education' to the Selling of the Marshall Plan in Italy," in *The Political Re-Education of Germany and her Allies After World War II*, ed. Nicholas Pronay and Keith Wilson (Totowa, NJ: Barnes & Noble Books, 1985), 219–239.

14 Martin Clarke, *Modern Italy, 1871–1995* (London: Longmans, 1996).

15 James E. Miller, "Taking Off the Gloves: the United States and the Italian Elections of 1948," *Diplomatic History* 7, no. 1 (winter 1983): 35–55. According to Ellwood, the Vatican played a far greater role in mobilizing the people against the Communists than did US efforts; see his "Italian Modernisation," 26.

16 Norman Kogan, *A Political History of Italy: The Postwar Years* (New York: Praeger, 1983), 5–6.

17 Ibid., 3 and forward.

18 John C. Campbell, ed., *Successful Negotiation, Trieste 1954: An Appraisal by the Five Participants* (Princeton, NJ: Princeton University Press, 1976). The final boundary differed little from the one adopted in 1947.

19 I remind readers that this too is a fictional, but historically plausible character.

20 De Greiff, "International Centre," 187. See also de Greiff, "Two Peripheries."

21 NARA, AEC Records, RG 326, E67A, Box 46, Folder 3, "Reports of Foreign Shipments of Isotopes," Shipment 30, 1-22-48, 1 mc of ^{14}C for the "Biosynthesis of organic molecules to be used to study oxidative metabolism of marine invertebrate eggs...".

22 On this network built around leading lights in the University of Trieste, see de Greiff, "International Centre," 180–193.

23 Gerin would later serve as Italy's representative to the International Atomic Energy Agency and headed the UNESCO-affiliated International Institute for Human Rights in Trieste, http://www.sibi.org/ingles/cv.htm#g_gerin; accessed on February 4, 2004 and de Greiff, "Two Peripheries."

24 NARA, AEC Records, RG 326, E67A, Box 46, Folder 3, *Radioisotopes for International Distribution. Catalog and Price List. September 1947.* (Isotopes Branch, United States Atomic Energy Commission, P.O. Box E, Oak Ridge, Tennessee), 4. This paragraph and the next are based on this document.

25 Ibid., 15.

26 The details of the packaging are from *Radioisotopes for International Distribution*, 13.

27 Subsequently the AEC actually revised the wordings of these undertakings in line with the less demanding requirements made by Britain and Canada. The new version for allowing on-site inspections actually took over the British clause almost verbatim: the client agreed "To facilitate exchange of information and visits relative to work with radioisotopes between qualified scientists, in accordance with normal scientific practice." NARA, AEC Records, RG 326, E67A, Box 46, Folder 3, "AEC. International Distribution of Radioisotopes. Report by Director of Research," 15 pp, attached to AEC 231/12, "Note by the Secretary," May 17, 1950.

28 The onerous conditions of inspection and reporting required by the Americans were similar to those that they had implemented in their zone of Occupied Germany in order to control scientific research there. The British were less intrusive in their zone of the country: see David Cassidy, "Controlling German Science, I. U.S. and Allied Forces in Germany, 1945–1947; II. Bizonal Occupation and the Struggle over West German Science Policy," *Historical Studies in the Physical and Biological Sciences* I. 24, no. 1 (1994), 197–235; II. 26, no. 2 (1996), 197–239.

29 Quoted by Creager, "Export," 375.

30 NARA, AEC Records, RG 326, E67A, Box 46, Folder 3, Majority submission to the State Department, "Foreign Distribution of Radioisotopes," undated, but following on the meeting of the Commissioners on August 19, 1947, 1–2.

31 NARA, AEC Records, RG 326, E67A, Box 46, Folder 3, Majority submission to the State Department, "Foreign Distribution of Radioisotopes," undated, but following on the meeting of the Commissioners on August 19, 1947, 2.

32 Unless otherwise stated, the quotations in this and the following two paragraphs are derived from NARA, AEC Records, RG 326, E67A, Box 46, Folder 3, Majority submission to the State Department, "Foreign Distribution of Radioisotopes," and Atomic Energy Commission, Minutes of Meeting No. 95 at Bohemian Grove, August 19, 1947.

33 Quoted by Creager, "Export," 376.

34 NARA, AEC Records, RG 326, E67A, Box 46, Folder 3, "AEC. International Distribution of Radioisotopes. Report by Director of Research," 15 pp, attached to AEC 231/12, "Note by the Secretary," May 17, 1950, cites the letter.

35 NARA, AEC Records, RG 326, E67A, Box 46, Folder 3, Memo from Carroll L. Wilson to The Commissioners, September 25, 1947.

36 NARA, AEC Records, RG 326, E67A, Box 46, Folder 3, Letter Richard Lovett to David E. Lilienthal, August 28, 1947.

37 NARA, AEC Records, RG 326, E67A, Box 46, Folder 3, gives details in the Press Release "Radioisotopes for Medical and Biological Research Available to Users Outside United States."

38 NARA, AEC Records, RG 326, E67A, Box 46, Folder 3, Memo from Carroll L. Wilson to The Commissioners, September 25, 1947.

39 NARA, AEC Records, RG 326, E67A, Box 46, Folder 3, Letter from the Legation of Finland to the Department of State, August 31, 1948 included as an enclosure with a memo by Roy B. Snapp, December 9, 1948, "Program for Foreign Distribution of Isotopes—Inclusion of Finland."

40 NARA, AEC Records, RG 326, E67A, Box 46, Folder 3, "Policy Statement. Finland," Department of State, September 2, 1948. For the quotations, see 7, 1, and 5 respectively.

41 NARA, AEC Records, RG 326, E67A, Box 46, Folder 3, Minutes of AEC Meeting on December 15, 1948, section 11, "Program for Foreign Distribution of Isotopes."

42 NARA, AEC Records, RG 326, E67A, Box 46, Folder 3, "Inclusion of Finland in the Program for Foreign Distribution of Isotopes. Report by the Deputy Director of Research," undated, but around February 1949.

43 Ibid.

44 NARA, AEC Records, RG 326, E67A, Box 46, Folder 3, Minutes for meetings in January and February, 1950, as approved at AEC meeting on January 3, 1950 refer to this issue. The memo K. S. Pitzer to R.W. Cook, "Review of foreign requests for radioisotopes," January 19, 1950, immediately tightened up the procedures for the automatic approval of requests without consultation. Prior authorization was required if "the facts [or] the appearances would suggest that the radioisotope requested will be used by or on behalf of a foreign military establishment."

45 Creager, "Export," 381; Peter Goodchild, *J. Robert Oppenheimer: Shatterer of Worlds* (Boston, MA: Houghton Mifflin, 1981), 193–196.

46 This statement was used repeatedly by the AEC to justify the policy, see for example, NARA, AEC Records, RG 326, E67A, Box 46, Folder 3, letter Sumner Pike to Senator McMahon, June 28, 1951.

47 Quoted in Goodchild, *Oppenheimer*, 195.

48 The phrase is from Patrick J. McGrath, *Scientists, Business and the State, 1890–1960* (Chapel Hill, NC: University of North Carolina Press, 2002), 145–146.

49 NARA, AEC Records, RG 326, E67A, Box 46, Folder 3, (draft) letter David E. Lilienthal to Senator McMahon, February 2, 1950.

50 NARA, AEC Records, RG 326, E67A, Box 46, Folder 3, "AEC. International Distribution of Radioisotopes. Report by Director of Research," 15 pp, attached to AEC 231/12, "Note by the Secretary," May 17, 1950. On tritium see Minutes of the 525th Meeting of the AEC, 2-14-51, item AEC 231/13. For a full list of the 99 isotopes made available see Proposed Press Release, Appendix "E" to "Note by the Secretary," AEC 231/15, May 11, 1951. See also R. G. Hewlett and F. Duncan, *Atomic Shield: A History of the United States Atomic Energy Commission, Vol. 2: 1947–1952* (Berkeley, CA: University of California Press, 1962), 442–484. Creager, "Export," provides a fine analysis of the use of the radioisotopes in research.

51 NARA, AEC Records, RG 326, E67A, Box 46, Folder 3, "Information for the Press and Radio," 2-1-50. This was also an invitation to press, radio, and periodical representatives to visit the site.

52 NARA, AEC Records, RG 326, E67A, Box 46, Folder 3, "Proposed Press Release," Appendix "E" to "Note by the Secretary," AEC231/15, May 11, 1951.

53 NARA, AEC Records, RG 326, E67A, Box 46, Folder 3, "Reports of Foreign Shipments of Isotopes," Shipment 718, 1-19-50, 40 mc of Phosphorus-32 to the Provincial Neurological Sanatorium in Trieste for "Continuation of treatment of a brain tumor case which responded to previous administration of Phosphorus 32."

54 Campbell, *Successful Negotiation*, Appendix D gives the Press Release that accompanied the signature.

55 De Greiff, "International Centre," 182.

56 An anonymous Italian commentator remarked in June 1953 that "95% of all Europeans—friends and enemies of America—judge American society by what they see at the cinema." Hollywood "...was useful above all in reinforcing the European admiration for the American standard of living, for American technique...Undoubtedly film has given the U.S. a propaganda triumph, to the extent that it has reminded Europeans of their traditionally optimistic vision of the 'American paradise'"—cited in Ellwood, "Propaganda," 225. For the libratory place of the cinema in women's lives, albeit in a different country, see Victoria de Grazia, "Mass Culture and Sovereignty: The American Challenge to European Cinemas, 1920–1960," *Journal of Modern History* 61 (March 1989): 53–87. De Grazia points out that British working-class women found American movies a source for a "new women" peer culture. The cinema was a place where women could go unescorted, often with female family members or friends, and movies became a major subject of discussion and memory, influencing mannerisms and fashions.

57 On the celebrations in Trieste see "Triestenes Cheer Return to Italy as New Accord is Announced: Tricolor Flown, Bells Toll While Thousands Parade—Dissenters Stay Away From Demonstration," *New York Times*, October 6, 1954, 5; "Trieste Acclaims Italian General as he Arrives to Assume Control," *New York Times*, October 7, 1954, 6. The flag hoisted in the Piazza dell'Unita on the 5th measured 65 feet by 35 feet. The Basilica of San Giusto (Saint Just) overlooks Trieste and has stood for centuries as an emblem of the Italian character of the city.

58 The letter is reproduced in Campbell, *Successful Negotiations*, Appendix B. Dated September 10, 1954, it asked Tito in "this friendly fashion to intervene personally in the Trieste negotiations to settle the exceedingly small differences remaining." At that time the US was considering a Yugoslav request for wheat and economic aid. Eisenhower suggested that this would be forthcoming if Tito cooperated. The US President signed the letter "With my warm personal greetings and good wishes."

59 Barton J. Bernstein, "In the Matter of J. Robert Oppenheimer," *Historical Studies in the Physical Sciences* 12 (1982): 195–252. See also Richard Polenberg, *In the Matter of J. Robert Oppenheimer: The Security Clearance Hearing* (Ithaca, NY: Cornell University Press, 2002).

60 Quoted in Goodchild, *Oppenheimer*, 196. Lilienthal was the first chairman of the USAEC.

61 To move the field forward in the direction taken in this paper, it is essential that historians of science and technology engage with the significance of American techno-scientific power after 1945. This enabled her scientists and science administrators to try to shape and control the research agendas of America's scientific "partners" in line with US interests. More than ever before international scientific exchange did not take place on a level playing field: there was a massive imbalance in every respect in favor of the US after the war. In the context of Cold War rivalry, a ruthless struggle ensued for the "hearts and minds of men" that was targeted at the general population in Western Europe but which required the active cooperation of the European scientific elite for it to succeed. To understand the full extent of this struggle, and the role of science and technology in it, we do not need to engage simply with political and diplomatic history, but also with cultural history. This paper is the first step along such a path. It is a tall order, but an exciting challenge.

10 Secrecy and science revisited

From politics to historical practice and back

Michael Aaron Dennis

If conventional understandings of science were accurate representations of our world, the conjunction of science and secrecy might serve as a powerful example of an oxymoron. Writing in *Scientific American*, Jeffrey Richelson, a student of secret government intelligence programs, explained that the major source of difficulty in having cooperation of scientists with the US intelligence establishment was that such

> cooperation will require an accommodation between two cultures, those of science and of intelligence, that have essentially opposite methods of handling information. In science, the unrestricted dissemination of data is accepted as being necessary for progress, whereas in intelligence, the flow of information is tightly restricted by a "need to know" policy; only those who have the proper security clearances and who cannot carry out their assigned responsibilities without certain knowledge or information are given access to it.[1]

For Richelson and countless others, the distinctive character of science is manifested in its openness, that is, the unrestricted exchange of information and knowledge without regard for the race, creed, sex, or national origin of those involved in the exchange. Secrecy, however, is far from unknown within the world of science. All of us are familiar with the existence of a classified world of research, containing its own journals, meetings, and professional organizations. That world exists both within and apart from the world we experience on a daily basis. Even the materials, Richelson is addressing—the use of national intelligence databases to understand global environmental change, Project Medea—is predicated on the existence of a secret world where researchers, more often than not academics, produced the knowledge that we might now harvest.

Science and secrecy were not, and are not, the polar opposites of common understanding. Timothy Ferris, a regular *New Yorker* science writer, declared that

> real science is a white hole that gushes information; scientists (astronomers especially) prefer to tell one another almost everything, because if they don't they can't build on each other's results. (The gravest concern of those who do classified work is that if they are cut off from such constant exchange their careers will wither).[2]

Given that the history of science is littered with examples of willful and deliberate secrecy, whether on the part of individuals or institutions, including states, such a claim is patently false. Furthermore, despite his invocation of Soviet science as an example of what happens when science is kept secret, Ferris does not address David Holloway's remarkable claim: that researchers in the secret cities of the Soviet atomic bomb project (such as Sakharov) were the bearers of democratic values and practices during the long Cold War.[3] If one accepts Holloway's claim, secrecy is not simply part of science but essential for democracy.

What then is the relation between science and secrecy? Is there a single, necessary relationship between the production of knowledge and the technologies through which that knowledge is made and disseminated? Technologies in this sense also include the systems of classification and secrecy that surround much contemporary knowledge whether for reasons of national security or for corporate market position. This chapter is more assay than essay: an attempt to chart the terrain of understanding secrecy and/in the production of knowledge. At the same time, understanding secrecy is no longer simply a historical concern. Contemporary research with respect to bioterrorism and biological warfare is now in roughly the same place as the historical work in the physical sciences examined in this essay. History might provide a useful cautionary tale before either scientists or the federal government in the United States decides to impose a secrecy regime in the biomedical disciplines.[4] What follows is a discussion of the foundations of much of the existing work on secrecy. I argue that much, if not all, of this work views secrecy as being identical to questions of access; that is, questions of who can know specific pieces of information.[5] In this literature, arguments against secrecy are cast in the language of economic rationality—it is inefficient to keep knowledge from others who might needlessly duplicate work already done. Almost all discussions of secrecy and science take place in a context where secrecy is viewed as obviously necessary—a nuclear weapons laboratory, for example—or where such restrictions are viewed as absurd and hence inimical to the "advancement of science." In response to this literature, I suggest that we might read some other accounts of secrecy, like Edward Shils 1956 text, *The Torment of Secrecy*, or Norbert Wiener's autobiographical writings, as steps toward the development of a radically different view of secrecy.[6] Specifically, such works observe that access is but one aspect of understanding secrecy and science; another often-ignored dimension is the effect of such practices upon the content of knowledge developed under particular secrecy regimes. Such a perspective might draw upon much work in science and technology studies to render secrecy comprehensible, if not transparent.

Normal science?

Robert K. Merton's famous norms of science—communism, universalism, disinterestedness, and organized skepticism (CUDOS)—are the *locus classicus* for most understandings of the inimical and unnatural relation of science and secrecy. Drawing upon his pioneering study of Puritanism and the rise of the "new science" of the seventeenth century, Merton extracted what he identified as the guiding norms of the scientific community. In an influential 1942 article, "Science and Technology in a Democratic Order," Merton articulated his famous norms as a direct defense of the necessary relation of progress in science with democratic politics.[7] As David Hollinger has observed, Merton made it clear that science could only flourish under a democratic regime, not the fascist regime of Nazi Germany.[8] Merton clearly stated that secrecy was the antithesis of his norm of communism, the belief that scientific knowledge was the common property of all people. My point here is not to claim that Merton invented the idea that science and secrecy are anathema. After all, his claim was that he had identified this practice through his study of the history of science. Central figures in the so-called Scientific Revolution distinguished themselves from other knowledge producers because of their emphasis on the public, and published, character of their knowledge claims. He was merely making clear to social scientists what natural scientists took as a self-evident truth, one that was visible from the emergence of the Royal Society in seventeenth-century England.

For Merton the problem with secrecy in science was twofold. First, secret science could not provide the researcher with the appropriate credit for his discoveries. Given that the only recognition in Merton's universe came to those who established their priority in making discoveries or breakthroughs, secrecy was clearly not in a researcher's self-interest. While working on a particular problem, researchers might choose not to communicate with others about their work, but when the work was completed they would race to publish their findings. Priority was the means to a reputation, to greater credibility, and to the rewards of science—prizes, grants, and status.[9] Second, secret knowledge was not open to the scrutiny of others who might point out errors and problems related to both the production and interpretation of the knowledge claims. If, as Merton and others believed, science "worked" through the rigorous self-policing of knowledge claims, then secrecy or restricting the dissemination of information might lead to the production of false knowledge. Finally, note that Merton's norms also create an autonomous social space for science, since only other scientists could credibly discuss the veracity of specific technical knowledge claims. Those untrained in the ways of science were incapable of adjudicating intellectual matters.

If Merton and his students, especially Bernard Barber,[10] were among the prime intellectual sources for the post-Second World War understanding of the relationship between science and secrecy, then we must look to the war itself and the subsequent militarization of American science for the institutional context in which such discussions began. Here we must make a historical point. We may think of the war, especially the Manhattan Project, as the modern occasion for our discussions of science and secrecy, but that would be a profound mistake. Discussions about secrecy were endemic with the establishment of the first industrial research laboratories in early twentieth-century America and the great expansion of such laboratories in the post-First World War context, what one observer called "a fever of commercial science."[11] Similarly, the fear that corporate monopolies might control the production of scientific and technological knowledge, as presented in the Temporary National Economic Condition (TNEC) Hearings of 1939, was an early analogue of postwar fears of the military control of science.[12] To an extent we are largely unaware of; wartime discussions of secrecy drew upon these earlier debates as well as the recognition that for many researchers industry had not affected science in a negative manner. On the contrary, many commentators began to conceive of industrial research laboratories as universities in exile, a view that had little relation to corporate reality. With this caveat, let us turn to the war.

Pick up any memoir of the Manhattan Project and one will find a ringing denunciation of General Leslie Groves and his policy of compartmentalization. Even Richard Rhodes, our contemporary chronicler of nuclear history, accepts the seemingly universal condemnation of Groves' apparent obsession with security and restricting the flow of information.[13] Oppenheimer's creation of the Los Alamos seminar series is viewed by both participants and historians as a triumph of the values of science over military paranoia. Los Alamos might have been isolated, but on the Mesa science ruled. Alas, such a perspective is seriously defective. First, while some researchers such as Szilard clearly fought the classification and compartmentalization system, others accepted security as a necessary wartime evil. Far from chafing under the demands of security, these researchers flourished and relished knowing that they were responsible for only one aspect of a larger project. Second, all such accounts view secrecy and the military as the "enemy." Unfortunately, this ignores another view of secrecy that is quite important. Secrecy and the ability to keep secrets were important ways in which the researchers might gain the confidence of their military colleagues and paymasters. Vannevar Bush, the leader of the wartime research and development establishment made this clear when he said to his colleague, Karl T. Compton, the president of MIT, "you and I are responsible for rather serious things, and the maintenance of our relations with the Army and Navy depends upon an orderly handling that inspires confidence."[14] Keeping secrets was essential to establishing and maintaining the credibility of the civilian researchers.[15] This is a definition or function of secrecy that we often forget. The relationship of academic researchers and the armed forces was new; building the connections that we accept as a historical given was an accomplishment in its own time. Undergirding Bush's statement was his recognition that only by properly handling the security issues would he and his organization acquire the trust of the military officers who were actually planning and fighting the war. Individuals like Leo Szilard, who bridled under the security regulations and complained unceasingly about compartmentalization, were either ignored by the military or released from the government service. Playing by the military's rules about information distribution allowed one the possibility of actually having an effect upon their actions.

Another problem with our over-reliance upon the Manhattan project for our understanding of wartime secrecy is that we seldom look at the other research and development programs. Take the case of the proximity fuse, which Bush believed even more technically difficult than the atomic bomb. In this case, the development of a sophisticated electronic device demanded the creation of new laboratories and new forms of industrial–military–academic cooperation. Merle Tuve, the leader of the project, instituted a compartmentalization policy that extended into the workers' eating habits. Researchers often ate lunch at a local "Hot Shoppes." Tuve learned that laboratory workers sometimes discussed their classified work at lunch. This led to a wonderful memo, which was posted throughout the laboratory explaining that the

Hot Shoppes was not a secure site and hence any discussion of the fuse project inside the restaurant would result in the arrest of all the members of a conversational group. Tuve's staff got the message, loud and clear, but they did not understand Tuve's intentions. Of course, Tuve was concerned that enemy agents might be serving the meat loaf, but more pressing was the possibility that staff members might learn about work unrelated to their own specific job assignments. Compartmentalization was a form of management as well as a security precaution. For Tuve, controlling the flow of information among the researchers was as important as, if not more important than, controlling the possible loss of information to an enemy.[16] Localized secrecy was the means to an end, but not an end in itself.

Secrecy might also be considered an essential element of the design process regardless of a technology's military value. The design and development of new technologies are marked by initial periods of contestation and struggle over goals, methods, and even the very possibility of the goal. Hence, if one is developing a new technology—such as a proximity fuse, an atomic bomb, or an inertial guidance system—it can prove beneficial to restrict the sheer number of voices until the group working on the project has produced what they believe is a stable vision or version of the technology. In other words, secrecy might reduce the stress of interpretive flexibility—the inherent plastic meaning of any technology. For example, take the case of inertial guidance for aircraft and ballistic missiles. For this technology to "work" it was essential that the inertial apparatus separate the acceleration of the plane from the acceleration of gravity. For many people, including George Gamow, the famous physicist, such a separation was impossible since it would violate Einstein's relativity theory. Those involved in developing the technology were of a rather different opinion, but the multiplicity of groups working on the problem aggravated the task of responding to Gamow's criticism since there was far from one solution to his objection.[17] Had the managers of the inertial projects kept their work a better secret they might not have had to deal with Gamow's critique until after they had stabilized their devices and methods. Once again, secrecy acts as a management technique, one that is quite powerful but easily abused. One can easily imagine researchers working on a device that shows little promise, but where the secret status of the project allows the work to persist. While we have several examples of this, including the Navy's canceled stealth A-12, secrecy need not necessarily breed corruption.[18]

Understanding the range of ways in which secrecy was part of the wartime research effort is important, but we are forced to return to the atomic bomb. Certainly the bomb was among the best kept secrets of the war; on August 5, 1945 fewer than 100 people knew the full scale and scope of the project.[19] Furthermore, all knowledge relating to the bomb was secret; any public discussion required an active decision to declassify particular pieces of information. Even the Smythe Report, perhaps the oddest press release in American history, did not present technical details but only a general discussion of the project and its work.[20] More precisely, we might note that the report revealed the physics, but not the physical chemistry necessary to making a working weapon. However, the report's final paragraph contains the fundamental idea behind the report: an informed citizenry, with the tutelage of physicists, can make an informed set of decisions about the future of nuclear weapons. The interesting point here was that the government censors were the adjudicators of what the American people needed to know about the Manhattan Project—the autonomy of science had already been breached.

The postwar debate over the legislation establishing the Atomic Energy Commission dealt extensively with the issue of secrecy but largely in terms of the punishments for revealing America's atomic secrets. During the course of the congressional discussion emphasis shifted from the dissemination of Manhattan Project's knowledge to a policy of restricting and finally controlling the flow of information. Congress created a new taxonomy of secret, the category of "restricted data" defined as follows:

> All data concerning the manufacture or utilization of atomic weapons, the production of fissionable material, or the use of fissionable material in the production of power, but shall not include any data which the commission from time to time determines may be published without adversely affecting the common defense and security.[21]

What does the invention of a new level of secrecy do? First, it creates an additional class of individuals who have access to restricted data. Although this might be of interest to those studying the mixing of individuals with different clearances, or how particular organizations work, it is unclear how the taxonomy affects the issues with which we are concerned.[22] Is this not simply another example of access being the rationale and meaning of secrecy? Second, the invention of restricted data reminds us that during the immediate postwar period many people spoke and acted as if the revelation of a particular piece of information might "give away" the "secret" of the bomb.[23]

For students of this period, the growth of restricted data is both a problem and a blessing. If we view secrecy as a problem in access, then we are mainly concerned with acquiring that access for ourselves. In other words, we operate under the belief that whatever is classified should be declassified or removed from the penumbra of secrecy; in turn, we will have a better idea of what actually happened. Among the many assumptions present in our call for access is the belief that the classified and the unclassified are linked in some direct and unmediated fashion, as if the light of inquiry would make the past clearer. More than likely the opposite is true—the relation of the classified and unclassified is problematic and highly mediated. Knowing the contents of restricted data might not help us reconstruct events and processes; if I learn that beryllium is an important ingredient in thermonuclear weapons have I learned something important? This type of information is only important if I am attempting to understand the growth and development of the beryllium machining industry or the growth in incidences of complaints of beryllium poisoning or a related inquiry.[24] In other words, restricted data in and of itself might prove more meaningless than meaningful. Hence, if access is why we are interested in secrecy we really don't have much to say other than on a case-by-case basis. It is one thing to know what actually took place at the Gulf of Tonkin by reading the previously classified cables from the region; it is another thing to know that element X is used in technology Y. Knowing secrets may be exciting, but it may not be intellectually interesting for students of science and technology studies. Of course, knowing what actually transpired in the Gulf of Tonkin might have kept the US Congress from granting Lyndon Johnson the ability to wage war in Vietnam. As Daniel Ellsberg's career makes clear, secrets are corrosive in a democracy when they affect how a polity is to understand its role in the world.[25]

So, what is interesting about secrecy?

Open the newspaper nearly any day of the week: secrecy is on display. New products, like Gillette's three-blade razor, are the result of industrial processes so guarded that they make the Manhattan Project look like a sieve.[26] Secrets are only known when they are no longer secrets, but the power to unveil and display a secret is what makes secrets useful and dangerous. These types of events and practices do not figure in our understandings of secrecy and science, despite the way in which the atomic bomb's use at Hiroshima might be likened to the unveiling of a new and powerful product.[27]

Return to our earlier ideas about why access is not what is interesting about secrecy. What is interesting is how practitioners discuss secrecy. The most common belief appears to be that secrecy is a necessary evil, but one that ultimately undermines the development of science. It is one thing to keep secrets in wartime, another to do so under the conditions of peace. Yet researchers keep secrets all the time, sometimes quite inadvertently. In his study of Toshiba's management of intellectual capital, David Fruin tells us that Toshiba had a great deal of trouble in setting up Knowledge Works factories overseas; indeed, the skills and knowledge necessary to make a Knowledge Works factory operate are so site- and person-specific that there is no way to capture this know-how short of exporting the people from a successful factory. As Fruin makes clear "the nature of factory know-how is not contained in manuals but is found instead in practice and experience."[28] For students of science and technology studies, it is clear that Fruin is talking about tacit knowledge—that knowledge which is practice-specific and often incapable of being articulated in any formal way.[29] Although this is not an intentional secret, like restricted data, it has a similar effect. Restricted data is also about slowing the spread of a technology; an inability to transmit tacit knowledge slows the ability of Toshiba to grow and compete with other Japanese and American firms. Clearly, tacit knowledge doesn't count as secrecy; rather it is part of the "tricks of the trade."

Another reason researchers argue against secrecy is the claim embodied in the Smythe report: secrecy denies the public the ability to learn about issues vital to the survival of the polity. This is a contested and contestable thesis. During the debate over the H-Bomb, Leo Szilard believed the American public incapable of making the right decision with respect to the weapons' development.[30] More information was not going to help the public; the decision had to be made by those who knew best: physicists. Restricted data created a community of inquirers capable of making the best possible decision. Szilard's world was far from democratic. Accountability was a problem for everyone but scientists. Despite his obsession with secrecy, Szilard accepted a political ideal that was a pure technocracy; a point made clear in his seminal story, "The Voice of the Dolphins."[31] Readers will recall that the story's underlying narrative, that intelligent dolphins rather than politicians were capable of ending the nuclear arms race, rested upon keeping the dolphins' actual work practices secret. In turn, after the story's happy ending, Szilard reveals the possibility that the dolphins were simply a cover for scientists imposing their rational vision upon international politics. In Szilard's universe secrecy prevented the uninformed from playing an authoritative role in politics. Ignorance was more than bliss; it was the basis upon which one might erect a rational political order.

If, as Yaron Ezrahi argues, science plays an authoritative and constitutive role in liberal democratic polities because it is transparent, then secrecy—as at the Gulf of Tonkin—might undermine democracy.[32] Transparency refers to the public's ability to see the process through which authoritative claims are made; conceivably, anyone with enough time and patience might gather "the facts" and understand how a decision was made or a policy developed. Diane Vaughan's account of the Challenger disaster is an example of the belief in transparency; Vaughan's meticulous reconstruction of the cultures of NASA and Morton Thiokol as well as the conversations leading to the launch decision exemplify transparency's political value.[33] Vaughan as both scholar and citizen wades through the documents and pieces together what she believes is the actual story. The alleged transparency of technical processes, the belief that with enough time and resources we might understand any given decision, appears at odds with secrecy. Transparency, however, might equally rest upon the credibility of researchers who vouch for the truth of what takes place in the classified world. Individual researchers become spokespeople for the government's massive investment in secret research. In turn, the credibility of individuals becomes a surrogate for the credibility of the state. In this sense, secrecy and democratic politics do not appear as diametrically opposed as researchers and analysts might believe.

Reading accounts of secrecy in science from the postwar era written by researchers or those involved in the loyalty and security programs reveals a common strand: a belief that secrecy was a new evil.[34] That is, whether it is Shils' *The Torment of Secrecy* or Wiener's *Invention* or his autobiography *I am a Mathematician*, one is struck by the overwhelming sense of nostalgia for a time when secrecy did not affect science. Read as Wiener discusses the state of science in 1956:

> There is not doubt that the present age, particularly in America, is one in which more men and women are devoting themselves to a formally scientific career than ever before in history. This does not mean that the intellectual environment of science received a proportionate increment. Many of today's American scientists are working in government laboratories, where secrecy is the order of the day, and they are protected by the deliberate subdivision of problems to the extent that no man can be fully aware of the bearing of his own work. These laboratories, as well as the great industrial laboratories are so aware of the importance of the scientist that he is forced to punch the time clock and to give an accounting of the last minute of his research. Vacations are cut down to a dead minimum, but consultations and reports and visits to other plants are encouraged without limit, so that the scientist, and the young scientist in particular, has not the leisure to ripen his own ideas.[35]

The poignant character of Wiener's lament should not be lost on us, but it is important that this is a complaint about two important issues. First, losing control over the direction of research. Second, losing control over the actual content of the knowledge produced by the researcher. Secrecy was an imposition from those who did not understand the Mertonian ethos that scientists took for granted. In other words, the scientist always possessed dual citizenship: first, in what Michael Polanyi called the "republic of science" and next in a particular nation-state.[36] Implicit in the Mertonian formulation that Wiener and researchers embraced was the very possibility of divided loyalties. Choosing between science and country became something akin to choosing between a friend and country. Research problem choice could be seen as a way of assessing loyalty to a government; even if a researcher did not find the work interesting s/he would have to work on the project or risk being labeled as disloyal. The professed norms of science and the norms of secrecy were not merely antithetical, they were mutually exclusive. However, the knowledge produced in secret was "science," at least to those privy to its production and use.

Wiener recognized that secrecy, citizenship, and knowledge production were of a piece; secrecy affected the very content of knowledge. This is certainly a far more controversial point since we are leaving the realm of access behind. Wiener's point, and that of Edward Shils, was not simply the question of economic rationality, that is, secret science forced the unnecessary duplication of work that had already been completed. Rather, it was a qualitative point more difficult to address. Put simply, Wiener is arguing that one gets a certain type of knowledge from a particular social organization, in this case a secret organization or research that is secret. This knowledge is different from what might be produced in a more open space. The argument is not that secrecy allows "bad" or incompetent science to flourish, although that was certainly a possibility if one believed in the scientific community's homeostatic propensities. Instead, it was an argument about the constraints and conditioning of the imagination. Secret knowledge produced a different map of intellectual geography, a different sense of the horizons of possibility. Pursued over time, such knowledge would produce an entirely different and separate world, one in which access would be the least of an outsider's problems. Even with access, the outsider would find themselves as visitors in a foreign country without any sense of the nation's language or grammar. Obviously, translation would prove possible over time, but such a scheme undermined the possibility of claims to universalism, let alone the claim that scientific knowledge was public property. Secrecy eroded the extent to which scientific knowledge, and concomitantly the world explained with that knowledge, might serve as a common currency for culture across boundaries.[37] However, Wiener's worldview could not account for the dilemma posed by actual espionage or the appearance of espionage. It may be that such secret worlds are essential and that the ability to create knowledge so separate from the world at large is a political tactic necessary within a competitive international order.

We might also read these discussions of secrecy as versions of Paul Forman's belief that knowledge is made to order; you get what you pay for.[38] That is, secrecy is at one with the idea that scientists are employees following orders. As employees why should we expect that they would control the content and direction of their research? While such a perspective is provocative and appealing, it does not appear to connect with the ways that scientists present themselves; indeed, we might read Forman as being more like Wiener and Shils insofar as he laments the transformation of physics into its secret and corporate present.

Finally, we need to note that a new source of secrecy emerges from the tangled webs created by university–industry relationships in the age of biotechnology. Scholars such as Roger Geiger are beginning to address this issue through substantive empirical research rather than diatribes about the threat posed by corporate interests.[39] What is becoming clear is that secrecy and restricted access to research results are increasingly common as the university becomes a corporation in the marketplace. Universities, such as Cornell, have an active interest in developing a sizeable patent portfolio with which they might fund research. Ironically, during the first biotechnology bubble of the 1980s, commentators feared that researchers might take their profitable research off campus and create corporations capable of siphoning the university's potential income. Today, universities have made it clear that researchers are employees whose inventions belong to the university, not the faculty member. In this sense universities and industrial research facilities are now alike.

On the matter of conclusions?

Far from being straightforward, the relationship of secrecy with the production of knowledge opens up a hermeneutic can of worms that science and technology studies must address. Part of the problem is that conventional understandings of science are inadequate to the task since they are implicated in the problem. In her work on research subpoenas, Sheila Jasanoff makes it clear that simply acquiring access to the raw materials that an investigator uses to write a scientific paper does not provide one with a road map to the construction of any particular paper. Instead, such access transforms those demanding the data into interpreters who must provide their own story about the materials or explain why the materials cannot be used to make the claims that are at issue.[40] Lawyers have an advantage generally not available to historians or sociologists: the discovery process. More recently, discovery has acquired a new meaning. At MIT, students working for startup companies established by individual professors are required to sign nondisclosure agreements, contracts that forbid the student from discussing the product under development. Professors working on related products have designed their course homework assignments to determine the nature and status of a competitor's work. Student employees are caught in a bind, violate their nondisclosure agreement or fail the homework assignment.[41] Industrial espionage masked as pedagogy has brought the marketplace squarely into the classroom, but it also raises the issues of secrecy in a powerful and palpable form.

We cannot acquire all the relevant materials, no matter how much we desire to do so. At the same time we need to think of ways to discuss the ways in which the classified world relates to the world to which we do have access. How are we to imagine the relations between realms that have very different reciprocal relations? Once again, we are back to questions of access, but with a difference. The question is not how to access this world, but how to assess that world's impact on what is visible.[42] How is the hand that stamps the security seal on a document linked to the hands that write the document? Is our situation reminiscent of the physicist studying a black hole: how can we find out what happens in a black hole if nothing can escape from it?

Or, is it that some things do move from the classified to the unclassified worlds—people, for example, and information. By studying the shape and form of what we can see, might we not make inferences about our visible world as well as the secret world?[43] Or is it, as Wiener suggested, utterly outside the scope of our imaginations?

Notes

1   Jeffrey T. Richelson, "Scientists in Black," *Scientific American* 278, no. 2 (1998): 48–55, 48.
2   Timothy Ferris, "Not Rocket Science," *The New Yorker* 74 (July 20, 1998): 4–5.
3   David Holloway, *Stalin and the Bomb* (New Haven: Yale University Press, 1994).
4   For more on this see Susan Wright with David Wallace, "Varieties of Secrets and Secret Varieties: Secrecy and the Biotechnology Industry," in *Secrecy and Knowledge Production*, ed. Judith Reppy, Cornell University Peace Studies Program, Occasional Paper #23 (October 1999); revised in *Politics and the Life Sciences* 19, no. 1 (March 2000), 45–57.
5   Some examples of this work are Sissela Bok, *Secrets: On the Ethics of Concealment and Revelation* (New York: Vintage, 1989); Herbert Foerstel, *Secret Science: Federal Control of American Science and Technology* (Westport, CT: Praeger, 1993); and the collection edited by Marcel La Follette, "Secrecy in University-based Research: Who Controls? Who Tells?" *Science, Technology and Human Values* 10, no. 2 (1985): 3–119.
6   Edward A. Shils, *The Torment of Secrecy* (Chicago, IL: Ivan R. Dee, 1956; reprint 1996); Norbert Wiener, *I am a Mathematician: The Latter Life of a Prodigy*. (Cambridge: The MIT Press, 1956); Norbert Wiener, *Invention: The Care and Feeding of Ideas*. (Cambridge: The MIT Press, 1993).
7   Reprinted as Robert K. Merton, "The Normative Structure of Science," in *The Sociology of Science: Theoretical and Empirical Investigations*, ed. Norman W. Storer (Chicago, IL: University of Chicago Press, 1973), 267–278. It is also important to realize that Merton's norms were subject to a powerful and devastating critique that is largely forgotten: Ian I. Mitroff, *The Subjective Side of Science: A Philosophical Inquiry into the Psychology of the Apollo Moon Scientists* (Amsterdam: Elsevier, 1974). Mitroff convincingly demonstrated that whatever activity might be explained by a set of norms might also be explained by a set of counter-norms. Hence, it is possible to understand the entire process described by Merton with a set of norms articulating the opposite set of values—private property, local understanding, interestedness, and organized credulity. Unfortunately, it does not lend itself to a neat acronym.
8   David A. Hollinger, "The Defense of Democracy and Robert K. Merton's Formulation of the Scientific Ethos," in *Knowledge and Society*, ed. Robert Alun Jones and Henrika Kuklick (Greenwich, CT: JAI Press, 1983), 1–15. Also of interest here is Everett Mendelsohn, "Robert K. Merton: The Celebration and Defense of Science," *Science in Context* 3 (1989): 269–290.
9   Certainly I don't mean this to be an exhaustive list, merely evocative. It is altogether too easy to translate Merton's norms into a framework for the acquisition of social capital. If we do that, secrecy might become both an asset and a liability.
10   Bernard Barber, *Science and the Social Order* (New York: Collier Books, 1962) is an especially good source for the antithetical relationship of science and secrecy.
11   I am embarrassed to do this, but some discussion of this issue can be found in Michael Aaron Dennis, "Accounting for Research: New Histories of Corporate Laboratories and the Social History of American Science," *Social Studies of Science* 17 (1987): 479–518.
12   On this point, see Larry Owens, "Patents, the 'Frontiers' of American Invention, and the Monopoly Committee of 1939: Anatomy of a Discourse," *Technology and Culture* 32, no. 4 (1991): 1076–1093. For a specific example of the fear of industrial control, see Peter Galison, Bruce Hevly, and Rebecca Lowen. "Controlling the Monster: Stanford and the Growth of Physics Research, 1935–1962," in *Big Science: The Growth of Large Scale Research*, ed. Peter Galison and Bruce Hevly (Stanford, CA: Stanford University Press, 1992), 46–77.
13   Given that so much information went to the Soviet Union, one might wonder if Groves' obsession was really so unwarranted. For Richard Rhodes, see *The Making of the Atomic Bomb* (New York: Simon and Schuster, 1986).

14 See April 1, 1941, VB to KTC, Box 26, Folder 609 (KTC '39–'42), Vannevar Bush Papers, Library of Congress.

15 Clearly this bears some relation to the discussion of credibility in Steven Shapin, *A Social History of Truth: Civility and Science in Seventeenth-Century England* (Chicago, IL: University of Chicago Press, 1994).

16 On these points, see Michael Aaron Dennis, "Technologies of War: The Proximity Fuse and the Applied Physics Laboratory," in *A Change of State: Political Culture, Technical Practice and the Origins of Cold War America* (Baltimore, MD: The Johns Hopkins University Press, forthcoming).

17 For this specific example, see Donald MacKenzie, *Inventing Accuracy: A Historical Sociology of Nuclear Missile Guidance* (Cambridge: The MIT Press, 1990); and Michael Aaron Dennis. " 'Our First Line of Defense': Two University Laboratories in the Postwar American State," *Isis* 85, no. 3 (1994): 427–455. Given the complexity of this particular example, it may be a poor choice. Gamow probably came to his knowledge of inertial techniques through his membership on the Air Force Science Advisory Board. Conceivably, one might argue that as a board member Gamow was doing his job by expressing his beliefs about the untenable character of the research. What is striking is that Gamow does not appear to have visited or contacted any of the groups trying to develop this technology before he produced his critique.

18 See Robert Holzer, "DOD Secrecy Drives Up Weapons Cost, Development Time," *Defense News*, Oct. 21, 1991, 10.

19 Richard Hewlett, " 'Born Classified' in the AEC: A Historian's View," *Bulletin of the Atomic Scientists* 37 (December 1981): 20–27.

20 Henry DeWolfe Smythe, *Atomic Energy for Military Purposes* (Washington, DC: Government Printing Office, 1945; reprint Stanford, CA: Stanford University Press, 1989).

21 Hewlett, "Born Classified."

22 For an interesting discussion of these very issues, see Hugh Gusterson, *Nuclear Rites: A Weapons Laboratory at the End of the Cold War* (Berkeley, CA: University of California Press, 1996), 8–100.

23 On the idea that there was a single secret and its consequences see Gregg Herken, *The Winning Weapon: The Atomic Bomb in the Cold War, 1945–1950* (New York: Vintage, 1981). Shils' work, cited in Note 6, also addresses this particular conception of an "atomic secret."

24 Or if I am trying to build my own bomb. However, even if I learn this particular fact and others, I still need to do a great deal of work if I want my own nuke. As recent events make clear, even impoverished nations are willing to use scarce resources to build the infrastructure necessary for a nuclear arsenal. My point is simply that individual factoids are not going to teach anyone how to build a bomb. Many Iraqi physicists knew how to build a bomb, but the Iraqi state lacked the resources to actually build one after the first Gulf War. Perhaps even more striking is just how difficult it is to accurately gauge what a nation is doing in secret. The United States grossly underestimated Iraq's progress toward a nuclear weapon before the first Gulf War and grossly overestimated the same nation's ability to build a weapon before the invasion of Iraq under President George W. Bush.

25 In particular, see David Ellsberg, *Secrets: A Memoir of Vietnam and the Pentagon Papers* (New York: Viking, 2002); also interesting in this respect is the late Senator Daniel Patrick Moynihan's remarkably candid and thoughtful volume on the problem of secrecy and American democracy in Daniel Patrick Moynihan, *Secrecy: The American Experience* (New Haven, CT: Yale University Press, 1998).

26 See Mark Maremont, "Gillette Finally Reveals Its Vision Of The Future, And It Has 3 Blades," *Wall Street Journal*, April 14, 1998, sec. A, p. 1.

27 This is one theme of an audacious and little-cited article, M. L. Smith, "Selling the Moon: The U.S. Manned Space Program and the Triumph of Commodity Scientism," in *The Culture of Consumption: Critical Essays in American History, 1880–1980*, ed. R. W. Fox and T. J. J. Lears (New York: Pantheon, 1983), 175–236.

28 W. Mark Fruin, *Knowledge Works: Managing Intellectual Capital at Toshiba* (New York: Oxford University Press, 1997), 162. Another excellent example of this is Intel's "Copy Exactly" program. Faced with the massive costs of chip fabrication facilities Intel came to believe that

the way to make a new manufacturing facility was to duplicate an existing facility down to a rather dramatic level of detail. For more on this program see C. J. McDonald, "The Evolution of Intel's Copy EXACTLY! Technology Transfer Method," *Intel Technology Journal* (4th Quarter, 1998), http://developer.intel.com/technology/itj/q41998/articles/art_2.htm, accessed on May 17, 2004.

29 On tacit knowledge, see H. M. Collins and R. G. Harrison, "Building a TEA Laser: The Caprices of Communication," *Social Studies of Science* 5 (1975): 441–450.

30 On Szilard's undemocratic perspective, see Peter Galison and Barton Bernstein, "In Any Light: Scientists and the Decision to Build the Superbomb, 1952–1954," *Historical Studies in the Physical and Biological Sciences* 19 (1989): 267–347.

31 Leo Szilard, *The Voice of the Dolphins and Other Stories*, expanded ed. (Stanford, CA: Stanford University Press, 1992).

32 Yaron Ezrahi, *The Descent of Icarus: Science and the Transformation of Contemporary Democracy* (Cambridge: Harvard University Press, 1990).

33 Diane Vaughan, *The Challenger Launch Decision: Risky Technology, Culture, and Deviance at NASA* (Chicago: University of Chicago Press, 1996). I owe this insight, even in this mangled form, to my former colleague, Sheila Jasanoff.

34 One wonders what scientists would have done had the following works been available to them. See Pamela O. Long, *Openness, Secrecy, Authorship: Technical Arts and the Culture of Knowledge from Antiquity to the Renaissance* (Baltimore, MD: Johns Hopkins University Press, 2001); and W. Eamon, *Science and the Secrets of Nature: Books of Secrets in Medieval and Early Modern Culture* (Princeton, NJ: Princeton University Press, 1994).

35 Wiener, *Mathematician*, 361.

36 Michael Polanyi, *Science, Faith and Society* (Chicago, IL: University of Chicago Press, 1946).

37 I think that this constraining of possibilities is what Ian Hacking is going on about in "Weapons Research and the Form of Scientific Knowledge," in *Nuclear Weapons, Deterrence, and Disarmament*, ed. David Copp (Calgary: The University of Calgary Press, 1986).

38 See Paul Forman, "Behind Quantum Electronics: National Security as Basis for Physical Research in the United States, 1940–1960," *Historical Studies in the Physical Sciences* 18 (1987): 149–229; and Paul Forman, "Inventing the Maser in Postwar America," *Osiris (2nd Series)* 7 (1992): 105–134.

39 Roger L. Geiger, *Knowledge and Money: Research Universities and the Paradox of the Marketplace* (Stanford, CA: Stanford University Press, 2004).

40 Sheila Jasanoff, "Research Subpoenas and the Sociology of Knowledge," *Law and Contemporary Problems* 59 (Summer 1996): 95–118. Obviously this point is also related to the historian's use of laboratory notebooks in reconstructing scientific and technical practices. How is what is in the notebook related to what is in the published document? A fascinating example of this is found in Gerald L. Geison, *The Private Science of Louis Pasteur* (Princeton, NJ: Princeton University Press, 1995). Note that I have not discussed what Merton and others take for granted: the need for some secrecy in the quest for priority since such claims rest on an assumption of eventual openness.

41 See Amy Dockser Marcus, "MIT Students, Lured To New Tech Firms, Get Caught In A Bind," *Wall Street Journal*, June 24, 1999, sec. A, p. 1.

42 Ron Doel is getting at a related idea near the end of his essay, "Scientists as Policymakers, Advisors, and Intelligence Agents: Linking Contemporary Diplomatic History with the History of Contemporary Science," in *The Historiography of Contemporary Science and Technology*, ed. Thomas Söderqvist (Amsterdam: Harwood Academic Publishers, 1997), 215–244.

43 For example, could we not argue that the International Geophysical Year (IGY) of 1957 was simply arms control by other means? That is, by measuring the earth's gravitational field and producing sophisticated maps of the Arctic, Russia and the United States acquired the information necessary to allow inertial guidance systems to fly to their targets with a greater degree of accuracy. More information allowed for greater claims of inevitable destruction.

Part V

History detectives

New ways of approaching modern
science, technology, and medicine

11 The conflict of memories and documents

Dilemmas and pragmatics of oral history

Lillian Hoddeson

Oral history interviews can breathe life into the research and writing of recent history. By filling in the nerves and connective tissue (motives, inspirations, fears, obsessions, etc.) that link actions with each other, participant accounts can animate a narrative with details not found in documents, shade it with nuances, and vitalize interpretations with insights into diverse points of view.

Interviews can also help the historian go about studying an institutional or theoretical development.[1] They may provide information that is unrecorded anywhere else or guide the historian through archival collections, adding annotations, and revealing documents, such as letters or notebooks, which had been squirreled away in an attic or cellar, or clippings which were interspersed in personal files. Interviewees sometimes contribute references that the historian could not possibly have found unaided, or help the historian decode agendas that lurk behind documents. It is not uncommon for the interview process to restore lost information by tickling the memories of "living sources," so they are able to recall events that they had effectively forgotten.

One of the most dramatic uses of oral history—about which I will have much to say here—occurs when memories and documents come into conflict. Such conflict remains at the core of the most devastating criticism that has been leveled against using interviews in historical research. Yet in the hands of a skilled oral historian, this conflict can become a powerful tool in a methodology that leads to deeper histories having subtler overtones than can be achieved without the use of interviews.

Before presenting the methodology, I need to define what is meant by oral history and explain the major objections that historians have raised to it.

Definitions and distinctions

The term oral history is commonly used in three different ways. It can refer to the interviews themselves, to the methodology for conducting them, or to histories that were written *using* the interviews. Most of the literature on oral history deals with the first two uses. This chapter will consider as well the histories that interviews enrich.[2] These histories divide into two basic types: those that are based almost entirely on interviews (e.g. Studs Terkel's *Hard Times: An Oral History of the Great Depression*) and those (of more interest to me) that depend as much on other sources as they do on interviews.

An important distinction separates "passive" and "interactive" interviews. This distinction turns on whether the historian is a significant influence. Historians differ radically over what they consider the correct degree of influence. The well-known fact that an interviewer's leading questions can cause distortions of memory, or in the worst case implant false memories,[3] suggests to many that passive interviews are more reliable than interactive ones and that historians should play as small a role as possible in conducting the interviews. Yet a completely passive interview is impossible, for even in the extreme case when an interviewee speaks directly into a tape recorder, without an interviewer present, the historian still plays a role in shaping the interview, if only in arranging it—as Ron Grele, one of the founding fathers of modern oral history, has noted.[4] While many guides to oral history emphasize the case for doing passive interviews, I will concern myself mainly with the argument for conducting fully interactive interviews whose content is self-consciously tailored by the historian with the help of questions based on considerable research.[5] The resulting dynamic dialogues between the historian and interviewee often yield deeper insights.

The case against oral history

Although historians have been using oral sources to help create and present their narratives since ancient times, the methodology of doing oral history is not fully established, even today.[6] Perhaps it was the invention of the wire and then the tape recorder that caused historians to begin worrying about oral history. For in the very era when these technologies began to upgrade the quality and usefulness of interviews, many historians began raising sharp objections to them.

Historians who dismiss oral history usually focus on one or more of the following: oral histories are inherently biased; human memory is unreliable; oral history deals with present reflections on past events; and oral histories evolve and change, even while they are in the process of being created.[7] These are all real problems, but none of them invalidate the use of interviews in the writing of history. Indeed the primary objection of bias applies just as well to documents, which may be as distorted as interviews by the agendas that are in place at the time of their creation.

All people rewrite their memories constantly, recalling some sections in detail, others more vaguely, deleting portions, blowing some parts up while shrinking others. While many researchers adhere to the time-honored picture of memories stored like pictures in a photo album, a growing research literature deals with the fact that memory is selective, distorted, and changing.[8] Most researchers of memory in the fields of cognitive psychology and neuropsychology have come to believe that "we do not store judgment-free snapshots of our past experiences but rather hold on to the meaning, sense, and emotions these experiences provided us."[9]

The fact that interviews suffer from the full range of the distortions that human memory creates is not news to the experienced oral historian. Any scholar who conducts interviews soon learns what the anthropologist and historian of African societies Jan Vansina once wrote, that "memory is not an inert storage system like a tape recorder or a computer. Remembering is an activity, a re-creation of what once was. It uses for this purpose not just this or that bit of information but everything available in the information pool that is needed in this circumstance, reshaped as needed for this particular re-creation."[10]

Human memory may reorder events temporally or move remembered events into other situations or time frames. In a dissertation about the discourse of oral history, Jean Robertson pointed out that "straightforward chronological progression from earliest to latest event is not the dominant temporal form employed within an oral history and is never maintained continuously over the entire account." A person typically reflects on the past associatively, allowing "one memory to trigger another by a process of association, [and] he often will join events that are historically separate into a single topic of discussion."[11]

The oral historian also learns that the memories expressed in interviews draw on information from the present as well as the past in its constructive process and that memories are colored by the interests, images, and values of the period in which the memories are expressed.[12] In a perceptive essay about oral history, the historian Michael Frisch noted that Terkel's *Hard Times* is as much about people "trying to live in and understand the 1960s" as it is about life in the 1930s.[13] Memories adapt and can shift their content to fit the prevailing values of communities.[14]

Interviews are affected by the interests, knowledge, and experiences of the historian who will shape her narrative to fit the interests of her audience. She has the power (and often draws on it) to pattern the historical fabric she weaves. Even the way the historian selects and arranges materials from the oral and written accounts is affected by her ideas, interests, and accumulating knowledge derived from many sources and studies.

In this essay I will explain why, for the case of oral history, I feel this is as it should be. The injected feedback of historians can bring interviewees to usefully rethink, refine, and, in some cases, alter their recollections when they conflict with documents.[15] The process of historical research using interviews may be compared with studying the kind of system (ranging from the quantum-mechanical to the psychological) in which the process of observation changes what is being observed, thus injecting a degree of indeterminism. Interviews can be seen as dialectics that operate between historian and interviewee, between present and past, and between interviews and every other kind of source (Figure 11.1).

The usefulness of suspicion: the mask

> "I have done that," says my memory. "I cannot have done that," says my pride, and remains inexorable. Eventually—memory yields.
>
> (Friedrich Nietzsche)[16]

To use interviews to best advantage, the historian needs to be aware of how a person's interests, beliefs, values, aspirations, and full range of emotions (including pride, shame, fear, pity, and love) affect their memories, and how a person's memories are

Figure 11.1 Historian Lillian Hoddeson with Hans Bethe at Los Alamos in 1979, preparing for an interview about its Theoretical Division during the Second World War.

Source: Los Alamos photo.

shaped and reshaped to support his present self-image. In studying the transmission of traditions in African culture, Vansina developed a useful way to model this process. He identified the public account, the one people reveal readily, as a cover story which he calls the "mask." This mask is "built up in terms of roles and statuses, values and principles—the noble mask of oneself."[17] Behind this mask is the hidden portrait, or "face," the authentic account. The face is "much less often limned and reveals traces of doubt and fear as quite contradictory experiences are remembered." The distinction between mask and face, Vansina explains, varies from culture to culture, from one discipline to the next, and from one profession to another. Both mask and face are important objects of study for the oral historian; exploring their relationship helps her to understand how interviewees see themselves in relation to their culture. But to construct a deeper history, she must go farther and dislodge the mask, to discover the face.

The masks of executives, politicians, and other public figures tend to be more firmly attached than those of scientists or craftsmen; for the glue that attaches mask to face hardens through repeated tellings of the stories.[18] My least successful interviews have been with individuals who deal regularly with the public. I have also found that mask-related problems intensify when interviews are conducted in places that reinforce the masks. Thus my 1970 interview with James Fisk, a former Bell Labs president, suffered from my mistake of interviewing him in his executive office. The result was indistinguishable in tone and content from Fisk's articles in the *Bell Labs Record*. I might have had better luck had I interviewed him in a taxi or a pub— an environment that could weaken, rather than support, the hold of his mask.

The historian's principal tool for examining the mask is her set of questions. The questions must be incisive and compassionate at the same time. Good interviews have a tension that derives from the uncomfortable probings of both the historian and interviewee, who are each trying "to understand the historical position from which the other speaks," as Grele described it.[19] The historian keeps the interview tense with efforts to explore and elucidate the story. "When did that happen?" "Were you aware of his work on that subject?" "Why have you left this part of the story out?" "Who else was there?" "What was her response?" "Why does your account differ from X's in this way?" Such questioning can be disturbing, even embarrassing. But the historian cannot allow the tension to amplify so much that it threatens the continuation of the interview.

In my interviews I routinely use documents as additional tools. I typically bring along quite a collection. The trick is to bring out relevant materials at just the right time. Consider my interviews in May 1978 with Robert R. Wilson, the founding director of the high-energy physics laboratory in Illinois known as Fermilab (Fermi National Accelerator Laboratory).[20] As the facility's creator and chief executive, Wilson gave many public lectures and press conferences. In the process, he developed his standard account of why he built the laboratory. This was a mask I needed to probe. He was happy to present it to me during our first session, which was fairly relaxed and revealed little beyond the written literature, but it allowed us to make contact. Now I could prepare my strategy.

I came to the next session equipped with a pile of carefully selected documents about Fermilab's origins. As Wilson told his story, I used the materials to gently correct him, adding dates, names, and other details. The interview went smoothly until I produced a particular document that conflicted with one of his favorite stories—as it happened, just as he was about to tell it. He became visibly upset and we were both embarrassed. But not for long, for the session was interrupted by an urgent phone call. He ran out and I was left wondering whether I ought to pack up.

The call upset him further, although I did not learn why until later. It was the notification that Wilson's resignation as Director of Fermilab, provisionally accepted in February 1978 by the board of Fermilab's managing organization (the Universities Research Association), was now official. Wilson had not expected this turn; his resignation had been but a gesture in his attempts to increase funding for the laboratory.

When Wilson returned to the interview, he continued his story, but there was a noticeable difference. The conflict that I had engineered, between memory and documents, had loosened his mask. The phone call's further blow to his public persona had pried his mask wide open. He offered a revised account that not only fit the documents but was also much richer in detail. The two jolts—one planned, the other coincidental—had helped him to expand his memory. I also believe that our interview, which was based on a developing friendship and genuine interest in his Fermilab career, helped Wilson adjust to his own cold reality. For the historical era we were probing in the interview had suddenly ended.

The encounter had an additional important consequence for my work. Wilson enlisted as a helpmate in my historical study of Fermilab. From that point on, he and I could act as a team in reconstructing Fermilab's history. When an historian can achieve this kind of collaboration with an interviewee, the possibility opens of reaching new levels of historical re-creation that are not possible using memory or documents alone.

Horace Judson tells a related story about the adaptation of memory to fit historical inquiries and the subsequent opening of research opportunities. The story comes from Judson's experience interviewing Francis Crick, with papers and correspondence in place before them. About ten months after a particular interview, Judson gave Crick a manuscript to read based in part on the interview. Crick noticed an inconsistency. "Did I really tell you that? Well it wasn't quite like that at all." At this point, Crick offered a refreshed memory that fit the full body of evidence.[21]

Let me now turn to some examples from my research on the wartime and postwar history of Los Alamos National Laboratory. Just before an interview on building the hydrogen bomb (H-bomb), Edward Teller confided in me: "Lillian, I have a terrible problem." He was working on his memoirs.[22] A year or so earlier, I had helped him to gain access to his Los Alamos papers from the 1940s and 1950s. "What's your problem, Edward?" "The problem is simple," he replied.

Then, throwing up his arms, he exclaimed with considerable emotion: "The documents don't agree with my memory!" He was referring to the story of the development of the Ulam-Teller design of the fusion bomb system. Teller's account was significantly shifted—in his favor—from the one told by Hans Bethe, an account that *did* fit the documents.[23] Having thus warned me, Teller then proceeded to offer his memories, which did not agree with the documents.

I am fully convinced, however, that Teller was not trying to invalidate or supercede the documents. His memories were what he *could* offer. By sharing his distress about their disparity with the documents, he had made *his* problem also become *my* problem. And by pointing to this problem openly and honestly at the time of our interview, Teller spared me the embarrassment of calling the disparity to his attention. His perception of the disparity as a "terrible" problem then opened the door to several highly productive interviews that we had subsequently about the history of the H-bomb—interviews in which Teller's mask was indeed partially dislodged.

How the conflict between documents and memory more typically operates is illustrated by an interview that Gordon Baym and I conducted in Pasadena in 1979 with the theoretical physicist Richard Feynman, then 61. I brought to that interview a large, recently declassified, stack of wartime theoretical documents that had been discovered at Los Alamos in an old safe. I thought that the documents, many in Feynman's own hand, might help him recall his wartime work. Unfortunately, Baym and I had scheduled this interview too soon after the first of Feynman's major surgeries; the operation clouded his memory. Blankly turning page after page, Feynman grew increasingly frustrated by his inability to recall anything about them. He blurted out angrily: "You know what you oughta do with these documents? Throw 'em in the garbage!" The interview remained uncomfortable—indeed anguished—for at least half an hour.

Then Feynman discovered in the stack several interesting blueprints that caused his memory barrier to lift. These blueprints had figured large in one of his favorite stories, told on many occasions and eventually published several times.[24] In his anecdote, "Little Richard" was on a wartime mission at the Oak Ridge laboratory. At one point, several Oak Ridge engineers were describing the safety precautions built into one of their chemical plants. They had gathered around a *"loooooong* table" covered with the enormous blueprints. As they spoke, Feynman realized that he didn't know the meaning of a certain symbol, a cross inside a circle.

As far as he could make out, the symbol referred either to a valve or to a window. But at that point in the discussion, to simply ask what the symbol meant would have been embarrassing. Feynman had waited too long. So he ventured a guess. Bracing himself, he put his finger on one of the symbols and asked, "What happens if this valve gets stuck?" He fully expected the Oak Ridge engineers to say, "That's not a valve, sir, that's a window!" Instead, according to his version of the tale, the engineers began to move their hands quickly up and down over the blueprints, which had been spread out over several tables. They continued for a time, only to end up staring at each other in shock. Feynman had pointed to the crucial weakness that could have caused the plant to blow up. The engineers concluded (as in many of Feynman's stories) that he was a genius.

Feynman was disappointed to observe that the real blueprints were much smaller than the supersize ones featured in his story. They had grown through many tellings. But he was delighted all the same to see the original documents after many years. And he was fascinated to read some of the other documents relating to this Oak Ridge visit, documents that had been filed along with the blueprints. They offered, among other things, the names of the engineers involved, and these names further tickled his memory.

The crisis was over. Now Feynman was eager to again tell his story of how he saved the Oak Ridge plant. And then to tell other stories. The blueprint and its related documents had jolted Feynman's memory back into action. It is of utmost importance for the oral historian to become skilled at using such jolts.

Jolts can be incorporated in the oral historian's manner of questioning. In 1987, the pioneering oral historian Charles Morrissey published a classic paper, which has become part of the standard literature on doing oral history. In it he described a method of questioning that employs the jolt. He called the method the "two-sentence format."[25] In the first of the two sentences (which can be a group of sentences), the historian lays out the agreed-upon knowledge, including some features of the mask. In the second, he probes the mask using a question based on the response of the interviewee to the first sentence. In further rounds of questioning, the historian probes deeper and deeper, interlacing further questions resulting from answers to earlier questions.

I can illustrate Morrissey's technique by appealing again to my Los Alamos interviews with Teller. In one interview I was trying to learn Teller's side of why Bethe, who was serving as head of the Theoretical Division, moved Teller out of the Division in June 1944. Bethe's account, substantiated by his colleagues Robert Bacher and Carson Mark, held that Teller had been moved out because he stopped working on his implosion bomb assignment and was spending his time instead studying the H-bomb.

Drawing on Morrissey's system, I began by outlining a few facts, saving the main question for later. I showed Teller an organization chart that revealed the new H-bomb group established in June 1944 with Teller in charge. Discussing this well-documented, and noncontroversial, event brought Teller back into the period and gave him the chance to tell his story—that is, to articulate his mask. I encouraged him to develop the picture in detail. "Does this indicate that there was an increased effort at Los Alamos to study the hydrogen bomb?" My question brought Teller to object and present his well-known view that Oppenheimer had placed far too little emphasis on the H-bomb during the Second World War.

My next step was to gently probe the mask: "Bethe told me that he moved Rudolf Peierls into your position as leader of the implosion theory group because you were working too much on the H-bomb." That gave Teller a space in which to counter Bethe's story with a more detailed version of his side. He explained that he was just not good at doing the intricate hydrodynamics calculations which were then required in the theoretical work on implosion. This point, suppressed in the standard account, made everything fit.

But was it merely a retrospective reinterpretation? I could answer this question only through more research. I would need to go back into the Los Alamos documents and see whether they shed any light on Teller's ability to do such calculations. I would probably need to speak with others familiar with the wartime implosion calculations and with the nature of Teller's particular talents. I would need to reinterview Bethe and anyone else I could find who had served in the Theoretical Division at Los Alamos during the Second World War. (Peierls and Mark were by then no longer available.) After that, I could return to Teller with more informed questions and more evidence with which to test his recollections. Doing repeated interviews interspersed with periods of document research typically generates accounts that become increasingly responsive to the existing evidence, especially as the historian or the interviewee add relevant points that had been left out of the discussion or resolve conflicts in the account. Interviews can become even more reflexive if budgets allow the transcription of interviews and editing of the transcripts by both the interviewee and interviewer.[26]

Whether I continue to probe this particular issue of why Teller left the implosion program will depend largely on whether the answer is important in my future research. A historian has to judge whether pursuing a question is worth the effort it takes to address it—and when to stop any particular inquiry.

The need to suspend suspicion: the dilemma of trust

In tension with the need for suspicion and confrontation is the need to establish a relationship of trust and collaboration between the historian and the interviewee. The oral historian must always keep in mind that the accounts she is offered depend on the willingness of her living sources to share their stories. We are not likely to dislodge the mask of an interviewee who mistrusts our sympathy with her role in the history.[27] Thus Teller would not have offered me his problem had I not previously established a trusting relationship with him. Nor would Crick have agreed to Judson's extensive interviewing had there not been trust between them. The point, as Vansina once put it, is that interviews are "social processes of mutual accommodation during which transfers of information occur." If there is no social relationship (as with the administration of questionnaires), the information transfer will be minimal.[28]

The dilemma the oral historian faces is to simultaneously maintain both critical detachment and a relation of trust. Solving this dilemma calls for compromise in which the subtlest part is negotiating between the opposing positions of confrontation and trust.[29] The oral historian is caught in a bind. To maneuver she must surrender part of her or his objectivity. She must bend her professional commitments just enough to establish a genuine, trusting relationship with the interviewee. In the interest of gathering the best information, she may have to (at least for a time) become a participant in the very system she is studying.

A similar bind arises for the historian who studies institutions or societies. In the 1970s, when I needed access to source materials about the invention of the transistor, it helped to have earlier held a student research position at Bell Laboratories. Without that "in," I do not believe that I would have been granted access to the relevant primary source materials. Analogous insider connections have also helped in my studies at Fermilab and Los Alamos. These links had costs, for I now bear the dueling responsibilities of respecting the interests of the laboratories while I struggle also to uphold the standards of my profession. Similarly, a historian who acquires a security clearance to gain access to classified materials is obliged (in fact legally required) to respect classification restrictions. She may be required to withhold from publication some of the information she uncovers.

This historian's dilemma of being both on the inside and outside of a system she or he is studying is similar to the situation faced by the social historian or anthropologist who enters a community she wants to study. Joining the community may be necessary for gaining access to materials, but it always entails a loss of objectivity, no matter how much anthropological distance (or "strangeness") one tries to preserve.[30]

If the historian succeeds in handling this insider–outsider dilemma, further difficulties are in store. A trusting interviewee may reveal historically valuable information that is potentially damaging to the interviewee or to someone else, perhaps another interviewee with whom the historian is in a relationship of trust.[31] Because oral historians work within an unwritten code that requires them to protect the words of interviewees who have confided in them, the historian cannot disclose this information. Furthermore, because an interview is jointly owned by interviewer and interviewee, neither are free to disclose material in an interview that both have not explicitly released. One way for the historian to get around an ethical problem that arises is to show the interviewee how he plans to use the sensitive information, explaining the importance and asking the interviewee for his opinion. However the interviewee responds, the historian is obliged to respect his wishes.

The nuts and bolts of oral history: using iteration to achieve convergence of a story

I have argued that doing oral history requires the historian to be at once confrontational and collaborative, objective and personal, and suspicious and trusting. She must also maintain her commitments to her professional community and to all her interviewees. How can this be managed?

Pragmatically. We must operate in whatever mode works at the time. Over the course of many related interviews, a balance between suspicion and trust can usually be reached and particular distortions will tend to average out. We move in stages toward convergence, to a stable and robust interpretation that is the end of our historical research.[32] While I can well imagine situations in which convergence is not possible, I have never in my own work encountered one.

I can best illustrate this pragmatic procedure by calling again on my experience of writing a technical history of the first atomic bombs.[33] When I began my Los Alamos study in 1978, I faced thousands of memoranda, technical reports, data books, program reports, and the like. It was impossible to read them all in the span of a decade or so and also write this history up.

Over the next several years I created a small research team with which to divide the work, a team that included two history graduate students, the Los Alamos archivist, and a group of other historians, another archivist, and a number of scientists who moved in and out as consultants.[34] We began by familiarizing ourselves with the received view, conducting about fifty initial interviews on various aspects. To prepare for the initial round, the team gathered a subset of documents on which to base the first stages of the work. But we did not initially know the history well enough to make the best selections.

The key to progress was interviews—many interviews. The first documents we gathered, most of which were not used or even referred to in our final book, helped us prepare questions for the first round of interviews. Many of them also served us during the interviews, by helping the interviewees step back into their past and by delivering the message that their stories would be carefully checked. Almost all the interviewees were curious to examine papers they had created years earlier and not been allowed to see since.[35] Most of the interviewees were careful to qualify which points in their recollections they remembered only vaguely.

Inviting our interviewees back to Los Alamos for their interviews proved simpler and less expensive than having large numbers of documents declassified. We found that it also helped them to trigger their memories. For Proust it was the madeleine that did it.[36] For the Los Alamos scientists, it was the mesas, the canyons, and even the security guards that revived memories our team could now capture with a tape recorder.[37]

The group we interviewed at Los Alamos clearly welcomed the chance to unpack the complex emotions that were still entwined with their memories of the intense wartime period. Typically in their seventies, these interviewees would speak with us or study documents in the archives for hours at a time. One interviewee—Bernard Waldman, who had measured the blast at the Trinity test and had flown on the observer plane into Hiroshima—made it a point to come to Los Alamos and be interviewed by our team, as he had only two weeks left to live. Dying of cancer, he told us that if he did not make this effort now, no one would ever know his unique story.

The rough account that emerged from the first round of interviews served as a guide to selecting further documents. Subsequent research exposed gaps and inconsistencies. Now we could dig deeper and refine the emerging picture through further interviews, further archival work, even more interviews, more studies in the archives, and so forth.

In many cases, the interviews offered clues that helped us interpret documents. For example, in one committee report I noticed a reference to the eminent French experimental physicist Frédéric Joliot. Initially I thought the reference indicated that the Los Alamos concern about spontaneous fission (a nuclear splitting that is not triggered by neutrons but occurs of itself) stemmed from findings by the Paris group. But when Emilio Segrè told me in an interview that no one at Los Alamos trusted Joliot's work, I deduced that Joliot was mentioned here simply to justify the experiment being proposed by the Los Alamos physicists.

As the team accumulated more interviews and studied more documents, the distortions of particular interviews came to matter less. With a stronger base of information we could cope better with the conflicting interpretations and better interpret the biased documents. And we could more effectively draw useful historical conclusions from our slanted sources. Over a period of many months (the whole project took fifteen years from start to publication), the dialogue between documents and interviews yielded a series of successively improved models of the history. They eventually converged on a story that remained stable during the past three years of our work, the story we published in 1993. The process we had undertaken is analogous to computer modeling—indeed to the methodology of any empirical project, including traditional historical research.

I have drawn on my experience of writing the Los Alamos history to illustrate how the historian can work to resolve the conflicts between memory and documents, between trust and suspicion, and how, with the use of many interviews and much research in documents, a point can eventually be reached where the composite account stabilizes. At this point, further interviews and further archival research produce only negligible change in the historical reconstruction. We have found the face behind the mask, the story on which all the partial accounts converge. At some future time this face may destablize in response to new materials or a more compelling interpretation, but for the time being it is the best we can do.

Acknowledgment

This chapter is based on two talks: the first was delivered in 1992 at "The Relation of Oral and Archival Sources to Writing History and Biography" workshop, a session of the History of Science Society, organized by Ronald E. Doel; the second was presented on April 28, 1994 at a Stanford University workshop on "Interviews in Writing the History of Recent Science," organized by Horace Freeland Judson to whom I'm grateful for much useful editorial help in shaping this piece.

Notes

1 Allan Nevins told Saul Benison, when the historian of medicine was embarking on his first oral history project, "The people who you interview are going to become your teachers..." Saul Benison, "It's Not the Song, It's the Singing: Panel Discussion on Oral History," in *Envelopes of Sound: The Art of Oral History*, 2nd edn, ed. Ronald J. Grele (New York: Praeger 1991), 54.

2 Many hard questions about oral histories as historical sources are raised critically by Soraya de Chadarevian, "Using Interviews to Write the History of Science," in *The Historiography*

of Contemporary Science and Technology, ed. Thomas Söderqvist (Amsterdam: Harwood Academic Publishers, 1997), 51–70. For more on the debate over whether interviews should count as "history," see the exchange between Dennis Tedlock and Alice Kessler Harris in "It's Not the Song," 50–105. According to Harris, interviews are raw data, not "history." To qualify as history, she feels, a piece of writing needs to address the question, "What does all this mean?" See p. 90.

3 For a psychologist's account of the implantation of false memories, see, for example, Elizabeth Loftus and Katherine Ketcham, "Lost in a Shopping Mall," in *The Myth of Repressed Memory* (New York: St. Martin's Griffin, 1994), 73–101. For an historian's acknowledgment of passive interviews, see, for example, Dennis Tedlock in "It's Not the Song," 91–92.

4 Remarks by Grele at "Interviews in Writing the History of Recent Science" workshop, Horace Freeland Judson, organizer, Stanford University, April 28, 1994 (hereafter cited as Stanford workshop).

5 This is, in fact, the kind of interview one most often encounters in research-oriented oral history collections, such as the one maintained by the American Institute of Physics Center for History of Physics, College Park, Maryland.

6 Spoken historical testimony dates back at least to Herodotus, Thucydides, and the writers of the Gospels. Recorded oral history became possible after the advent of the wire recorder. On that basis, Allan Nevins established the first modern oral history program at the Columbia University Oral History Research Office in 1948. Large numbers of historians began to employ interviews in the 1960s, when tape recorders became widely available.

7 Today the objections to oral history, while often voiced, are rarely committed to print. One article that includes a number of the objections is J. L. Heilbron, "An Historian's Interest in Particle Physics," in *Pions to Quarks*, ed. Laurie M. Brown, Max Dresden, and Lillian Hoddeson (Cambridge: Cambridge University Press, 1989), 47–56.

8 The literature on memory distortion has grown to enormous proportions in the last decades. A sampling of sources about memory distortions in the field of cognitive psychology can be found in Daniel L. Schachter, *Searching for Memory: The Brain, the Mind, and the Past* (New York: Basic Books, 1996). See, esp., the references for chapter 3, "Of Time and Autobiography"; chapter 4, "Reflections in a Curved Mirror: Memory Distortion"; and chapter 5, "Vanishing Traces: Amnesia and the Brain," 72–161.

9 Schachter, *Searching for Memory*, 5.

10 Jan Vansina, *Oral Tradition as History* (Madison: University of Wisconsin Press, 1985), 176, 147–148.

11 Jean Ellis Robertson, *Language in Oral Histories: The Shape of Discourse About the Past*, (Ann Arbor, MI: University Microfilms International, 1986), 14–16. See also Joseph C. Miller, "Listening for the African Past," in *The African Past Speaks: Essays on Oral Tradition and History* (Folkestone, England: Archon, 1980), 13–15; Elizabeth Tonkin, *Narrating Our Pasts: The Social Construction of Oral History* (Cambridge: Cambridge University Press, 1992), 103–121; and John P. Dean and William Foote Whyte, "How Do You Know if the Informant Is Telling the Truth?" in *Elite and Specialized Interviewing*, by Lewis Anthony Dexter (Evanston, IL: Northwestern University Press, 1970), 122.

12 See, for example, Ronald J. Grele, "Can Anyone Over Thirty Be Trusted? A Friendly Critique of Oral History," in *Envelopes of Sound*, 1991, 206–207.

13 Michael Frisch, "Oral History and *Hard Times*: A Review Essay," in *A Shared Authority: Essays on the Craft and Meaning of Oral and Public History* (Albany, NY: State University of New York Press, 1990), 7.

14 Vansina, *Oral Tradition*, 94.

15 At the 1994 Stanford workshop, William Provine was among those who argued strongly for extensive preparation before interviews; Judson countered that an interviewer's ignorance can sometimes help coax out a more complete (rather than telegraphic) description from the interviewee. For other points of view on this issue, see Susan Allen McGuire, "Expanding Information Sets," *Oral History Review* 15 (Spring 1987): 61; Mary Stuart,

"'And How Was it for You Mary?' Self, Identity and Meaning for Oral Historians," *Oral History* (Autumn 1983): 62; Frisch, *Shared Authority*, esp. 188–189; and Eva M. McMahan, *Elite Oral History Discourse: A Study of Cooperation and Coherence* (Tuscaloosa, AL and London: University of Alabama Press, 1989), 5.

16 Friedrich Nietzsche, *Beyond Good and Evil: Prelude to a Philosophy of the Future*, trans. W. Kaufmann (New York: Vintage, 1966), 80, par. 68.

17 Vansina, *Oral Tradition*, 8. This notion of the mask is referred to by many other scholars. We find it, for example, in Lorraine Daston's notion of "scientific personae" and in Leon Edel's compelling distinction between the public and private myth in biography, summarized by Nathaniel C. Comfort in *The Tangled Field: Barbara McClintock's Search for the Patterns of Genetic Control* (Cambridge: Harvard University Press, 2003), 2–4.

18 The psychoanalyst George Moraitis pointed out at the 1994 Stanford workshop that a person's mask can shift. For example, when a subject that had earlier been kept hidden is used later to help a person avoid a different subject, the mask becomes part of the interviewee's defense system.

19 Grele's remarks at the 1994 Stanford workshop.

20 These interviews were among the earliest in a series that led to Lillian Hoddeson, Adrienne Kolb, and Catherine Westfall, *The Ring of the Frontier: The Rise of Megascience at Fermilab 1967–1989* (Chicago: University of Chicago Press, forthcoming).

21 Horace Judson, private communication to Lillian Hoddeson, December 18, 1999.

22 These memoirs are now available as Edward Teller and Judith Shoolery, *Memoirs: A Twentieth-Century Journey in Science and Politics* (Cambridge, MA: Perseus, 2001).

23 These different accounts appear in interviews that Hoddeson conducted with Bethe and Teller at Los Alamos during July 1995. They are held by the Los Alamos National Laboratory Archives, Los Alamos, New Mexico.

24 See, for example, Richard Feynman, *"Surely You're Joking, Mr. Feynman": Adventures of a Curious Character* (New York: W. W. Norton, 1985): 120–125.

25 Charles Morrissey, "The Two-Sentence Format as an Interviewing Technique in Oral History Fieldwork," *Oral History Review* 15 (Spring 1987), 43–54, esp. 51; also Morrissey, "On Oral History Interviewing," in Dexter, *Elite*, 112.

26 At the 1994 Stanford workshop, Ron Grele made the distinction between the research interview and the archival interview; the two are aimed at different audiences. Obviously, editing has little value when the goal is to maintaining the "purity" of an interview. Most archival interviews were created, however, for research purposes, and for these, editing to clarify the content and sense intended by the interviewee can be very valuable.

27 Some eminent historians of science (e.g. Thomas S. Kuhn) have been spectacular failures as oral historians because they could not put aside their own programs and focus on what their interviewees were in the best position to tell them. See, for example, Kuhn's interview with Niels Bohr in the Archive for History of Quantum Physics, held at AIP Niels Bohr Library, College Park, Maryland, and elsewhere.

28 Vansina, *Oral Tradition*, 63.

29 Morrisey, "Two-Sentence," 47. Also Ronald J. Grele, "Private Memories and Public Presentations: The Art of Oral History," in *Envelopes of Sound*, 260–261.

30 The concept of strangeness in anthropological studies of scientific practice is discussed, for example, Bruno Latour and Steve Woolgar, *Laboratory Life: The Construction of Scientific Facts* (Princeton, NJ: Princeton University Press, 1986), chapter 2; and Sharon Traweek, *Beamtimes and Lifetimes: The World of High Energy Physics* (Cambridge: Harvard University Press, 1988), Prologue.

31 Vansina tells a story in which a young oral historian learns about a murder in one branch of a family committed by a member of the other branch. Jan Vansina in section II, "It's Not the Song," 68.

32 This concept of robustness has been discussed by a number of philosophers, for example, William Wimsatt, "Robustness, Reliability, and Overdeterminism," *Scientific Inquiry and the Social Sciences* (San Francisco, CA: Josey-Bass Publishers, 1981).

33 Lillian Hoddeson, Paul, W. Henriksen, Roger A. Meade and Catherine L. Westfall, *Critical Assembly: A Technical History of Los Alamos during the Oppenheimer Years, 1943–1945* (New York: Cambridge University Press, 1993).

34 The history graduate students were Catherine Westfall and Paul Henriksen; the Los Alamos archivist was Roger A. Meade; the consulting historians were Robert Seidel and Richard Hewlett; the consultant archivist was Alison Kerr; and the scientists were Gordon Baym, Robert Penneman, and Leslie Redman. Behind the scenes were strong administrative allies at the laboratory, including Gilbert Ortiz, L. M. Simmons, and David Sharp.

35 Donald Kerr, the director of the Los Alamos laboratory (and later Kerr's successor, Sigfried Hecker) signed a paper allowing former Los Alamos scientists under proper conditions to see their own classified wartime documents while they were being interviewed in secure quarters at Los Alamos.

36 Marcel Proust, *Swann's Way* (New York: Random House, 1934), 33–36.

37 The association of place with memory has been extensively employed over the ages as a mnemonic device. The "method of loci" is said to have been used in 477 BC by the Greek orator Simonides. See F. A. Yates, *The Art of Memory* (Chicago, IL: University of Chicago Press, 1966), 1–26.

12 Reading photographs

Photographs as evidence in writing the history of modern science

Ronald E. Doel and Pamela M. Henson

In the late 1940s, G. C. Heron sat at his desk, frustrated. The photograph archive he directed at New Zealand's National Library, the Alexander Turnbull Library, was actively soliciting photograph collections. At the same time individuals throughout the country, seeing little value in them, were tossing images away. Heron found this deeply troubling. Photographs "preserve scenes, impressions and faces of bygone days and form an historic record as surely as does any manuscript, diary or printed work," he wrote, but few historians were doing anything with them.[1]

During the last half-century photographic archives have burgeoned: over 8 million images fill the still photographs stacks of the US National Archives alone.[2] Yet historians remain hesitant about embracing photographs as evidence in the way Heron desired. Many only seek photographs at the end of research projects. Then they use them like potted plants, hoping to illuminate stories based on written archives. Keith McElroy has rightly observed that historians use photographs "primarily as illustrations, and frequently their content has contradicted the thesis of a publication that was derived from literary sources." Yet few historians pause to consider what images might teach them.[3]

Photographs are nevertheless becoming more important to scholars of modern history. The total number of books that directly address photographs as an aspect of historical methodology remains astonishingly small—just twenty-five have been published, all since 1973. But a growing number of historians are drawing on intense interest in contemporary visual culture to "read" photographs as evidence, as historians Alan Trachtenberg and Robert M. Levine have urged.[4] In recent years social and cultural historians have employed photographic collections to analyze the living standards and cultural expectations of lower-class citizens of late nineteenth century Brazil, urban life in turn-of-twentieth-century America, the practice and ideology of the British Empire, rural life in the New Deal era, the cultural world-view of editors at the National Geographic Society, the ambitions of planners and financiers who promoted the first world's fairs, and the practice of medicine in post-Civil War America.[5]

Historians of science have moved more slowly than colleagues in other fields to incorporate analysis of photographs of the social, technical, and institutional practices of science into their writings. Sustained by the rich textual and archival sources of a privileged and highly verbal elite, few historians of modern science have felt the pressures that inspired social historians to utilize novel methodological approaches.[6] There are nevertheless promising exceptions. Gregg Mitman has explored the relationship of the naturalist tradition and popular culture by studying nature films in *Reel Nature*, and two pairs of historians (Peter Galison with Emily Thompson, and Scott Knowles with Stuart W. Leslie), have employed photographs to study how architectural design has influenced the practice of science in twentieth-century America.[7] These works draw on the far larger volume of research by historians of science that addresses visualization in science, including the ground-breaking writings of Martin Rudwick and Barbara Marie Stafford.[8] But these writings are primarily about sketches, drawings, paintings, and artistic prints. Largely ignored are the hundreds of thousands of photographs of scientists, scientific institutions, science exhibits at county fairs and at world expositions, field camps, laboratories, university classrooms, remote expeditions, and museum dioramas produced since the invention of photography.[9]

Much can be asked of photographs, but it is not always easy to make them deliver the goods we seek. This chapter addresses the challenges and opportunities of using still photographs as historical *evidence*—alongside correspondence, diaries, memoirs, laboratory notebooks, published reports, newspaper accounts, and census reports—to enhance our understanding of the history of recent science.[10] Photographic images allow us to revisit what it meant to do science, how the practice of science differed among disciplines, how citizens learned about science, and who got to participate. They allow us to explore the visual iconography of science, the presentation of self, and the confidence we feel about our general understanding of the ecology of scientific institutions and the production of knowledge. They encourage us to revisit seemingly familiar topics and themes, including gender relations and national styles of science. By carefully "reading" photographs of modern science, we may gain the kind of new insights and questions that the application of oral history to history of science achieved a generation ago.[11]

Many approaches to the history of recent science can be limited to the Cold War and post-Cold War periods without doing injustice to their subjects. The task of analyzing photographs as historical evidence, however, is different. Visual tropes have exceedingly long lifetimes. Iconographic styles and fashions from nineteenth century print culture still resonate in portraits and expedition photographs composed more than a century later.[12] Moreover, comparisons *between* images, particularly those from different time periods, often allow historians to tease out valuable historical information from individual photographs. While we intend to offer a new approach to the history of recent science, reaching this aim requires us, along with our colleagues in the history of photography, to adopt a long view.

Challenges of interpreting historical photographs

Photographs of scientists and science-related activities are one slice of the much larger volume of photography documenting the human condition and the natural realm since the invention of photography in 1839. But why privilege photographs at all—why value them as a source of information? What do historians of modern science and technology need to know to read photographs as historical documents?

A long-standing argument for the significance of photographs is their verisimilitude, a result of the photographic process itself. "Not everybody trusts paintings," the venerable twentieth-century landscape photographer Ansel Adams boasted, "but people believe photographs."[13] Adams was voicing an old concept: already by 1859 the writer and jurist Oliver Wendell Holmes declared the photograph a "mirror with a memory," the most reliable witness to reality then available.[14] The idea that photographs are akin to a membrane peeled from nature, a transparent window into the past, has not lost its appeal in popular culture. Although skeptical modern readers quickly learn to spot slant in newspaper reporting, the "straight photographs" that accompany news stories are granted a higher evidentiary value, as they allow—or seem to allow—readers to gain direct insights about events and people.[15]

Nevertheless, historians of photography properly regard Holmes's optimistic "mirror with a memory" concept as naive. Individual photos are far from transparent, objective windows into historical events—and using images as evidence requires historians to test their authenticity, their accuracy, their representativeness, and their significance, just as with archival documents. One obvious concern: photographs can be doctored. Contemporary digital photography has made it easy to alter the content of photographs (even if few are as humorous as the addition of Groucho Marx and Sylvester Stallone to the Yalta Conference in 1945 in a parody published in 1990).[16] But the problem is hardly new. In the Stalinist and Khrushchev eras discredited public figures in the Soviet Union routinely vanished from reissued photographs, such as the hapless cosmonaut edited out from a Black Sea gathering in the famous 1961 "Sochi Forgery."[17] More than six decades earlier, a skilled photograph retoucher convincingly set British Prime Minister William Gladstone outside the entrance of a rakish London pub, hoping (in vain) that the photo forgery would create scandal. Outraged, a senior British jurist declared that any photographer now needed to "prove his photograph" to vouchsafe its authenticity.[18]

Historians of science might imagine that altered images belong to the realm of politics, remote from the laboratory or the field, but this is not the case. Photographer Frank Hurley violently scratched out from one glass negative the ship that ultimately rescued the crew of the ill-fated *Endurance* expedition of 1914, using that plate to instead portray Shackleton leading his rescue team *away* from their frozen refuge at the edge of the Antarctic.[19] A more notoriously altered photograph stemmed from a visit Albert Einstein made to General Electric's research headquarters in Schenectady, New York in 1921. A staff photographer recorded Einstein standing with such GE luminaries as David Sarnoff, Irving Langmuir, F. W. Alexanderson, and Charles P. Steinmetz, more than a dozen individuals in all. But when the photograph was reprinted in 1924, Einstein's sole companion had become Steinmetz, the brilliant mathematician, engineer, and corporate manager at GE, a cultural hero no less respected in his day than Thomas Edison and Henry Ford. A prominent socialist, Steinmetz also stood out by his serving as the chief engineer of a large capitalist corporation. GE publicists recognized the value of crafting a photograph implying a private meeting between Einstein and Steinmetz, despite their clumsy misstep of modifying an image already in wide circulation.[20]

But most alterations of photographic scenes are more subtle and difficult to spot. Professional photographers seeking images that conformed to their aesthetic standards or to the artistic trope of their era frequently improved on nature. In the late nineteenth century, the pioneering photographer William Henry Jackson routinely had assistants clean streets and public areas; they also sawed off tree branches that "disrupted the organization of a 'proper' view."[21] Ansel Adams eliminated the chalk-white letters "LP" from a foreground hill in the final print of his 1944 "Winter Sunrise, Sierra Nevada, From Lone Pine, California." By so doing, Adams later wrote, "the image [retained the appearance of] a sublime rather than an 'altered landscape'."[22] His contemporary Arthur Rothstein, a celebrated member of the Farm Security Administration's famous army of image-makers, faced withering criticism from Midwestern farmers and officials for moving a cattle skull several feet to compose his well-known 1936 portrait of extreme drought in the Badlands of South Dakota.[23] Yet far fewer people then (or now) realized that Rothstein's most famous photograph, "Dust Storm, Cimarron County, 1936"—the icon of the Dust Bowl—was not spontaneous, but also carefully staged and rehearsed. The farmhouse towards which the farmer and his hapless sons are running was in fact a crumbling shed behind a sturdy plains house, and the dust, while swirling, was comparatively thin on that day.[24] Professional documentary or "straight" photographs, to use Beaumont Newhall's helpful term, sometimes fall short of faithful representation.[25]

Historians of photography also recognize that photographers alter the meaning of their images simply by how they choose to compose them. "Merely to sight through the viewfinder," as James Davidson and Mark Lytle have written, "reminds us that every photograph creates its own frame, including some objects and excluding others."[26] A classic though little-known example comes from the history of science. In 1869 Timothy O'Sullivan, the famed nineteenth century photographer selected to accompany Clarence King's Fortieth Parallel survey, photographed a geological formation in present-day Utah called Witches Rocks. O'Sullivan's photograph was reproduced in the Expedition's official report and circulated widely as an example of conglomerate rock from the Tertiary era.[27] Many of O'Sullivan's compelling exposures of the then unfamiliar American West are now classic nineteenth century American icons, and are often praised as quintessential straight photographs from an era when daguerreotypes and the wet plate (collodion) process limited darkroom experimentation. But in the late 1970s members of the Rephotographic Survey Project retraced the steps of the Hayden, King, Powell, and Wheeler explorations of the American West. Photographic historian Rick Dingus, seeking to recapture O'Sullivan's "Witches Rock," made an important discovery: in composing his photograph O'Sullivan had tilted his camera about fifteen degrees from the horizontal, thereby exaggerating the steepness and instability of the formation. Dingus reasonably speculated that O'Sullivan was influenced by King's catastrophist theory of geology, a topic aired during campfire discussions.[28] The discovery of O'Sullivan's distorted framing is important for historians of photography, but no less so for historians of science: despite the intense scrutiny devoted in recent years to visual traditions in modern geology, we have only begun to explore the impact of photography on this discipline.[29]

To interpret photographs as evidence, historians of science also need to consider what can happen to images when they leave the hands of photographers and are handled by publishers, promoters, and editors. William Henry Jackson lost artistic control of the photographs he made of the Columbian Exposition in Chicago in 1893—and at least one Jackson image of the deserted grounds reproduced in the official fair volume, *The White City*, gained flags, gondolas, boats, ornamental gardens, open concession tents, and even spectators through an expert retoucher's brushstrokes. Moreover, publishers and authors can shift meaning by cropping photographs, altering the frames selected by photographers themselves. For his 1938 *Land of the Free*, the poet and writer Archibald MacLeish reproduced a Dorothea Lange photograph of a white plantation owner in Clarksdale, Mississippi. MacLeish had sought to illustrate the themes of self-reliance and independence. But MacLeish's edited version of this poignant Farm Security Administration photograph cropped out four gaunt farmhands, all black, grouped behind the plantation owner. Lange had sought to portray an oppressive southern social hierarchy, not a symbol of independence.[30] While the famous Clarksdale edit is an obvious instance, historians of science ought to recognize that published photographic essays from scientific expeditions and institutional yearbooks often include cropped images as well—and photographs selected for publication within them sometimes reflect a narrow sampling of those actually available to their editors.

Finally, historians must be concerned with the relation between photographs and accompanying documentation, including captions. While written texts are instantly interpretable through linguistic analysis, photographs have a different epistemic value than letters and documents. They typically require linked written information—a date, an indication of geographic place, field notes, a listing of individuals in the scene—before photographs can serve as reliable historical evidence. Captions and images are thus interdependent, as Carol Armstrong has noted. But historians need to ascertain the validity of accompanying texts even as the photographs are scrutinized.[31] A photograph of Seminole Indians gathered in formal native dress before a stone building facade, for instance, may depict a ceremonial gathering—although our impression of the scene changes in reading the caption, "A Picture of ten Indian Cut Throats and Scalpers, who were confined in the Old Spanish Fort as prisoners of war by the United States Government." In this instance, as the American West specialist Lee Clark Mitchell discerned, the label is clearly at odds with the photograph. Americans at the turn of the twentieth century were not sure what stories to tell about Native Americans, and the caption on this albumen stereograph print was erroneous—or deliberately deceptive.[32] But even in less jarring instances the ways that captions influence how we read photographs remains profound. When *Time* magazine published a wire service photograph of US soldiers ordered to take part in the first tactical test involving nuclear weapons at the Nevada Test Site in 1951, their caption emphasized that this "hand-picked audience of 5,000 G.I.s [watched from] seven safe miles away." Reproduced in the National Archives' 1999 retrospective exhibit "Picturing the Century," with *Time*'s misleading caption replaced by the simpler and more descriptive "Exercise Desert Rock," this same photograph clearly illustrated not only a central issue in the early Cold War, but the shifting historical contexts in which all photographs (and documents) must be viewed.[33]

Other challenges face historians of science seeking to examine photographs as evidence. We must decide if photographs are staged or record spontaneous scenes. We need to analyze the motives of photographers in making them, and those of publishers in reproducing them. Furthermore, as archivist James C. Anderson has written, the content and organization of large photographic collections in museums and libraries "have been shaped by art-historical or other equally narrow perspectives which severely restrict the scholar in his search for our visual history."[34] Photographic archives are less easy to use than manuscript collections.

Finally, as digital images begin to replace standard photograph prints—in early 2004, Eastman Kodak announced that it would stop selling traditional film cameras in Western Europe, Canada and the US—historians will face new challenges in inter-preting photographs. In contrast to film negatives (which often survive alongside proof sheets), digital camera users typically preserve only selected individual images, erasing unwanted exposures. This means that it is harder to reconstruct the larger sequence of exposures taken at any one time. While many digital photographs are printed on paper, thus increasing their chances of surviving alongside traditional photographic prints, digital images residing primarily on computer hard drives may be lost to posterity faster than film negatives dispatched to storage cartons. Of greater concern to historians is the ability of photographers employing image-processing software such as Photoshop to alter the very content of their photographs. While image manipulation is hardly new, the ease with which digital images can be retouched has further undermined the expected verisimilitude of photographic prints.[35]

Yet these challenges are not insurmountable. Historians routinely employ similar skills in reading historical documents (and, in recent years, in reading oral history interviews).[36] Although reading photographs requires extending these skills, the rewards of doing so are great.

Photographs as sources for history of science: history and methods

What makes the abundance of photographic collections especially valuable for historians of science is their timing on the world stage. The rise of professional photography coincided with three critical developments in the second half of the nineteenth century: first, the emergence of professional science; second, the tremendous expan-sion of the modern state; and finally, the start of the second Industrial Revolution. In the United States, O'Sullivan, Jackson, John K. Hillers, and other determined entre-preneurs rode the wave of expanding federal power after 1865 as professional photog-raphers for the major exploring expeditions in the American West. By then the American public, like their counterparts overseas, were fascinated by photography. Photographic studios mushroomed across America after the first daguerrotypes were made. In the first decade after the Civil War many citizens bought stereoscopes to view stunning vistas of Paris, New York City, and the American West.[37] Widely accepted because of the perspective realism they inherited from landscape paintings, these photographs also embodied the dominant political ideologies of their age.[38] The archives of state-supported photographers in the nineteenth century clearly document the rise of the modern nation-state no less than they do the growth of science and technology.[39]

The desire of the state to document its burgeoning bureaucratic resources and demonstrate their achievements domestically and abroad also reflected a rising nationalism, and the identification of science and technology as symbols of progress. By the turn of the twentieth century (in some cases well before), photographers were chronicling activities in the US Department of Agriculture, the US Naval Observatory, and the Army Corps of Engineers. One can peer at scientists and shop foremen working at spacious lab benches inside the sprawling complex of the Fixed Nitrogen Research Laboratory, established in Washington in 1919 as part of the War Department's Chemical Warfare Service; one can likewise witness researchers, including Dr George E. Ward, working to produce synthetic rubber at the Department of Agriculture's Northern Regional Research Laboratory in Peoria, Illinois, in 1942.[40] These efforts accelerated during the Second World War. Photographers from the Office of War Information—many holdovers from the famous Farm Security Administration photo-documentary project—began recording scenes of college and university training in the sciences. Their cameras recorded women attending science classes at Iowa State and young black scientists at Atlanta University, images later used in war propaganda and to shape public opinion.[41] During the Cold War, leaders of the US Information Agency, created in 1953, sent photographers across America to document activities helpful in waging "public diplomacy." USIA "picture stories" documented scientists and technological systems, from science exhibits at the Brussels Universal Exposition of 1958 to the launch of satellites and spacecraft from Cape Canaveral.[42] Federal photographers simultaneously recorded the construction of US research facilities in Antarctica during the International Geophysical Year of 1957–1958, and their counterparts in the Department of Agriculture sought to highlight the scientific and social advantages of pesticides.[43]

Commercial photographers not employed by the state also began putting scientific subjects before their lenses as the medium of photography gained popularity. An important inspiration came from the documentary impulse of the Progressive Era, where the camera's unflinching gaze gave it status as a truthful witness.[44] More than many historians have realized, leaders of the documentary tradition also occasionally turned their attention to scientists and to research. Lewis Wickes Hine's 1935 photograph "Finding Out What Caused That Recent Eclipse," showing an ethnically diverse group of attentive middle-schoolers with their teacher in a cheery, well-lit New York classroom, revealed his faith in education, and provides a powerful counterpoint to Hine's better-known photographs of gaunt spinning girls and bobbin boys in the dark textile mills of Alabama.[45] Photographers employed by Science News Service, founded in 1920, recorded thousands of images of scientists, many of which made their way into major newspapers and in what became the periodical

Science News.[46] While scientific subjects were only occasionally reported in mainstream American periodicals prior to the Second World War, the use of the atomic bomb against Japan in 1945 and the unprecedented identification of science and technology with military campaigns made scientists and research a special fascination. The eminent photographers Alfred Eisenstadt and Arnold Newman used their cameras to do character studies of such public figures as Albert Einstein and J. Robert Oppenheimer. Their images reached millions of viewers through *Life* and *Look.*[47]

By the turn of the twentieth century, professional photographers were hardly alone in capturing images of science. Amateur photography blossomed a generation after the birth of the medium. Some 13,000 Kodak cameras were sold within a year after they first went on sale in 1889, and Kodak's one-dollar Brownie camera, introduced in 1900, led to the proliferation of amateur photography clubs, magazines, and exhibitions.[48] Scientists themselves were among the earliest camera enthusiasts. Grove Karl Gilbert, the geologist, was among those whose cameras documented the Harriman Expedition to Alaska in 1899, while the paleontologist Samuel H. Knight at the nascent University of Wyoming documented expeditions to unearth dinosaur bones in the arid hills beyond the campus.[49] Later in the twentieth century major research institutions came to employ in-house documentarists, including John MacFall at the Scripps Institution of Oceanography in La Jolla, California, and Alfred F. Huettler at the Woods Hole Oceanographic Institution in Massachusetts. They are among many thousands of individual photographers whose work now fills university archives.[50] More than 5 million photographs, some in scrapbooks and albums and many concerned with science activities, are housed at the Steenbock Library of the University of Wisconsin and a similar number reside at the University of Illinois archives.[51] Cameras were ubiquitous throughout the entire twentieth century, and the photographic prints they produced have proven as durable as letters.[52]

Nor do these examples exhaust the range of photographic collections available to historians of science. The archives of the Rockefeller Foundation—like those of many other foundations that supported scientific programs—bulge with photographs documenting their funded programs, including fields of hybrid plant strains at the heart of the Green Revolution.[53] Major photograph repositories, first among them the photography collection at the George Eastman House in Rochester, New York, hold rich selections of images documenting the social practices of science. Anthropological photographic projects are another helpful resource.[54] So are the archives of major geographic institutions and societies—a point Catherine Lutz and Jane Lou Collins demonstrated in their 1993 *Reading National Geographic.*[55] And sometimes historians discover unique photographic archives by accident. While searching for materials on Dwight Moody's 1893 evangelical campaign in the archives of the Moody Bible Institute, James Gilbert came across images from a 1940s US Air Force-funded program called "Sermons for Science." Finding this unexpected military connection led Gilbert to write a new study on religion and science.[56]

In recent years, the Internet has offered historians yet another means of gaining access to photographs and photograph collections. Several major libraries and archives have begun to digitize samples of their visual holdings and have placed these online. In the United States, for instance, a large (and still-increasing) sampling of Science Service photographs, together with original captions, appears on the Smithsonian Institution's National Museum of American History website.[57] The American Museum of Natural History in New York is creating an online collection of photographs documenting its field science expeditions, while the University of Washington has digitized all photographs contained in a commemorative album of the Harriman Alaska Expedition of 1899—providing Internet access to 253 images, or 5 percent of the 5,000 photographs taken on this trip. On the European continent, international programs such as the European Visual Archive provide access to numerous individual repositories through a single search engine.[58] Digital libraries seem certain to expand as archives (including those at smaller colleges, historical societies, and institutions worldwide) strive to make their collections more accessible to virtual users.

How to *read* this vast warehouse of information is often a greater challenge than locating relevant images. Reading photographs, like reading documents, requires professional savvy. We must be aware of the provenance of photographs so that we can better judge how particular images were framed and posed. We must also be aware of their intended use. Just as we would not interpret a fund-raising letter to potential patrons that described a desultory, poorly equipped, overcrowded laboratory as an unbiased and accurate depiction of actual conditions, we must determine the motivations of individuals behind the viewfinders. It is sometimes easiest to do this when looking at the work of a particular prolific photographer. Just as with traditional archival collections, the attitudes and viewpoints of individuals are most clearly seen when a significant fraction of that person's work is open to inspection.[59] Another important skill that historians must bring to bear when analyzing photographs is a very familiar one: knowledge of the existing historiography. We must know what historians of science have already discerned from traditional archival evidence to spot what is truly surprising and unexpected in the photographs we inspect: initially perplexing images have the most to teach us, for they challenge us to raise new questions and rewrite current interpretations. Are laboratories arranged in unconventional ways? Are women researchers present when we do not expect them? A 1954 photograph of the laboratory of Smithsonian biologist Gustav Arthur Cooper offers just this kind of insight (Figure 12.1). Cooper (1902–2000) was a noted American paleontologist. His wife, the linguist Josephine Cooper, is far less known. Josephine Cooper learned to prepare paleontological specimens to aid her husband, and translated many scientific articles in foreign languages. They formed a family industry, not unlike that of Sir Joseph Hooker—director of Britain's Royal Botanical Gardens at Kew and the most eminent botanist of the nineteenth century—and his wife, Frances Harriett, daughter of the Cambridge professor of botany who had trained Charles Darwin. While Josephine Cooper held no Smithsonian post, the camera unambiguously reveals her sharing her husband's laboratory and career. This single image expands the relatively small number of known creative couples in the sciences.[60]

But photographs are more complex documents than letters, and historians of science must ask additional questions of them to fully appreciate what stories they

Figure 12.1 Gustav Arthur Cooper and his wife, Josephine Cooper, at work in the Division of
Vertebrate Paleontology, United States National Museum, Washington, DC, June
1954.

Source: Smithsonian Institution Archives.

can tell. In contrast to personal correspondence—largely written with expectations of
privacy and hence candid windows into personal feelings—we often encounter
photographs as *published* documents rather than as plates or film negatives in archival
folders. The universe of published photographs is smaller than that of candidate
images photographers submitted to publishers. These candidate photographs in turn
are a small fraction of the number of photographs made at any one time—and only
this large *portfolio* of individual photographs is truly analogous to individual letters
in archival collections. What criteria did editors use to select images for publication?
What stylistic conventions or tropes did photographers use, consciously or unconsciously, in composing their photographs? Answering these questions requires historians of science to be familiar not only with their particular fields but also with the
history of photography and cultural history. While we cannot view earlier
photographs precisely in the ways they were seen at the time, historical geographer
James R. Ryan has declared, "it is possible to situate photographs in the historical
and cultural contexts in which they were made and displayed, to show how their
meanings were framed by wider discourses."[61]

Historians of science are coming to appreciate this point. In a provocative reading
of the now famous photograph of James B. Watson and Francis Crick peering at a
mechanical representation of DNA in April 1953, made shortly after their discovery
of the structure of the double helix, Soroya de Chadarevian has demonstrated that this
image gained status rapidly only after the 1968 publication of Watson's best-selling

The Double Helix. By examining the original contact print containing eight separate images of Watson and Crick made by the photographer Antony Barrington Brown, de Chadarevian makes clear that this image became a cultural icon only after it was hitched to Watson's blunt and ruthless narrative of scientific practice. The remaining images (capturing a less jaunty Watson) would remain largely unknown. How one particular photograph came to represent Watson's moment of discovery, de Chadarevian argues, gives us "insight into the retrospective construction of the event the image itself represents." Equally important is a second point she makes: "both the photographs and the events they supposedly portray have acquired [new meanings] retrospectively." Indeed, portraits—formal and informal—can tell us a great deal about the subject's place in the world, and photographic portraits reflect the social construction of personal identity.[62]

While Watson (born in 1928) did not shy from publicity, this was not necessarily the case for scientists born one or more generations earlier. Writing to Charles B. Davenport, the long-time director of the Eugenics Record Office of the Carnegie Institution of Washington, the writer and editor Waldemar Kaemppfert complained in 1915 that American scientists were far less willing than their European counterparts to allow images of themselves in their labs to be published. "It is almost impossible for me to obtain the type of scientific picture I want," lamented Kaemppfert, later an influential science writer for the *New York Times*.[63] While Science Service photographers often succeeded in obtaining captivating images of scientists and scientific activities by the late 1920s, we must remain aware that published photographs in this era may be quite distinct from images that stayed in institutional file drawers or photo albums.[64]

Nor can we assume that published sets of photographs inevitably reflect the full interests of their creators. Photographers occasionally were stymied when they sought to place their favored images in print. Herbert Ponting, the noted expedition photographer who accompanied Captain Robert Scott to the Antarctic from 1910 to 1913, frequently photographed scientists at work on the expedition's ship *Terra Nova*.[65] In several dramatic images, Ponting's lens captured biologist Edward L. Nelson drawing water samples from below the sea ice in the chill of the Antarctic night. Ponting intended to sell these images to the London *Daily Mirror*, with accompanying stories about what these scientists had achieved. But Scott, learning of his plan, would hear nothing of it, and forbade Ponting from making what he regarded as unseemly financial gain.[66]

After Scott died in March 1912 on his failed effort to return from the South Pole, Ponting initially respected Scott's wishes. Then, after the First World War, Ponting again tried to market his photographs of scientists. But he discovered British audiences were more interested in images of adventure than serious research. "I found it was absolutely necessary to cut out all the Science" in public lectures, Ponting lamented in 1921 to Frank Debenham, a fellow member of the *Terra Nova* expedition: "The people would simply not stand for it." Ponting remained convinced that his photographs of scientists would have sold better in the US, believing Americans were science-crazed. One reason for his optimism was enthusiasm from one of the most prominent Americans of his time. After visiting a Ponting photograph exhibit in London in 1914 that retained his selection of science images, former President Theodore Roosevelt penned him a note: "I came within an ace of asking my host to bring me behind the scenes and present me to you yesterday, but I thought you would probably be so busy ... I do not know when I have seen any exhibition which impressed me more than yours did. The pictures were wonderful ..."[67] But Ponting remained in England, and by the 1920s primarily marketed heroic images of exploration. Only in recent years, as interest in turn-of-the-century Antarctic exploration has again peaked, have historians benefited from published reproductions of Ponting's documentary studies of British scientists.[68]

The best question may not be whether photographs can help historians sharpen their interpretations. Clearly they do. We might better ask: can they deliver the goods? Are there sufficient numbers of clear, well-documented photographs that provide significant new information to challenge and enlarge existing historical accounts, and raise provocative new questions? Is it worth the added effort to analyze photographs?

New interpretations: revisiting the history of recent science

Historians of photography have already made a convincing case: careful analysis of photographs can influence how we write the history of science. Rick Dingus's discovery that O'Sullivan tilted his camera to dramatize nature makes clear that visual evidence in geology was far from transparent but instead susceptible to prior theoretical commitments.[69] We ought expect further scrutiny of photographs to tell us more about big-picture themes in the history of recent science. They certainly will give us more insight about how women and minorities took part in scientific institutions and field research, how instruments were actually used, how scientists chose to present themselves to their publics, how citizens learned about science, and about the fields of science that dominated the public and land-grant universities. (Our perspective on the history of science in modern America is primarily derived from the twenty most productive elite institutions, a list that includes relatively few state universities and land-grant institutions, even though these schools enrolled a larger proportion of students.)[70] But examining photographs will not simply contribute to an improved social history of recent science: it should improve our understanding of how elite science was done.

In preparing a book on photographs as evidence in the history of science, we have visited more than twenty-five photographic archives in the United States, Britain, and Western Europe. We have examined thousands of individual photographs, as well as runs of magazines and newspapers that routinely included photographs of scientists, scientific institutions, and the teaching and practice of science. We are now comparing the stories told in these photographs to those familiar to us in the current historiography of American science in the nineteenth and twentieth centuries. What have we learned thus far?

Several benefits and new insights emerge immediately on studying photographs carefully. One great advantage of photographic evidence is that it allows immediate comparisons across varied institutions and cultures. We have found that women were far better integrated into science classrooms and field camps in early twentieth-century

America than many past accounts have suggested: at least a third of participants in upper-division geology classes at the University of Wyoming between 1911 and 1918 were female, and similar numbers of women participated in this university's spring term field classes.[71] Half the members of the University of Colorado's Mountain Laboratory for Biology and Botany summer study group *ca* 1911 were women, and visiting female students outnumbered men at the University of Washington's Puget Sound Marine Station (later Friday Harbor Laboratories) in 1911.[72] We have discovered that virtually all biology classrooms—from Ithaca, New York, to Porto Allegre, Brazil, in the late nineteenth century through the mid-twentieth century—featured abundant visual displays. This indicates that visual thinking was characteristic of this discipline and not an isolated instance of national style.[73] Photographs of astronomical observatories—more than traditional archival sources—remind us that even remote astronomical facilities such as Lick and Palomar were not isolated monasteries of research but also on occasion places of public witnessing of science.[74]

Many photographs provide insight into the history of particular scientific disciplines and the social history of women and minorities within the scientific community. But others call attention to big-picture themes. Clean-scrubbed young boys attending "Corn Clubs" events in the southeastern US states from 1908 to 1913 peer anxiously—sometimes proudly—into the lenses of photographers attached to the Rockefeller Foundation. They reveal the intense interest of Foundation officials in agricultural science in the Deep South at the same time its famous hookworm eradication program was underway.[75] Indeed, photographs of county fairs in America in the first third of the twentieth century remind us of the deep importance agriculture still held then for US citizens. Fair-goers often viewed demonstrations of applied science and were enticed into eugenics fitness exhibits. As important as physics was to the rapidly industrializing United States, these photographs remind us how little we know about the history of American agriculture compared to the physical sciences.[76] These photographs also remind us of another dark corner in our historical map: historians of science have largely overlooked primary and secondary schools, which shaped the attitudes of future scientists and administrators of scientific institutions and agencies, as well as the broader mass of citizens who were called on to fund public scientific work.[77]

Sometimes it is easiest to spot challenges to existing historical accounts by looking at photographs made a century earlier. After all, the historiography of this period is more richly developed, the traditional archives are better explored, and these images are often sufficiently unfamiliar to encourage us to examine them with special care.

But in this book we are particularly concerned with understanding *recent* science. How can photograph collections better inform us about the history of science since 1940? Let us address this challenge through the following questions, and case studies.

What can we learn about science and propaganda efforts during the Second World War?

Consider this photograph of an unnamed black zoology student at Atlanta University, made by Arthur Rothstein in March 1942 for the Office of War Information (OWI) (Figure 12.2). Rothstein's photograph favors the subject, who is

Figure 12.2 Arthur Rothstein, "Chemistry Student," Atlanta University, Atlanta, Georgia, March 1942.

Source: Prints and Photographs Division, Library of Congress.

cleanly dressed, studious, and handsomely posed. The photograph suggests the calm confidence of this young investigator; it also suggests his home institution, Atlanta University, possesses advanced research or instructional facilities. We know Rothstein was one of the better-known Farm Security Administration (FSA) photographers, whose 1930s work helped to shape American documentary photography. But this elite yet struggling black university has not received much attention from historians, and the photograph is not well known. What can this photograph tell us?

Rothstein's photograph in fact opens an important door on the experience of minorities (and women) in American science during the Second World War. A bit of detective work reveals more than a hundred images documenting scientific training and scientific work in the early 1940s in the OWI files. It is a significant part of the FSA/OWI collection—and, astonishingly, until now, almost entirely unknown. All of them were made by noted FSA photographers, including Rothstein, Jack Delano, Russell Lee, Marjory Collins, and Esther Bubley, and continued to reflect the rigorously documentarist tradition this group pioneered during the 1930s. Their concern with documenting scientific practices first emerged in 1942. Except for Russell Lee (a trained chemist who in the 1930s had photographed agricultural specialists and the shiny dome of the new McDonald Observatory etched against the Davis Mountains in Texas), FSA photographers had not seen scientific activities relevant to their mandate to document the social transformations of the rural American South during the Great Depression.[78] This raises new questions. Why did these leading photo-documentarians now consider science a worthy study, and what viewpoints guided their photographic work?

Answers begin to emerge from the archives of the OWI—a source thus far explored by few historians of modern science. Created in June 1942, the OWI was the US government's leading propaganda agency during the Second World War. Of the photographic units it contained, the most famous was that headed by Roy Stryker, the self-driven former Columbia University economics instructor who had led the famous FSA photography project in the 1930s.[79] Before his distinguished unit was transferred from the Department of Agriculture to OWI in late 1942, Stryker grew anxious that mounting Congressional opposition would shut down his operation. In April 1941, Stryker declared it necessary to "keep our finger in defense activities the way the whole world is moving now"; by early 1942 he reached an agreement with the office of the Coordinator of Information (COI), created by President Franklin D. Roosevelt in 1940, to provide photographs useful to the COI's foreign propaganda campaign. The COI was deeply involved in secret activities—it was a direct antecedent to the Central Intelligence Agency—and FSA photographers were required to undergo security checks from the FBI.[80] While Stryker insisted that the project's legendary dedication to objective photojournalism would not be altered by its evolving mission, he certainly appreciated that such topics as the increased employment of women, the "new curricula in schools and colleges," and scientific work in general, were now high priorities.[81] Indeed, Rothstein's photograph of the Atlanta University chemistry researcher was apparently composed for the COI, for it was reproduced in an OWI pamphlet *Negroes and the War*, highlighting the contributions of blacks to wartime efforts.[82]

The OWI photograph archives also provide intriguing new information on undergraduate training in science in the Second World War. In May 1942, a buoyant Jack Delano drove his car through Ames, Iowa, reaching the Iowa State

College campus on threadbare tires. This former FSA photographer, hustling between shooting assignments of Swedish-Americans in Minneapolis and bustling American railroad yards, visited Iowa State on orders from Stryker to document food production in the American heartland and education at agricultural colleges.[83] Delano spent several days on campus, photographing numerous individual students as well as shuttering views of classrooms in zoology, veterinary medicine, forestry, and chemistry.[84] A dedicated New Dealer who shared Stryker's populist faith in "the essential goodness of ordinary people," Delano focused primarily on students rather than on faculty or facilities. He wrote detailed notes on individual students he photographed and studied Iowa State research programs designed to increase agricultural promotion. The experience impressed him; Delano later recalled that "For a city boy like me, the exposure to such an environment was an education in itself." But a particularly fascinating detail in Delano's photographs— one that escaped comment in any of his surviving notes—is the number of women filling Iowa State's undergraduate classrooms, as men left for military service. In his image of one large chemistry lecture, not a single young man can be found (Figure 12.3). In this single document, Delano recorded a fact of social history that might otherwise not be teased out without examining dozens or hundreds of university records. The sheer number of female science students at Iowa State raises important questions about enrollments in other land-grant institutions throughout country, and extends Margaret Rossiter's pioneering analysis of women scientists in America.[85]

Figure 12.3 Jack Delano, "A Class in Chemistry: In the Lecture Room of the Chemistry Building," Iowa State College, Ames, Iowa, May 1942.

Source: Prints and Photographs Division, Library of Congress.

These FSA-OWI photographs are significant in their own right—but they also suggest a new interpretation of twentieth century American documentary photography. A prevailing view among scholars of FSA photography is that the demise of Stryker's extraordinary project marked the end of an important photo-documentary tradition.[86] Yet a better argument may be that OWI photographs of scientists, universities, and related technological systems (ignored by scholars examining more traditional FSA images) instead marked the *beginning* of a new documentary tradition that placed increased emphasis on the technological and scientific products of the state. US Information Agency photographers in the early Cold War often highlighted American technological know-how and scientific achievements.[87] Rockefeller Foundation leaders similarly sought to document their contributions to advances in plant genetics that underlay the Green Revolution, producing scores of photographs beginning in the 1960s to document higher-yield crop varieties in less developed countries.[88] Like Stryker's OWI team and later USIA photographers, Rockefeller Foundation documentarists created dramatic images that recorded food production and the role of modern science in making this so. Photographs of ecstatic third-world farmers contemplating 10-foot high corn stalks (Figure 12.4) celebrated not only improved productivity but democracy in action: a marriage between modern science and Thomas Jefferson's ideal of the yeoman farmer, an unmistakable suggestion to Cold War-era audiences of links between democratization and free scientific inquiry.[89] These photographs challenge our understanding of the origins of psychological warfare in US foreign policy. Diplomatic historian Kenneth A. Osgood correctly argues that Eisenhower Administration officials were especially eager to escalate the use of psychological strategies during the Cold War, particularly involving science. But these previously unexamined photographs point to earlier influences in the Roosevelt Administration, and hint at a much richer history by including photographs created by private foundations and nongovernmental organizations.[90]

How can photographs help us understand the practices and values of scientific communities and expeditions?

Photographs can also help resolve conflicts in interpreting the practices of modern scientific expeditions, suggesting how new questions can be raised about the field sciences. For instance, seemingly conflicting accounts exist of the Norwegian-British-Swedish Expedition to Antarctica in 1949–1952. This important multinational effort sought to address claims from the noted Swedish glaciologist Hans Ahlmann, employing data from the German Antarctic Expedition of 1938–1939, that high-latitude glaciers were receding, suggesting potentially rapid climate change with profound political, economic, and national security implications.[91] The expedition's youngest member, the Oxford-trained Charles Swithinbank, later observed that personal relations between foreign members of this first-ever international expedition to Antarctica were rarely strained: "If ever national rivalries had developed," he wrote, "most of us would have concluded that international expeditions were doomed." He quoted John Giæver, the Norwegian leader of the NBS expedition, as saying: "I do not think there has ever been a polar expedition with so little friction

Figure 12.4 Rockefeller Foundation, "Application of nitrogen and phosphorous often produce
fantastic results in the tropics..." Agricultural Station, Peru, ca. 1963 or 1964.
Photographer unknown.

Source: Courtesy the Rockefeller Archives Center, Sleepy Hollow, NY.

between members." Yet Brian B. Roberts, a senior Antarctic scientist representing
the British Committee to the expedition, confided to his private diary very different
impressions, at least of the expedition's leadership. Encounters with the Norwegian
oceanographer Harald Sverdrup, the taciturn leader of the Norwegian Committee

and nominal chair of the Expedition's international committee, had left him deeply discouraged: "The withholding of information causes suspicion and rumours. Once nationalism appears, it engenders more nationalism in those who are irritated by it or from whom information is withheld."[92] Can photographs made during the expedition itself help resolve these apparently incongruous views?

It seems that they can. As luck would have it, Swithinbank was a shutterbug, and the photographic record he made of his experiences in Antarctica has survived.[93] Swithinbank's camera recorded scenes on the Expedition ship *Norsel* as well as various encampments on the Antarctic continent. A number of photographs do give evidence of nationalistic displays: Norwegian and British flags flutter from amphibious weasel transports and from campsites.[94] Yet a larger number of photographs show Expedition members from these three nations adapting to one another's presence in their cramped shelters during the long Antarctic night. Swithinbank captured images of fellow researchers repeatedly in communal story times and in impromptu concerts (Figure 12.5); he also documented emergency surgeries and communal celebrations. Such scenes are familiar to historians of scientific expeditions from the nineteenth-century forward. They suggest that, at the lower rungs of the NBS Expedition, nationalism likely was not a major source of friction. Likely both Roberts and Swithinbank were accurate in their interpretations: nationalism was a far greater concern for decision-makers more closely connected with state interests and policy-making than for junior scientists wintering in Antarctica.

Because expeditions are transient institutions, traditional archival materials documenting them are often scarce. In such instances, surviving photograph collections can provide valuable insights into the practice of field science.

How can photographs help us understand the production of knowledge in recent science?

Let us raise one last example: how can historians of science employ photographs to help us better understand the factors that influence how science is actually done? In his classic *The Presentation of Self in Everyday Life* (1959), sociologist Erving Goffman analyzed how individuals put on a show "for the benefit of other people": what people wear, with whom they associate, how they arrange themselves in group photographs, all help reveal their identities and world-views. Goffman's ideas are a rich instrument with many applications, helping us see the social relations of individuals within particular cultural contexts.[95] Candid individual photographs, interpreted in this light, can suggest new and unexpected questions to historians of science. A lone female chemist appears for the first time in a group photograph of the 1940 meeting of the American Institute of Chemists held in Atlantic City, New Jersey; grim-faced and tight-lipped, she stands isolated from her colleagues (Figure 12.6). Existing studies of twentieth-century American chemistry are not yet sufficiently fine-grained to indicate how significant this photograph might be.[96] A far warmer personal relationship is evident among members of the Department of Chemistry at the University of Illinois, photographed in front of Farwell's Restaurant, home to daily faculty meetings (Figure 12.7); this camaraderie ought

Figure 12.5 Stig Hallgren, "Story-telling during the Norwegian–British–Swedish Expedition," Antarctica, 1951.

Source: Courtesy of Charles Swithinbank.

be counted among reasons that Illinois's work in physical chemistry led the nation in the mid-twentieth century.[97] Snapshots of Smithsonian Institution Secretary Alexander Wetmore, an ornithologist, and Watson Perrygo preparing skins in Panama in 1951 provides a rich window into their mini-laboratory in the tropics (Figure 12.8); they also show that Wetmore, a skilled Washington inside player and science admin- istrator, comfortable in a three-piece suit in a Congressional hearing room, was happiest as seen here, in the field, working with his hands surrounded by scientific

Figure 12.6 "Eighteenth Annual Meeting of the American Institute of Chemists," 1940.

Source: Chemical Heritage Foundation, Philadelphia, PA.

Figure 12.7 "Faculty members, Department of Chemistry, University of Illinois, before Farwell's Restaurant," 1956.

Source: Chemical Heritage Foundation, Philadelphia, PA.

Figure 12.8 Smithsonian Secretary and ornithologist Alexander Wetmore and field
collector Watson M. Perrygo prepare bird specimens in El Valle, Panama,
March 31, 1951.

Source: Smithsonian Institution Archives.

specimens.[98] An informal photograph of three US Congressmen accompanying
Yevgeni P. Velikov, a renowned physicist and leading science advisor to the Kremlin,
during a trip to inspect a Soviet radar facility in September 1987, reveals close bonds
between US and Soviet officials during the early years of glasnost.[99] Similarly,
"Breakout at Ainslie Genome Meeting" at the Cold Spring Harbor Laboratory (1998)
reminds historians that some of the important scientific discussions at scientific
conferences are informal, individual, and often poorly documented (Figure 12.9).

We can also gain insights by comparing group photographs side by side. Several
distinct research groups emerged within Columbia University's Lamont Geological
Observatory, founded in 1949, including those surrounding the physical chemist
J. Laurence Kulp and the German physical oceanographer Georg Wüst. Members of
Kulp's team appear conservatively dressed in white shirts and socks; they wear their
hair short and none apparently sport beards or moustaches. By contrast, the clothing
and physical arrangement of individuals around Wüst suggest a distinctly European
origin. These differing characteristics actually matter: oral history interviews have
made clear that many in Kulp's team were bound by a fundamentalist Christian faith
and unlike Wüst's group were involved in controversial classified research projects;
cultural frictions did emerge with other Lamont groups.[100]

Photographs can also provide valuable insights into the experiences of women in
public education and in major scientific organizations—issues often absent from
archival documents. In 1943, the OWI photographer Esther Bubley—soon one of the

Figure 12.9 Break at Ainslie Genome Meeting, Cold Spring Harbor Laboratory, 1998.
Source: Cold Spring Harbor Laboratory Archives.

best-known women photographers of her day—visited the highly regarded Woodrow
Wilson High School in Washington, DC. There Bubley chronicled well-dressed white
students, primarily female, completing chemistry and biology experiments in comfort-
ably equipped school laboratories.[101] The previous year her OWI colleague, Marjory
Collins, took her camera to a segregated black high school in Washington. Kneeling
down to capture wide-eyed black elementary and middle-schoolers witnessing chemistry demonstrations at eye level, Collins, later a social activist, documented a little-
known world distant from that of Wilson High School, with overcrowded facilities and
rudimentary equipment (Figure 12.10).[102] Also at this time, Ann Rosener, yet another
woman photographer attached to the OWI, took time from photographing women
assembling aircraft and heavy munitions to visit the US Geological Survey headquarters. There, her camera captured images of women producing rectified maps of Alaska,
then a vulnerable American territory, working alongside USGS Director William
Wrather. In contrast to Wrather—faintly smiling in his tight-fitting suit—his female
engineering aides appear comfortable in their skirts and blouses. Indeed, in one
provocative 1943 image, Rosener shows these aides sprawled bare-footed, skirts
billowing, across a vast warehouse-floor workspace where map reference points are
being carefully plotted (Figure 12.11). Contemporary audiences would hardly have
missed this teasing, playful Betty Grable allusion and suggestion of sexuality. But
pinup girls they were not: Rosener's subjects were lively and attractive, took delight in
their work, *and* were actively involved in serious wartime pursuits.[103]

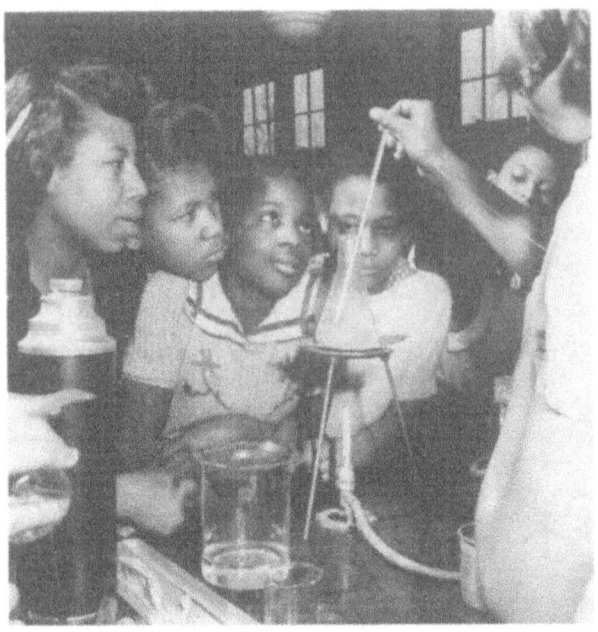

Figure 12.10 Marjory Collins, "Science Class in a Negro High School," March 1942.
Source: Prints and Photographs Division, Library of Congress.

Figure 12.11 Ann Rosener, "Mrs. Marvel Barenkov, engineering aid, Joel Langjofer, chief engineering aid, employees of the U.S. Geological Survey, Alaska Branch, and Dr. William Wrather, director of the Geological Survey," June 1943.
Source: Prints and Photographs Division, Library of Congress.

Each of these individual photographs suggests potentially important insights into the role of women in science in America during the Second World War. But what may be most significant is that all these images were produced by *women* photographers. As Andrea Fisher has noted in *Let Us Now Praise Famous Women: Women Photographers for the US Government 1935 to 1944*, the work of women photographers has largely "remained invisible in the writing of history." Their perspective on the role of women in the modern workforce, Fisher argues, differs from that of their male counterparts, for their images "raise questions about our concept of documentary truth and the relationship of women to it."[104] Yet historians thus far have left unexamined their photographs of scientific training or scientific practice. For instance, Fisher's path-breaking analysis of Bubley's photographs includes just two images from her Wilson High School shoot, both made *after* school has let out.[105] Here, perhaps, the unfortunate disciplinary divide between social history and the history of science may have made images of science seem foreign. But our discovery that these photographs do exist may well now inspire historians of modern science to pose new questions and scour these sources.

* * *

Locating significant photographs—then interpreting them—is not easy. But locating and interpreting documents—just like conducting good oral history interviews—is not easy either. These pursuits are simply more familiar to us. And we ignore photographic evidence at our peril.

We realize this argument may not be widely heeded in the near future. Analyzing photographs as historical evidence within the history of science probably has the status now that oral history did in the early 1960s—a suspicious enterprise, but given grudging acknowledgment. We fully realize that the method will flourish only when archivists gain the resources to better catalog photographic collections. But we are in a visual culture, and future historians of science, we predict—certainly we hope—will wonder why it took so long for the significance of photographs to sink in.

Acknowledgments

This research was aided by NSF grants DIR-9112304 and SBR-9511867 (Doel). We are grateful for comments and criticisms from Finn Aaserud, Daniel Alexandrov, Dee Baer, Lowell Berentsen, Joan Bromberg, Pete Daniel, David DeVorkin, Rick Dingus, Michael Edwards, Brandon Brame Fortune, James B. Gilbert, Bill Grant, Kris Harper, Laura Henderson, Melissa A. Johnson, Freida Knobloch, Larry Landis, Mark Largent, Ellen Miles, Nick Natanson, Ben Primer, Ron Rainger, Bill Robbins, Terri A. Schorzman, and Charles Swithinbank. Much earlier, when this project was little more than an inchoate idea, Doel received very helpful encouragement from both Ansel Adams and Beaumont Newhall.

Notes

1 G. C. Heron, "Photograph Collections and the Alexander Turnbull Library: An Appeal for the Preservation of Photograph Collections on Account of their Historic Value," n.d. (*ca* 1948), pamphlet, Cambridge University Library.

2 Bruce I. Bustard / U.S. National Archives and Records Administration, *Picturing the Century: One Hundred Years of Photography from the National Archives* (Washington, DC: NARA, in Association with the University of Washington Press, 1999), 2.

3 Keith McElroy, review of Robert M. Levine, *Images of History: Nineteenth and Twentieth Century Latin American Photographs as Documents* (Durham, NC: Duke University Press, 1989), *American Historical Review* 96, no. 3 (1991): 642. A staff member at the Library of Congress's Prints and Photographs Division in 2001 recalled one confident young PhD historian intent on finding photographs to illustrate his already completed dissertation who left distraught, aware that the images he found contradicted his manuscript-based thesis.

4 Alan Trachtenberg, *Reading American Photographs: Images as History, Mathew Brady to Walker Evans*, 1st ed. (New York: Hill and Wang, 1989); Robert M. Levine, ed., *Windows on Latin America: Understanding Society through Photographs* (Coral Gables, FL: North-South Center, University of Miami, 1987); Peter Burke, *Eyewitnessing: The Uses of Images as Historical Evidence* (Ithaca, NY: Cornell University Press, 2001); and Robert M. Levine, *Insights into American History: Photographs as Documents* (Upper Saddle River, NJ: Prentice Hall, 2004). We searched the Library of Congress's online catalog [1898 through the present] for "photography in historiography"; our search of HOLLIS (Harvard University libraries) yielded a remarkably similar list.

5 Levine, *Images of History*; James R. Ryan, *Picturing Empire: Photography and the Visualization of the British Empire* (Chicago, IL: University of Chicago Press, 1997); Peter B. Hales, *William Henry Jackson and the Transformation of the American Landscape* (Philadelphia, PA: Temple University Press, 1988); Michael Lesy, *Wisconsin Death Trip*, 1st Anchor Books ed. (New York: Anchor Books, 1991); Pete Daniel, *Official Images: New Deal Photography* (Washington, DC: Smithsonian Institution Press, 1987); and Catherine Lutz and Jane Lou Collins, *Reading National Geographic* (Chicago, IL: University of Chicago Press, 1993). Case studies abound; see Keith McElroy, *Early Peruvian Photography: A Critical Case Study* (Ann Arbor, MI: UMI Research Press, 1985); Eric Margolis, "Mining Photographs: Unearthing the Meaning of Historical Photos," *Radical History Review* 40 (1988): 33–49; and Walter Rundell, "Photographs as Historical Evidence: Early Texas Oil," *The American Archivist* 41 (1978): 373–398.

6 An important exception is Martin Kemp, "Seeing and Picturing: Visual Representation in Twentieth-Century Science," in *Science in the Twentieth Century*, ed. John Krige and Dominique Pestre (London: Harwood Academic Press, 1977), 361–390; existing studies that draw on photographs as well as more traditional resources include William H. Goetzmann, *Exploration and Empire: The Explorer and the Scientist in the Winning of the American West*, 1st ed. (New York: Knopf, 1966); P. Thomas Carroll, "American Science Transformed," *American Scientist* 74 (1986): 466–485; David E. Nye, *Image Worlds: Corporate Identities at General Electric, 1890–1930* (Cambridge: MIT Press, 1985); and Margaret Rossiter, *Women Scientists in America: Volume 1: Struggles and Strategies to 1940* (Baltimore, MD: Johns Hopkins University Press, 1982).

7 Gregg Mitman, *Reel Nature: America's Romance with Wildlife on Films* (Cambridge: Harvard University Press, 1999); Peter Louis Galison and Emily Ann Thompson, *The Architecture of Science* (Cambridge: MIT Press, 1999); and Scott Knowles and Stuart W. Leslie, " 'Industrial Versailles': Eero Saarinen's Corporate Campuses for GM, IBM, and AT&T," *Isis* 92, no. 1 (2001): 1–33.

8 Martin J. S. Rudwick, *Scenes from Deep Time: Early Pictorial Representations of the Prehistoric World* (Chicago, IL: University of Chicago Press, 1992); Barbara Marie Stafford, *Voyage into Substance: Art, Science, and the Illustrated Travel Account, 1760–1840* (Cambridge: Harvard University Press, 1984); Barbara Marie Stafford, "Think Again: The Intellectual Side of Images," *Chronicle of Higher Education*, June 20, 1997, B6-B7; and L. J. Jordanova, *Sexual Visions: Images of Gender in Science and Medicine between the Eighteenth and Nineteenth Centuries* (New York ; London: Harvester Wheatsheaf, 1989). Science, objectivity, and images are explored in Lorraine Daston and Peter Galison, "The Image of Objectivity," *Representations* 40 (1992): 81–128; Jennifer Tucker, "Photography as Witness, Detective,

and Imposter: Visual representation in Victorian Science," in *Victorian Science in Context*, ed. Bernard Lightman (Chicago, IL: University of Chicago Press, 1997), 378–408, esp. 380–381; Brian S. Baigre, ed., *Picturing Knowledge: Historical and Philosophical Problems Concerning the Use of Art in Science, Toronto Studies in Philosophy* (Toronto: University of Toronto Press, 1996); and Alex Soojung-Kim Pang, "Visual Representation and Post-Constructivist History of Science," *Historical Studies in the Physical and Biological Sciences* 28, no. 1 (1997): 139–172.

 9 Our aim is not to analyze photography's applications as a *scientific* research tool, a distinct subject explored in a burgeoning literature; see Ann Thomas, *Beauty of Another Order: Photography in Science* (New Haven, CT: Yale University Press, 1997); Harry Robin, *The Scientific Image: From Cave to Computer* (New York: Abrams, 1992); Bettyann Kevles, *Naked to the Bone: Medical Imaging in the Twentieth Century* (New Brunswick, NJ: Rutgers University Press, 1997); Alex Pang, " 'Stars Should Henceforth Register Themselves': The Rhetoric and Reality of Early Astrophotography," *British Journal for the History of Science* 31 (1997): 177–201; and Peter Geimer, ed., *Ordnungen der Sichtbarkeit: Fotografie in Wissenschaft, Kunst und Technologie* (Frankfurt am Main: Suhrkamp, 2002).

10 The presentation of science in popular film and video, while important, lies beyond the scope of this essay; for an introduction see Gregg Mitman, "Cinematic Nature: Hollywood Technology, Popular Culture, and the American Museum of Natural History," *Isis* 84 (1993): 637–661; and Nathan Reingold, "Metro Goldwyn Mayer Meets the Atom Bomb," *Science, American Style: Selected Writings of Nathan Reingold* (New Brunswick, NJ: Rutgers University Press, 1991), 334–350.

11 Charles Weiner, "Oral History of Science: A Mushrooming Cloud?" *Journal of American History* 75, no. 2 (1988): 548–559; Pamela M. Henson and Terri A. Schorzman, "Videohistory: Focusing on the American Past," *Journal of American History* 78, no. 2 (1991): 618–627, and reprinted in the *Oral History Association Newsletter* 25 (winter 1992): 3–5, and 26 (spring 1992): 8; Spencer R. Weart and David H. DeVorkin, "Interviews as Sources for History of Modern Astrophysics," *Isis* 72, no. 263 (1981): 471–476; and Alex Soojung-Kim Pang, "Oral History and the History of Science," *International Journal of Oral History* 10, no. 3 (1989): 270–285; for an overview see Ronald E. Doel, "Oral History of American Science: A Forty Year Review," *History of Science* 41 (2003): 349–378.

12 National Museum of Art, Merry A. Foresta, and Charles Issacs, *American Photographs: The First Century* (Washington, DC: Smithsonian Institution Press, 1997); Estelle Jussim and Elizabeth Lindquist-Cock, *Landscape as Photograph* (New Haven, CT: Yale University Press, 1982); and Anne Farrar Hyde, *An American Vision: Far Western Landscape and National Culture*, 1820–1920 (New York: New York University Press, 1990).

13 Ansel Adams, *Simpson's Quotations*, http://www.bartleby.com/63/6/5806.html, accessed on January 8, 2005.

14 Oliver Wendell Holmes, "The Stereoscope and the Stereograph," *Atlantic Monthly* 3 (June 1859), 738–748; Sally Stein, "In Pursuit of the Proximate: A Biographical Introduction," in *Photographic Memories*, by Jack Delano (Washington, DC: Smithsonian Institution Press, 1997), iii–iv; Jerry L. Thompson, "Truth and Photography," *Yale Review* 90, no. 1 (2002): 25–53. Such claims were common in the decades following the introduction of photography; for instance, Civil War photographer Matthew Brady also declared photographs a transparent "eye of history"; for further background see Ryan, *Picturing Empire*; David A. Hollinger, *Science, Jews, and Secular Culture: Studies in Mid-twentieth-century American Intellectual History* (Princeton, NJ: Princeton University Press, 1996), 42–59; and Oscar Handlin, *Truth in History* (Cambridge, MA: Belknap Press, 1979).

15 Beaumont Newhall defined the "straight photograph" in his classic *The History of Photography: From 1839 to the Present* (New York: Museum of Modern Art; Distributed by New York Graphic Society Books, 1982), 167–197. A wide body of literature examines the relationship between photographs and other forms of visual evidence; for an entry see David Freedberg, *The Power of Images: Studies in the History and Theory of Response* (Chicago, IL: University of Chicago Press, 1989); and Gordon Fyfe and John Law,

Picturing Power: Visual Depiction and Social Relations, Sociological Review Monograph, 35 (London: Routledge, 1988).

16 Andy Grundberg, "Ask It No Questions: The Camera Can Lie," *New York Times*, August 12, 1990, sec. H, p. 1.

17 David King, *The Commissar Vanishes: The Falsification of Photographs and Art in Stalin's Russia* (New York: Metropolitan Books, 1997); see also Dino A. Brugioni, *Photo Fakery: The History and Techniques of Photographic Deception and Manipulation* (Dulles, VA: Brassey's, 1999); and William J. Mitchell, *The Reconfigured Eye: Visual Truth in the Post-photographic Era* (Cambridge: MIT Press, 1992), 218–220.

18 Tucker, "Photography as Witness," 379, referring to Ernest Arthur Jeff. The doctored photograph was printed in 1894. As photo-historians Chris Bruce and Andy Grundberg have pointed out, "The belief in the veracity of photographic information is a cultural convention, not a fact"; see Chris Bruce, Andy Grundberg, and Henry Art Gallery, *After Art: Rethinking 150 Years of Photography: Selections from the Joseph and Elaine Monsen Collection: With Essays* (Seattle, WA: Henry Art Gallery University of Washington, Distributed by The University of Washington Press, 1994), 16.

19 An insightful analysis of this clumsy manipulation appears in Caroline Alexander and American Museum of Natural History, *The Endurance: Shackleton's Legendary Antarctic Expedition* (New York: Knopf, 1998), see 202–203. Alexander herself terms Hurley's act a "grave indiscretion."

20 Ronald R. Kline, *Steinmetz: Engineer and Socialist* (Baltimore. MD: Johns Hopkins University Press, 1992), ix–x, 295, and 297. The *Schenectady Worker News* published the original photograph on May 20, 1921 and the doctored version on July 18, 1924. We thank Stuart W. Leslie for calling this incident to our attention, and Ron Kline for further discussions.

21 Peter B. Hales, "Photography and the World's Columbian Exposition: A Case Study," *Journal of Urban History* 15 (1989): 247–273, on 264.

22 Nevada Museum of Art, *The Altered Landscape* (Reno, NV: The Nevada Museum of Art in Association with the University of Nevada Press, 1999), xviii.

23 James Curtis, *Mind's Eye. Mind's Truth: FSA Photography Reconsidered* (Philadelphia, PA: Temple University Press, 1989), 75; and F. Jack Hurley, *Portrait of A Decade: Roy Stryker and the Development of Documentary Photography in the Thirties* (New York: De Capo Press, 1977), 90.

24 The conventional interpretation of this photograph is well-described in Penelope Dixon, *Photography of the Farm Security Administration: An Annotated Bibliography. 1930–1980* (New York: Garland, 1983), 119, but a careful reading of the sequence of photographs that Rothstein made at the farm in preparation for this image belies its apparent spontaneity; see Curtis, *Mind's Eye*, 82–87. Yet this photograph, while not literally accurate, did illustrate a truth about the severity of the Dust Bowl, and like inaccurate but often-repeated tales of key events told in oral history accounts, is historically illuminating. For more on this issue see Lillian Hoddeson in this volume.

25 Newhall, *History of Photography*.

26 James West Davidson and Mark H. Lytle, *After the Fact: The Art of Historical Detection*, 2nd ed. (New York: McGraw-Hill, 1986), 217.

27 Clarence King, *Report of the Geological Exploration of the Fortieth Parallel* (Washington, DC: Government Printing Office, 1870); and Arnold Hague and Samuel Franklin Emmons, *Descriptive Geology* (Washington, DC: Government Printing. Office, 1877), facing 338.

28 Rick Dingus, *The Photographic Artifacts of Timothy O'Sullivan* (Albuquerque, NM: University of New Mexico Press, 1982), 60–64. King's career is recounted in Goetzmann, *Exploration and Empire*, chapters 10, 12, and 16, and in Michele Aldrich, "King, Clarence Rivers," *Dictionary of Scientific Biography*, vol. 7 (1973): 370–371.

29 A significant start is Mark Hineline, "The Visual Culture of the Earth Sciences, 1863–1970" (PhD dissertation, University of California, San Diego, 1992).

30 Hales, "Columbian Exposition," 266–267; Carl Fleischhauer and Beverly W. Brannen, *Documenting America. 1935–1943* (Berkeley, CA: University of California Press in Association with the Library of Congress, 1988), 25.

31 Carol Armstrong, *Scenes in a Library: Reading the Photograph in the Book. 1843–1875* (Cambridge: MIT Press, 1998).

32 Lee Clark Mitchell, "The Photograph and the American Indian," in *The Photograph and the American Indian*, ed. Alfred L. Bush and Lee Clark Mitchell, (Princeton, NJ: Princeton University Press, 1994), xiii–xiv, 120; Vicki Goldberg, *The Power of Photography: How Photographs Changed Our Lives* (New York: Abbeville Press, 1991), 95; see also Armstrong, *Scenes*; Ryan, *Picturing Empire*; and Jefferson Hunter, *Image and Word: The Intersection of Twentieth Century Photographs and Text* (Cambridge: Harvard University Press, 1987).

33 "News in Pictures," *Time*, November 12, 1951, 26; Bustard, *Picturing the Century*, 106. The caption included in the National Archives exhibit reads in full, " 'Exercise Desert Rock.' Troops of the Battalion Combat Team, US Army 11th Airborne Division, watch a plume of radioactive smoke rise after a D-Day blast at Yucca Flats, as the much prepared Exercise Desert Rock reaches its peak." [Photograph by Corporal McCaughey, Las Vegas, Nevada, November 1, 1951.]

34 James Cletus Anderson, *Roy Stryker. the Humane Propagandist* (Louisville, KY: Photographic Archives of the University of Louisville, 1977), 3.

35 "Kodak to Stop Selling Traditional Cameras in U.S.," *USA Today*, January 13, 2004, http://www.usatoday.com/tech/techinvestor/techcorporatenews/2004–01–13-kodak-all-digital_x.htm (January 21, 2004), and "The Fate of Photography in a Digital Age," (US) National Public Radio, "All Things Considered," January 25, 2004, http://www.npr.org/features/feature.php?wfId=1616953. On the emerging crisis in contemporary records management (and proposed remedies) see Ann J. Gilliland-Swetland, *Enduring Paradigm. New Opportunities: The Value of the Archival Perspective in the Digital Environment* (Washington, DC: 2000), http://www.clir.org/pubs/reports/pub89/ contents.html, accessed on January 8, 2005. Peter Lyman and Howar Besser, "Defining the Problem of our Vanishing Memory: Background, Current Status, Models for Resolution," in *Time and Bits: Managing Digital Continuity*, ed. Margaret MacLean and Ben H. Davis (Los Angeles, CA: Getty Research Institute, 1999); and Seamus Ross, *Changing Trains at Wigan: Digital Preservation and the Future of Scholarship* (London, 2000), http://www.leeds.ac.uk/ cedars/OTHER/SRoss.htm (January 28, 2004). For a related discussion see Arne Hessenbruch, this volume.

36 See Hoddeson and Tansey, this volume.

37 Bustard, *Picturing the Century*, 2; on the history of nineteenth century photography see Peter B. Hales, *Silver Cities: The Photography of American Urbanization. 1839–1915*, *American Civilization* (Philadelphia, PA: Temple University Press, 1984), 4–5; and Foresta and Isaacs, *American Photographs*, 8. Photographs from nineteenth century federal expeditions to the US West are among the best-known of all images from this era; see Hales, *William Henry Jackson*; Don D. Fowler, *The Western Photographs of John K. Hillers: "Myself in the Water"* (Washington, DC: Smithsonian Institution Press, 1989); Eugene Ostroff, *Photographing the Frontier* (Washington, DC: Smithsonian Institution Press, 1976); and Hal G. Stephens, Eugene Merle Shoemaker, and John Wesley Powell, *In the Footsteps of John Wesley Powell: An Album of Comparative Photographs of the Green and Colorado Rivers, 1871–72 and 1968* (Boulder, Denver, CO: Johnson Books, 1987).

38 National Museum of American Art (US); Merry A. Foresta, *American Photographs: The First Century from the Isaacs Collection in the National Museum of American Art* (Washington, DC: National Museum of American Art, Smithsonian Institution, and Smithsonian Institution Press, 1996), 19; Ryan, *Picturing Empire*, 13, 17.

39 Robert V. Bruce, *The Launching of Modern American Science, 1846–1876* (New York: Knopf, 1987).

40 For two examples among many, see the Fixed Nitrogen Research Laboratory Collection, P/B Ch432.83, Box File #1, Chemical Heritage Foundation, Philadelphia, PA; and

Collection no. 2141, Prints and Photographs Division, Library of Congress, Washington, DC (hereafter PP-LC).

41 For example, Jack Delano, "A Class in Chemistry in the Lecture Room of the Chemistry Building" (Iowa State College), May 1942, negative no. 2758-D, and Arthur Rothstein, "Chemistry Student" (Atlanta University, GA.), March 1942, negative no. 269-D, Farm Security Administration photograph collection, PP-LC.

42 Bustard, *Picturing the Century*, 22; Robert W. Rydell, *World of Fairs: The Century-of-Progress Expositions* (Chicago, IL: University of Chicago Press, 1993); and Robert W. Rydell, John E. Findling, and Kimberly D. Pelle, *Fair America: World's Fairs in the United States* (Washington, DC: Smithsonian Institution Press, 2000).

43 Pete Daniel, *Lost Revolutions: The South in the 1950s* (Chapel Hill, NC: University of North Carolina Press for Smithsonian National Museum of American History, Washington DC, 2000).

44 *Photography from 1839 to Today: George Eastman House, Rochester, New York* (Cologne: Taschen, 1999), 574–575.

45 That Hine took this assignment late in his career, when in need of money, does not diminish its value for expanding what we know about early twentieth-century science; on Hine's later career see Judith Mara Gutman, *Lewis W. Hine, and the American Social Conscience* (New York: Walker, 1967).

46 Marcel C. LaFollette, *Making Science Our Own: Public Images of Science, 1910–1955* (Chicago, IL: University of Chicago Press, 1990); an entry to these photographs is Watson Davis, *Science Picture Parade* (New York: Duell, Sloan and Pearce, 1940).

47 See for instance Newman, "Dr. J. Robert Oppenheimer, 1948" reproduced in *Photography*.

48 Ian Jeffrey and National Museum of Photography Film and Television (Great Britain), *ReVisions: An Alternative History of Photography* (Bradford, England: National Museum of Photography Film & Television, National Museum of Science & Industry, 1999), 14, 63–65; Bustard, *Picturing the Century*, 2.

49 Edward Curtis served as the main photographer of the Harriman Expedition; see William H. Goetzmann and Kay Sloan, *Looking Far North: The Harriman Expedition to Alaska, 1899* (Princeton, NJ: Princeton University Press, 1982); on Knight, see Deborah Hardy, *Wyoming University: The First 100 Years, 1886–1986* (Laramie, WY: University of Wyoming Press, 1986).

50 Published collections of these photographs are useful windows into these archives; see John Perkins, *To The Ends of the Earth: Four Expeditions to the Arctic, the Congo, the Gobi, and Siberia* (New York: Pantheon Books, 1981); Margo Baumgartner Davis and Roxanne Nilan, *The Stanford Album: A Photographic History, 1885–1945* (Stanford, CA: Stanford University Press, 1989); Shaun J. Hardy, "The Earth and Space Sciences at Carnegie: A Pictoral Sampler from the First Six Decades," in *The Earth the Heavens and the Carnegie Institution of Washington*, ed. Gregory A. Good (Washington, DC: American Geophysical Union, 1994), 237–252; Elizabeth L. Watson, *Houses for Science: A Pictorial History of Cold Spring Harbor Laboratory* (Plainview, NY: Cold Spring Harbor Laboratory Press, 1991); and Charles V. P. Young, *Cornell in Pictures: The First Century* (Ithaca, NY: Quill and Dagger Alumni Association, 1965).

51 Personal communication with archivists at the Steenbock Memorial Library, University of Wisconsin Madison and at the University of Illinois at Urbana Champaign Library, 2000 and 2002 respectively.

52 The rise of digital photography makes the question of the survival of recent images problematic; see Grundberg, "Ask it No Questions." At the same time, archival photographs are increasingly available on the web; see for example the Emilio Segrè Visual Archives at the American Institute of Physics, http://www.aip.org/history/esva/ (January 8, 2005) and the Smithsonian Institution Archives History of the Smithsonian catalog, http://www. siris.si.edu (January 28, 2004). [*Note added in proof*: a large new online

collection, searchable by key words (for instance, "laboratory" or "museum") is the New York Public Library Digital Gallery, http://digitalgallery.nypl.org/nypldigital/index.cfm (March 14, 2005).]

53 See Arnold Pacey, *Technology in World Civilization: A Thousand-year History* (Cambridge: MIT Press, 1990); and John Perkins, *Geopolitics and the Green Revolution: Wheat, Genes, and the Cold War* (Oxford: Oxford University Press, 1997).

54 Historians of anthropology have begun to explore these collections in depth; for an introduction, see Elizabeth Edwards and the Royal Anthropological Institute of Great Britain and Ireland, eds, *Anthropology and Photography, 1860–1920* (New Haven, CT: Yale University Press in association with the Royal Anthropological Institute London, 1992); and Jon Wagner, ed, *Images of Information: Still Photography in the Social Sciences* (Beverly Hills, CA: Sage Publications, 1979).

55 Lutz and Collins, *Reading National Geographic*.

56 James B. Gilbert, *Redeeming Culture: American Religion in an Age of Science* (Chicago, IL: University of Chicago Press, 1997), 1–3.

57 "Science Service," Smithsonian Institution, http://americanhistory.si.edu/scienceservice/ (January 25, 2004).

58 "Over 100 Years of Fossil Collecting," American Museum of Natural History, http://paleo.amnh.org/fossil/FRC.xindex (January 23, 2004); "About the Harriman Alaska Expedition of 1899," University of Washington Libraries Digital Collections, http://content.lib.washington.edu/EHarriman/more-info.html (January 23, 2004). The European Visual Archive, http://192.87.107.12/eva/uk/search.asp (January 23, 2004).

59 See Levine, *Images of History*, 84–90

60 "Gustav Arthur Cooper and Josephine Cooper, Division of Invertebrate Paleontology, National Museum of American History, 1954," Gustav Arthur Cooper papers, RU 7318, Smithsonian Institution Archives. On the careers of women such as Josephine Cooper, see Roy MacLeod, "Fathers and Daughters: Reflections on Women, Science, and Victorian Cambridge," *History of Education* 8, no. 4 (1979): 321–333; Ann B. Shteir, "Botany in the Breakfast Room: Women and Early Nineteenth Century British Plant Study," in *Uneasy Careers and Intimate Lives: Women in Science, 1789–1979*, ed. Pnina G. Abir-Am and Dorinda Outram (New Brunswick, NJ: Rutgers University Press, 1987), 31–43; Ann B. Shteir, *Cultivating Women, Cultivating Science: Flora's Daughters and Botany in England, 1760 to 1860* (Baltimore, MD: The Johns Hopkins University Press, 1996), 173–182; and Helena M. Pycior, Nancy G. Slack, and Pnina G. Abir-Am, *Creative Couples in the Sciences: Lives of Women in Science* (New Brunswick, NJ: Rutgers University Press, 1996).

61 Ryan, Picturing Empire, 214.

62 Soroya de Chadarevian, "Portrait of a Discovery: Watson, Crick, and the Double Helix," *Isis* 94 (2003): 90–105, quoted on 105 and 93. Photographic portraits of individual scientists arose out of a well-developed tradition in art and photographers from the mid-nineteenth century on incorporated many traditions of portraiture in their work; for an introduction see Brandon Brame Fortune, Deborah Jean Warner, and National Portrait Gallery (Smithsonian Institution), *Franklin & his Friends: Portraying the Man of Science in Eighteenth-century America* (Washington, DC, PA: Smithsonian National Portrait Gallery; in association with the University of Pennsylvania Press, 1999).

63 William L. Laurence, "W. Kaempffert, Science Popularizer," *Science* 129 (April 19, 1957): 727–728; Kaempffert to Davenport, December 28, 1915, Kaempffert folder, Davenport papers, American Philosophical Society. We thank Mark Largent for providing us this citation.

64 On Science Service, see LaFollette, *Making Science Our Own*; see also Marcel C. LaFollette, "Eyes on the Stars," *Science, Technology, and Human Values* 13 (1988): 262–275.

65 Herbert George Ponting, *The Great White South; or, With Scott in the Antarctic, Being an Account of Experiences with Captain Scott's South Pole Expedition and of the Nature Life of the Antarctic* (New York: R. M. McBride & Company, 1923).

66 Ponting to R. F. Scott, November 4, 1912, in Papers of the British Antarctic Expedition 1910–1913, vol. 7 [Frank Debenham correspondence series], MS 280/28/7a; ER, Scott Polar Research Institute, Cambridge University [hereafter Debenham].

67 Ponting to Debenham, labeled "private," January 7, 1921, and T. R. Roosevelt to Ponting, June 14, 1914, both Debenham.

68 These works say little about the practice of science as revealed in Ponting's images; see Herbert George Ponting and Frank Hurley, *Antarctic Photographs, 1910–1916: Scott, Mawson and Shackleton expeditions* (London: Macmillan, 1979); and Beau Riffenburgh and Liz Cruwys, *The Photographs of Herbert G. Ponting*, ed. Jonathan Jeffes (London: Discovery Gallery, 1998).

69 Dingus, *Photographic Artifacts*; a related interpretation appears in Melissa A. Johnson, *Reflections of Photographing Princeton: An Essay to Accompany an Exhibit at the Seeley G. Mudd Manuscript Library, August 15, 1998–June 30, 1999* (Princeton, NJ: Princeton University Library, 1998), 24.

70 Given the questions that have dominated accounts of the history of science throughout the twentieth century, existing bias towards elite science is hardly surprising. It is especially evident in studies of foundation patronage; see Roger L. Geiger, *Research and Relevant Knowledge: American Research Universities since World War II* (New York: Oxford University Press, 1993); and Robert E. Kohler, *Partners in Science: Foundations and Natural Scientists, 1900–1945* (Chicago, IL: University of Chicago Press, 1991).

71 See the S. K. Knight photograph collection, American Heritage Center, University of Wyoming, including A21.625 [1911], A21.601 [1914–1915], and A21.602 [1918, all Samuel K. Knight collection. The large numbers of women participating in Wyoming paleontological expeditions are not noted in Tom Rea's *Bone Wars: The Excavation and Celebrity of Andrew Carnegie's Dinosaur* (Pittsburgh, PA: University of Pittsburgh Press, 2001) nor in histories of the University of Wyoming; see Wilson Ober Clough, *A History of the University of Wyoming, 1887–1937* (Laramie, Wyo.: Laramie Printing Co., 1937); and Deborah Hardy, *Wyoming University: The First 100 Years, 1886–1986* (Laramie, Wyo.: University of Wyoming Press, 1986). We are grateful to Frieda Knobloch for steering us to the Knight collection.

72 "Colorado Mountain Laboratory for Biology / Botany," [1908–1915, Image X-13855], Western History Division, Denver Public Library; and "Group of Students and Instructors," n.d. [summer 1911], Box 130, folder 10, W.U. President Collection 74–34, University of Washington Archives. Women were increasingly represented in natural history by the late nineteenth century, as collecting was not considered a "manly" activity, but the mix of men and women in this image collection raises important questions; see Philip J. Pauly, "Woods Hole, 1882–1925," in *The American Development of Biology*, ed. Ronald Rainger, Keith R. Benson, and Jane Maienschein (Philadelphia, PA: University of Pennsylvania Press, 1988), 121–150, see 126.

73 Photograph of the John Henry Comstock Laboratory at Cornell University, Department of Manuscripts and University Archives, Cornell University; "Agricultural Laboratory, Universidad do Rio Grande, Brazil," No. 22916-V, *ca* early 1950s, RAC; see Pamela M. Henson, "The Comstock Research School in Evolutionary Entomology," *Osiris* 8 (1993): 159–177.

74 "Visitors to Lick Observatory," n.d. [late nineteenth century], [Mary Lea Shane archives of the University of California]; "American Scientific Affiliation members visiting Palomar Mountain Observatory," 1949, reproduced in Gilbert, *Redeeming Culture*, 157. Such photographs remind us of the significance of witnessing scientific practices in recent times; the leading study on this issue is Steven Shapin and Simon Schaffer, *Leviathan and the Air-pump: Hobbes, Boyle, and the Experimental Life: Including a Translation of Thomas Hobbes, Dialogus physicus de natura aeris* by Simon Schaffer (Princeton, NJ: Princeton University Press, 1985).

75 "County Prize Winners, members of Boys' Corn Clubs," n.d.; "A Corn Club Boy (Frank G. Brockman, Amherst, Va.) and his 167 bushels of corn produced on one acre in 1912,"

and "Field Meeting of a Boys' Corn Club in Elbert County, Georgia," n.d., all in GEB 1054, folder 4000, Agricultural Demonstration Work, 1908–1913, RAC. On the hookworm campaign, see John Ettling, *The Germ of Laziness: Rockefeller Philanthropy and Public Health in the New South* (Cambridge: Harvard University Press, 1981).

76 On eugenics in America, see Daniel J. Kevles, *In the Name of Eugenics: Genetics and the Uses of Human Heredity* (New York: Knopf, 1985); Diane B. Paul, *Controlling Human Heredity, 1865 to the Present, The Control of Nature* (Atlantic Highlands, NJ: Humanities Press, 1995); Diane B. Paul and NetLibrary Inc., *The Politics of Heredity: Essays on Eugenics, Biomedicine, and the Nature-nurture Debate* (Albany, NY: State University of New York Press, 1998); and Rydell, Findling, and Pell, *Fair America*, 65; see also Garland E. Allen, "The Biological Basis of Crime: An Historical and Methodological Study," *Historical Studies in the Physical and Biological Sciences* 31, no. 2 (2001): 183–222; and Lene Koch, this volume. An important source for photographs of eugenics practice and education is the Eugenics Archive of the Cold Spring Harbor Laboratory, http://www.eugenicsarchive.org/eugenics/ (January 25, 2004).

77 While Philip J. Pauly's *Biologists and the Promise of American Life: From Meriwether Lewis to Alfred Kinsey* (Princeton, NJ: Princeton University Press, 2000), 171–193 addresses high school biology in the early twentieth century, this large topic remains woefully underexplored.

78 Fleischhauer, *Documenting America*, 25; and Roy Emerson Stryker and Nancy C. Wood, *In this Proud Land: America, 1935–1943, as seen in the FSA photographs* (London: Secker & Warburg, 1974), 14. Russell Lee's work is analyzed in Russell Lee and F. Jack Hurley, *Russell Lee, Photographer* (Dobbs Ferry, NY: Morgan & Morgan, 1978).

79 On Stryker's career and influence, see Curtis, *Mind's Eye*; Hurley, *Portrait*; Stryker and Wood, *Proud Land* ; and Jeanie Cooper, "Interpreting National Identity in Time of War: Competing Views in U.S. Office of War Information (OWI) Photography, 1940–1945" (PhD dissertation, Boston University, 1995).

80 Stryker to Russell Lee, March 18, 1942, Reel 6, Stryker microfilm. On the COI and its role within US intelligence, see Rhodri Jeffreys-Jones, *The CIA & American Democracy* (New Haven, CT: Yale University Press, 1989); and Arthur B. Darling, *The Central Intelligence Agency: An Instrument of Government, to 1950* (University Park, PA: Pennsylvania State University Press, 1990).

81 "Washington, Nerve Center for Global War," early 1942; Jack Delano and Ed Rosskam to Stryker, n.d. [likely March 1942]; see also Russell Lee to Stryker, September 7, 1942, all Reel 6, Stryker microfilm, PP-LC.

82 US Office of War Information, Bureau of Special Services, *Negroes and the War* (Washington, DC: Government Printing Office, 1942). Fleischhauer, Beverly W. Brannen, *Documenting America*, 5. On the significance of this pamphlet—attacked in Congress for promoting racial equality—see Allan M. Winkler, *The Politics of Propaganda: The Office of War Information, 1942–1945*, Yale Historical Publications: Miscellany, 118 (New Haven, CT: Yale University Press, 1978), 67; Nicholas Natanson, *The Black Image in the New Deal: The Politics of FSA Photography* (Knoxville, TN: University of Tennessee Press, 1992), 69–71, 258; and Cooper, "Interpreting National Identity," 58.

83 Jess Gorkin to Jack Delano, May 8, 1942, and Roy Stryker to Delano, both Box 2, folder 6, Jack Delano papers, Library of Congress; Delano, untitled notes on Iowa State College photo shoot, May 1942, lot 120, US Office of War Information, Overseas Picture Division, Written Records 1935–1948, microfilm, PP-LC. Writing Russell Lee, Stryker requested that he too "watch for and photograph things which dramatize the problem of Food and War—the idea of food as a way of life—as a part of the thing for which we are fighting"; see Stryker to Lee, January 26, 1942, Reel 6, Stryker microfilm, PP-LC.

84 Examples include "A class in chemistry in the lecture room of the chemistry building," No. 2758-D, "Anatomy lecture in veterinary school," No. 2672-D, and "Zoology class in the science building," No. 2654-D, all May 1942, Iowa State College series, PP-LC. FSA-OWI photographer John Vachon made a similar series of photographs for Stryker at the

University of Nebraska in this same month; see for instance "Jesse Younger studying in his room," May 1942, 3050-D, Lot 2 of 3, PP-LC.

85 Delano's career is explored in Delano, *Photographic Memories*; for a history of Iowa State College [now University] in this period—which by 1940 had the largest graduate enrollment of any land-grant college and was considered a "national school of science"—see Earle Dudley Ross, *A History of the Iowa State College of Agriculture and Mechanic Arts* (Ames, Iowa: The Iowa State College Press, 1942), 356. Rossiter used a photograph made of a classroom at the University of Michigan in the late nineteenth century—showing a black student integrated into a classroom split by gender—to explore social customs of this period; see Rossiter, *Women Scientists in America: Volume 1: Struggles and Strategies to 1940* (Baltimore, MD: Johns Hopkins University Press, 1982) 10, plate 1.

86 This point of view appears for instance in Anderson, *Roy Stryker*; and Stryker and Wood, *In this Proud Land*, esp. 16–17. While many of our colleagues in American Studies find persuasive the postmodernist critiques of Roland Barth, Susan Sonntag, and John Tagg (among others) that photographs are fundamentally unreliable sources, our experience from the history of recent science suggests this issue merits further discussion; see Roland Barthes and Stephen Heath, *Image. Music, Text* (New York: Hill and Wang, 1988); Susan Sonntag, *On Photography* (New York: Farrar, Straus and Giroux, 1977); and John Tagg, *The Burden of Representation: Essays on Photographies and Histories* (London: MacMillan Education, 1988). Cooper ("Interpreting National Identity," 270) provides support for our reinterpretation in arguing that "The theme common to both late war OWI materials and Cold War publications, I suspect, is US technological superiority"; her comments are also relevant to Roy Stryker's Standard Oil photo-documentary project, best explored in Steven W. Plattner, *Roy Stryker, U.S.A., 1943–1950: The Standard Oil (New Jersey) Photography Project* (Austin, TX: University of Texas Press, 1983).

87 Eric Sandeen, *Picturing an Exhibition: The Family of Man and 1950s America* (Albuquerque, NM: University of New Mexico Press, 1995).

88 See for instance "Mexican Wheat Shipment to India," 1966, "Collection [*sic*] of Corn on residual moisture prior to rainy season, Haxcala, Mexico, 1963–64," and "Appreciation of nitrogen and phosphorous often produce fantastic results in the Tropics," all 300D, Corn Improvement, RAC.

89 John Kasson, *Civilizing the Machine: Technology and Republican Values in America, 1770–1900* (New York: Grossman Publishers, 1976), 55–106.

90 Kenneth A. Osgood, "Form before Substance: Eisenhower's Commitment to Psychological Warfare and Negotiations with the Enemy," *Diplomatic History* 24, no. 3 (2000): 405–434. Sandeen is getting at this issue in *Picturing an Exhibition*, 99. We thank Nick Natanson and Eric Sandeen for discussions of this issue.

91 Histories of this expedition include John Giæver, *The White Desert: The Official Account of the Norwegian-British-Swedish Antarctic Expeditio*, trans. E.M. Huggard (London: Shatto & Windus, 1954); and G. De Q. Robin, "Norwegian-British-Swedish Antarctic Expedition, 1949–1952," *The Polar Record* 6, no. 45 (1953): 608–616. The expedition also sought to reinvigorate Scandinavian science after the Second World War and to reinforce Norwegian sovereignty claims in Antarctica; see Aant Elzinga, *Changing Trends in Antarctic Research, Environment & Assessment*, vol. 3 (Dordrecht; Boston, MA: Kluwer Academic Publishers, 1993), and Gerald S. Schatz and Antarctican Society, *Science, Technology, and Sovereignty in the Polar Regions* (Lexington, MA: Lexington Books, 1974).

92 Charles Swithinbank, *Foothold on Antarctica: The First International Expedition (1949–1952) through the Eyes of its Youngest Member* (Lewes, DE: Book Guild, 1999), quoted on 108, 226; Roberts "Antarctic, 1950–1951," MS 1308/7, SPRI, quoted on 103.

93 Some were published in Swithinbank's memoir, *Foothold on Antarctica* (cit. n. 92); a larger collection of his photographs are preserved at the Scott Polar Research Institute of the University of Cambridge [see Norwegian-British-Swedish Antarctic Expedition, 1949–1952, P53/24/1-10, SPRI].

94 See "Wintering party on the day that *Norsel* left us (20 February, 1950)," "Aftermath of a blizzard on the first Advance Base journey," and "My first sight of the mountains,"

reproduced in Swithinbank, *Foothold on Antarctica*, 68, 132, 133. [*Note added in proof:* reviewing this chapter, Swithinbank asserted, "*I* was the one who flew a Norwegian flag on my weasel throughout the expedition—recognizing the official British government position that Dronning Maud Land is Norwegian territory... Article 4 of the Antarctic Treaty accepts this. One good reason for flying flags was that they brightened up color photographs of an otherwise rather colorless snowy landscape" (Swithinbank to Doel, December 17, 2004, private communication). Oral historians have long known that photographs are invaluable for stimulating recall from participants, including those who, like Swithinbank, warn us that "one can learn a lot from photography but *be careful!*"]

95 Erving Goffman, *The Presentation of Self in Everyday Life* (Garden City, NY: Doubleday, 1959); see also Ronald J. Grele, "Movement without Aim: Methodological and Theoretical Problems in Oral History," in *The Oral History Reader*, ed. Robert Perks and Alistair Thomson (New York: Routledge, 1998), 38–52; we thank Daniel Alexandrov for discussions of this issue.

96 Arnold Thackray, *Chemistry in America, 1876–1976: Historical Indicators, Chemists and Chemistry* (Dordrecht and Boston, MA: Kluwer Academic Publishers, 1985) remains the best demographic guide to this discipline in America; see also John W. Servos, *Physical Chemistry from Ostwald to Pauling: The Making of a Science in America* (Princeton, NJ: Princeton University Press, 1990).

97 Untitled group photograph, Fall 1956, of Reynold C. Fuson, Carl Marvel, E. J. Corey, [Lenard?—handwriting unclear], J. C. Martin, and Ken Rinehart, Carl Marvel collection, Chemical Heritage Foundation, Philadelphia, PA. In an oral history interview on July 13, 1983, Marvel declared that twice-daily meetings at Farrell's substituted for faculty meetings; see Marvel Oral History Interview, p. 17, Chemical Heritage Foundation. In their history of DuPont (which drew scientific advisors from the University of Illinois), Hounshell and Smith get at this issue in discussing annual plays produced by Illinois chemistry faculty, but photographs in the Marvel collection are richly illuminating; see David A. Hounshell and John K. Smith, *Science and Corporate Strategy: Du Pont R&D, 1902–1980*, Studies in Economic History and Policy (Cambridge and New York: Cambridge University Press, 1988).

98 Smithsonian Secretary Alexander Wetmore and taxidermist Watson M. Perrygo preparing birds in El Valle, Cocle, Panama, 31 March, 1951, Wetmore image #4548, RU7006, B178, Album 1, Smithsonian Institution Archives. Naturalists who collect in the tropics must possess the craft skills to quickly dry and prepare their specimens to avoid rapid deterioration in the heat and humidity of the tropics. This image captures how scientists create a mini-laboratory in the midst of the tropics, with camp table, director's chairs, and scientific apparatus. Close examination reveals the many tools and supplies needed to prepare a bird specimen in the field. One in an extensive series, the image also shows the techniques of specimen preparation. The series of images also reveals Wetmore's close relationship with Watson M. Perrygo, his field assistant and surrogate son, as they travel and work side by side.

99 Thomas Cochran photograph of Congressmen Tom Downey and James Moody, Velikov, and former Congressman Bob Carr, September 1987, while visiting the Krasnoyarsk Radar site in Siberia as part of bilateral nuclear arms reductions talks; see William J. Broad, "Secrets of Krasnoyarsk—Soviet Eye on the Sky, Questions of Trust," *New York Times*, September 20, 1987, 3; John Krige and Kai-Henrik Barth, eds, *Global Power Knowledge: Science and Technology in International Affairs*, Osiris 21 (2006). We thank Kai-Henrik Barth for calling this photograph to our attention.

100 Tanya J. Levin and Ronald E. Doel, "The Lamont Doherty Earth Observatory Oral History Project: A Review of Preliminary Results," *Earth Sciences History* 19, no. 1 (2000): 26–32; on Kulp, see Mark Alan Kalthoff, "The New Evangelical Engagement with Science: The American Scientific Affiliation, Origin to 1963" (PhD dissertation, Indiana University, 1998); and J. Christopher Jolly, "Thresholds of Uncertainty: Science and Responsibility in the Fallout Controversy" (PhD dissertation, Oregon State University,

2003). In Laurence Lippsett, ed., *Lamont-Doherty Earth Observatory: Twelve Perspectives on the First Fifty Years. 1949–1999* (New York: Office of External Relations, LDEO, Columbia University, 1999) these scientist-authors, for the first time, addressed some of these issues in print. The ongoing oral history project may well have allowed senior Lamont scientists to write more openly about the social dynamics of their organization, not readily apparent in traditional archival documents.

101 Bubley, "A biology student at Wilson high school," No. 38649-D; "Chemistry students watching an experiment in class at Woodrow Wilson high school," 39378-D, both lot 944; "Chemistry student making notes on the apparatus used for an experiment at Woodrow Wilson high school," 39386-D, and "Chemistry students performing a class demonstration at Woodrow Wilson high school," 39406-D, both lot 942, all PP-LC.

102 Marjory Collins, unpublished manuscript autobiography, Cartons 1 and 2, Accession 90-M159, Radcliffe College Archives, Cambridge, Mass. See "Science class in a Negro high school, March 1942," 881-E, PP-LC.

103 Rosener, "Mrs. Marvel Barenkov, engineering aid, Joel Langjofer, chief engineering aid, employees of the US Geological Survey, Alaska Branch, and Dr. William Wrather, director of the Geological survey, adjusting templates to a base map manuscript," USGS, Washington, DC, June 1943, 31925-D, PP-LC. Photographs such as these raise new questions: what training did Wrather's assistants have, and to what extent did they contribute to research and mapping? Rosener's photographs suggest questions, not answers. Existing histories of the USGS do not address such issues in depth; see for example Mary C. Rabbitt, *Minerals, Lands. and Geology for the Common Defense and General Welfare: A History of Public Lands, Federal Science and Mapping Policy. and Development of Mineral Resources in the United States* (Washington, DC: US Geological Survey, 1979).

104 Andrea Fisher and National Museum of Photography Film and Television (Great Britain), *Let Us Now Praise Famous Women: Women Photographers for the U.S. Government, 1935 to 1944: Esther Bubley, Marjory Collins, Pauline Ehrlich, Dorothea Lange, Martha McMillan Roberts, Marion Post Wolcott, Ann Rosener, Louise Rosskam.* (London and New York: Pandora Press, 1987), 3; see also Cooper, "Interpreting National Identity."

105 Fisher, *Praise Famous Women*, 91, 94. Fisher's work nonetheless remains the most comprehensive review of New Deal and the Second World War-era women photographers in America, and limited biographical information exists for Rosener, Collins, and other women photographers in this period.

Part VI

New voices

Neglected and novel perspectives

13 What we still do not know about South–North technoscientific exchange

North-centrism, scientific diffusion, and the social studies of science

Alexis De Greiff A. and Mauricio Nieto Olarte

Introduction

In July 2001, the 21st Congress of History of Science took place in Mexico City. The title of this meeting was provocative and relevant: science and cultural diversity. It was particularly significant that, for the first time, this traditional gathering took place in Latin America. Mexico—cradle to some of the most complex and interesting American cultures, a vivid example of every excess of imperialism, and a proud example of the survival of an autochthonous culture—seemed an appropriate place to discuss cultural diversity. Fifty-two different countries and hundreds of historians attended the meeting. The opening lecture was delivered in the spectacular Palacio de Bellas Artes, whose enormous murals conveyed a clear sense of national pride. With the aid of simultaneous translation to various languages, we listened to the opening protocol remarks, all of which predicted that this was going to be a very special occasion.

The opening lecture was delivered by a philosopher and historian of science, the Cairo-born Professor Roshdi Rashed. His Southern origin was congruent with the spirit of the Congress and hinted that from the outset we were taking part in a stimulating renewal. Rashed, who had taught at the University of Mansoura in Egypt, highlighted the fact, "It is the first colloquium held in a country of ancient culture which is neither Mediterranean nor Asiatic" and "It is also the first colloquium on the history of science which is not hosted by an industrial country of the North."[1]

Yet soon we were dismayed. Instead of showing us the richness of questions about cultural diversity and science, Professor Rashed seemed to go out of his way to discourage anyone interested in such a problem. Our distinguished lecturer indeed expressed his concern about the diversity, "not to say dispersion," of the discipline of history of science, and his fear of the "flourishing temptation to extend social history to the conceptual tradition."[2] Throughout his presentation, he clearly stated what he conceived was history of science's true path: an internal history of ideas. According to Rashed, not only was it important to establish the difference between external social elements and genuinely scientific ones, it was also necessary "to ask ourselves what distinguishes it [science] from all other production of cultural works."[3] For Professor Rashed, the diffusion of knowledge is different from its production.

He went even further, telling us that external factors "may explain controversies when the facts are imperfectly established and proofs not rigorously carried out."[4]

We begin with this episode because it illustrates for us of the kind of obstacles we face in writing the history of South–North technoscientific exchange. Such claims made in Rashed's opening lecture not only discourage debates about science and cultural diversity but also create barriers that limit understanding science and technology as political practices. Interestingly, Professor Rashed was right about one thing: social studies of science, far from being a fully constituted discipline, are fields that still need to be shaped and developed. Indeed, almost a century after Ludwig Fleck published his celebrated book, and five decades after Thomas Kuhn started a new phase in the historical studies of science and technology—decades after various historians, sociologists, and philosophers showed once and again (through case studies) the historic and social roots of epistemological problems—and now that the writings of David Bloor, Steve Shapin, Michel Callon, Bruno Latour amongst others seem to be necessary readings for historians of science—even after all this, we can say that the history of contemporary science, technology, and medicine still has a long road to travel, and, possibly, the most interesting parts of the journey still lie ahead.[5]

Indeed, a comprehensive history of these fields is still missing. The origins of such a history can be traced back to the first stirrings of public and political concern about imperial and colonial interests, gender, race, the consequences of military technology, the Cold War, and our imperiled environment. The ideological foundations of this historiography include Marxism and the counterculture movement. Key contributors include Paul Forman in the United States, Hillary and Steven Rose in the United Kingdom, and Marcello Cini in Italy.[6] For numerous scholars of recent science, technology, and society writing today, the "Society for Social Responsibility in Science" and the "Science for the People" became spaces for political action and theoretical reflection about the role of science in contemporary society, especially in the Vietnam War. The heirs of J. D. Bernal's political commitment were apparent. Their interest in science and technology was stimulated by their wider concern about global politics. Indeed, their aim was to unveil the relation between capitalism, South–North asymmetrical relations, and science and technology. After all, the 1968 movement was inoculated by anti-imperialism conscience. Sadly, the political and intellectual commitment with South–North exchange almost disappeared from the professional agendas of historians and sociologists of science. Books on the topic are rare; there are very few courses that address the issue; scholars from the Third World are missing from most editorial boards.

How to remedy this situation? This essay introduces some of the central questions and topics in the study of South–North technoscientific exchange. Terms such as North, South, East, West, and technoscience were coined as political notions after the Second World War[7] Hence, South–North technoscientific exchange stands here for the international relations that entailed technoscience as a constituent part during the second half of the twentieth century. We argue that paying serious attention to analyzing South–North technoscientific exchange is essential to understand technoscience and society relations. Technoscience in the North does not develop independently of the South–North transactions. Furthermore, shifting our attention to these issues illuminates relationships and themes at the heart of the history of modern science, technology, and medicine.

In this chapter, we seek to encourage scholars to study the kind of problems, gaps, conceptual difficulties, and political issues related to South–North science and technology studies. We do not attempt to cover all potentially relevant topics. Rather, we offer examples that allow us to illustrate that—to talk about science and politics—we neither need to shift levels of analysis nor change the subject. In the first section, we explore how critics of traditional colonial historiography and postcolonial studies have contributed a critical view of science and technology. Then we offer a brief review of three different literatures: development, the Green Revolution, and the Cold War. All constitute powerful illustrations of the unbreakable bounds between science and politics. Simply put, they are good examples of science as politics (Figure 13.1).

We have chosen to review in some detail these subjects, because they are interesting and actually crucial fields of research.[8] But there are many others. Science and empire, voyages of discovery, natural history, geography, medicine, and anthropology are, among many others, fields of historical and sociological research where the relation between scientific practice and domination have been apparent.[9]

The topics reviewed here should help illuminate the importance of some of the arguments held by recent sociology of science. Despite all evidence, some historians of science still try to explain science as a human product different and independent from other cultural practices. This chapter aims to show that to reach a thorough understanding of South–North scientific exchange, we must consider seriously the

The World turned Upside Down

Figure 13.1 The missing perspective: south–north relations in the history of contemporary science, technology, and medicine.

Source: Hobo-Dyer Cylindrical Equal Area Projection, south on top, ©2004, www.odt.org (reproduced with permission).

fundamental lessons of sociology: that scientific knowledge and technology are inseparable from the exercise of authority, control, and domination.

Eurocentrism, postcoloniality, and the diffusion of science and technology

George Basalla's model of the diffusion of science in three stages—namely, a period in which the "nonscientific society provides a source of European science" (phase 1), followed by a period of "colonial science" (phase 2), and a third phase of "transplantation with a struggle to achieve an independent scientific tradition"—has been sufficiently criticized.[10] Some commentators have pointed out that Basalla's three stages might offer an adequate theoretical framework to discuss scientific development in countries such as the United States, Russia, or Japan, but that his proposals are insufficient to explain the history of science in nonindustrialized countries. We are not even certain that a situation like the one described by Basalla's third stage ever took place in less-developed countries. But if it did, such "national science" does not guarantee that the scientific practices of a Third World country are truly independent from foreign control.[11]

However, let us consider the assumptions that underlie this type of diffusion model. We could summarize them in the idea of modern science as a finished product that diffused without major "distortions" from a Center, Europe. One of the main contributions of the sociology of scientific knowledge is removing the traditional distinction between the contexts of discovery and the contexts of justification, showing that the production and the diffusion of knowledge are simultaneous processes. If we acknowledge that the creation or birth of what we call Western science is inseparable from its expansion, the study of its diffusion attains a fundamental meaning, one that is very different from the marginal and accessory place it usually occupies.

This means that the expansion of Western science cannot be explained in epistemological terms or by the rigor of its methods; on the contrary, its status is a consequence of its expansion. Criticism of the notion of a unique and superior Western science has also been advanced by intellectual traditions different from the social studies of science and technology. For postcolonial and gender studies are useful to examine the assumptions of a historiography that has centered its attention on Europe first and the United States later. In contrast, the historiography of technology still centers on innovation, rather than in the uses,[12] imposing serious limitations to the attempts to escape North-centrism.

Postcolonial historiography, which overlaps several questions raised by feminists, challenges the assumption that European expansion must be explained by the intrinsic superiority of such civilization. Western culture became meaningful as an identity marker when it confronted other cultures. Postcolonial studies have awakened interest in the explanation of the causal relationships between European expansion and the creation of modern science in Europe, drawing special attention to the notion of development. The expansion of scientific practices did not always have positive effects in the "peripheries." The "civilizatory" process has been repeatedly pointed out as a source of emancipation, being described as a cause of political revolutions. Local cultural traditions are depicted as opposing development and as obstacles to progress. Conversely, some have argued that the diffusion of scientific practices as geography, natural history, or medicine have been powerful mechanisms for establishing order and efficient ways of control and domination.[13] Such processes are also important inasmuch the idea of the unity of Western science entails the elimination of other ways of knowing. Critical examination of the processes of scientific diffusion opens a whole field for historical and political meditation in a constellation of issues formerly ignored by colonial historians devoted to legal, economic, and cultural aspects but oblivious to science. In this sense, Sandra Harding points out ways in which postcolonial studies of science and technology, a field that in fact still does not exist, might be conducted:

> the relationship between scientific and technological change and projects of European-American empire, anti-Eurocentric accounts of other cultures' scientific and technological traditions, and the implications of the now obvious failures of the North's attempts to increase the standard of living in the South— the failure of "development."[14]

Western science also has contributed to the deterioration of "others," namely women, ethnic minorities, nature, and Third World citizens.[15] The implantation of Western technoscience may also contribute to depriving natives of control over their own resources, taking power away from them.

Power, of course, is exercised through concrete social practices such as natural history, taxonomy, the manufacturing of maps, and sky and marine charts, searching for the cure for malaria, or building a nuclear plant. These practices constitute an active exercise of power and their dissemination must be seen as an attempt to gain control of new spaces. Our task—as Roy Macleod suggests—is to study science not within imperial history but as imperial history.[16]

Finally, there is an important group of scholars who, following the postcolonial movement, argue that the Third World can offer positive substitutes to Western science. By examining these ways of knowing (which they refuse to call "alternative"), they explore the possibilities of indigenous sciences.[17] The study of local knowledge may provide elements for a dialogue between the "experts in development" and the communities "to be developed," promoting cultural diversity. Social studies of science would benefit a great deal by exploring trading zones with studies on traditional ecological knowledge and environmental history.[18]

At this point, a few clarifications may be necessary. Our proposed aim—to escape Eurocentric and Americano-centric accounts of history, and to make other voices visible and perceptible—cannot be used to deny the importance of Europe or the United States in modern history or to neglect the central role of Western science. Instead, we must explain, historically, socially, culturally, and politically, its success and the consequences of such success. Also, the idea is not just to abandon our interest in Western science in order to rescue local knowledge. As Arif Dirlik has pointed out, "The distinguishing feature of Eurocentrism is not its exclusiveness, which is common to all ethnocentrisms, but rather the reverse: its inclusiveness."[19] Eurocentrism is not the result of ignoring others but rather the consequence of organizing the knowledge of the world, including other ways of knowing, into one single systematic whole. We must help to deconstruct this state of affairs. The temptation to identify ourselves with the excluded and to become the spokesmen of the subordinates brings about the high risk of presupposing that we are privileged and legitimate translators and spokesmen of the "other" and, therefore, of ratifying Western culture as the fundamental culture.[20]

The race for Third World hearts and minds: the seduction of development

During the 1950s and 1960s, large portions of the world's population strove to build a national identity within the context of international tension, national class struggles, and ideological debate. Although the state of revolt is a better-known face of the Third World, it is just one aspect of these nations' history. As Arturo Escobar and others point out, this period was marked by the construction of the "development discourse" as a new form of domination over the new nations and, more generally, over the so-called developing countries.[21] International institutions, notably the World Bank and the United Nations technical agencies, played a central role in nurturing the discourse and practice of development programs. The effect of these programs has meant the creation of ever-greater gaps between rich and poor countries, the widening of the internal social, cultural, and economic contrasts, the degradation of the environment, and other social ills. The workings of public rhetoric and practice of development are still largely unexplored in current historiography, although some efforts have been made to investigate the phenomenon in studies of discrimination through literary analysis, as well as in anthropological works on modernization and resistance, especially in Asia and Africa, and Latin America.[22] Curiously, even these works fail to tackle the issue of science and development to the extent that some historians have done in studying the close link between colonial domination, science, and technology. An important lesson to learn from these works is the necessity of shifting the object of study "from the people to be 'developed' to the institutional apparatus that is doing the 'developing'."[23] In other words, we need to abandon the idea that development and modernity are "unfinished" projects everywhere but in Western Europe and North America due to the cultural and/or structural obstacles of the peoples to be "developed." We need to start looking at the institutions for development as instruments of control and domination and realize that scientific programs are also political programs.

Escobar focuses on development as a practice—that is, the establishment and operation of institutions concerned with the implementation of programs, mobilization of resources, and creation of new spaces of representation based on the idea of "development." By contrast, Gilbert Rist, from a slightly different perspective, analyzes the history of the concept and how it shaped the views of world history during the twentieth century. He points out that "development" is a central element of the religion of modernity. It is therefore a set of beliefs deeply rooted in our conception of social, political, and economic relations in both national and international arenas. Development is a collective certainty, a dogmatic truth that is not debatable and, therefore, becomes a coercive force. "The action determined by the belief is obligatory," Rist notes, "and does not rest upon any choice."[24] In the name of development, mistakes are made and people are aware of them. In this sense the parallel with religion helps to understand the phenomenon. Religious believers realize the deep contradictions between the doctrine prescribed in the holy books and the practice of ecclesiastical institutions. However, they are tolerant. Thus, Rist defines "development" as "a belief and a series of practices which form a single whole in spite of contradictions between them."[25]

What is the role of science and technology in development theories? As John Agnew points out, all theories of development and social change contain within them position on the role and impact of science and technology on development.[26] Surprisingly, the dialogue between science and technology studies (including the history of contemporary science, technology, and medicine) and critical analyses of development is poor. For instance, Escobar is concerned with the negative impact caused by producing technology in one place to be "applied" in another. He thus advocates for "a policy of technological research and development in support of autono-mous peasant production system."[27] According to Escobar, the only way to tackle effectively the problem of poverty through useful knowledge is with peasants' self-understanding—then proceed to build a system of communication involving peasants, institutions, and researchers.

This is an all too familiar prescription to historians and sociologists of science and technology, who have insisted that science and technology are essentially local practices. While historians of recent science and technology have been virtually oblivious to South–North exchange for development, leading scholars of development continue to treat science and technology as black boxes. Rist, for instance, explains, "Belief is so made to tolerate contradictions—especially as, unlike scientific theories, it cannot be refuted. This is why science changes faster than belief, which has immunity against anything that might place it in question."[28] This pre-Kuhnian view of science contrasts with empirical studies that show scientists as conservative professionals, committed to their local research traditions. One should ask about the articulation between faith in development and faith in certain scientific theories and technical innovations. Indeed, Escobar has observed that science and technology act not just as "promises" (a word reminiscent of Kuhn's paradigm) but "makers of civilization." But in what way? Michael Adas' work—on the way science and technology acted as ideological instruments to establish colonial power—ends in the Great War.[29] However, the role of science and technology in international relations has intensified since then. After the Second World War international development programs translated into technical assistance and scientific manpower building. How new forms of domination (development discourse) worked back-to-back with science and technology is a subject that needs further exploration. While the workings of social scientists in the diagnoses and construction of social representations of the "developing" world are studied in detail, natural scientists are missing actors in most works. Several questions require study. For instance, what roles did technical personnel, engineers, scientific administrators, and scientists play in scientific and technological projects for development? What can we learn about their local negotiations and the use of local cultural resources to gain epistemological supremacy and, thus, access to resources? What image of science and technology did these agents try to establish? What kind of hybrid image resulted from that effort once joined with local knowledge?

There is an interesting asymmetry in the literature on technoscientific international relations. Whereas the works concerned with international relations between industrialized countries talk of "scientific exchange," the literature on South–North exchange is located in studies of "scientific and technical cooperation," namely analyses of assistance programs for development. It is as though scientific practices, not explicitly tied to development projects in the South, were marginal to political and scientific international relations. Such distinction between exchange and cooperation must be understood as a historical product in itself. Very little has been studied about scientific excellence in the South, to use Marcos Cueto's phrase, and horizontal exchanges with the North.[30] The available resources, professional practices, instruments, and impact are radically different. However, as in development theories, these differences are often perceived as defects, manifestations that southern nations are a step behind the "developed." Indeed, a sort of "sociology of obstacles" is still common in works on science, technology, and economic development. In his seminal book *The Social Function of Science*, J. D. Bernal stressed the constraints imposed by local cultures in Latin America, India, and the Islamic World.[31] Similarly, in *The Two Cultures* (1959), C. P. Snow called for sending an army of scientists and linguists to prevent the Third World fall into communist hands,[32] whereas Ziman and Moravcsik assumed and concluded that Paradisia, an imaginary country in the South, will never progress unless the Western model of science, technology, and its institutions is effectively transferred.[33]

There are, nonetheless, pockets of critical reflection that deserve attention. Critical voices, especially postcolonial studies, are heirs of *dependency* theory. The process of institutionalization and professionalization of science in Latin America preceded and was instrumental in the establishment of national scientific policies in the 1960s (and 1950s in the case of Argentina and Brazil).[34] Sociological and economic studies stimulated by the dependency theorists provided thoughtful analyses of the problems of science and technology in peripheral countries.[35] Geologist Amílcar Herrera, in his influential book *Ciencia y Política en América Latina*, developed a sociohistorical analysis of scientific research in Latin America. He criticized the contradiction between what he called the "explicit" and the "implicit" science policies operating in the South. The former, the rhetoric of science for development, was the façade covering the local elite's lack of commitment to national development, a disinterest that characterized the implicit policy.[36] Thus dependency theory, which can be placed among the "dissident voices" of development, led to a debate that was particularly fruitful in Latin America.[37] The central thesis was that "underdevelopment" in the periphery was inseparable from capitalist development in the metropolis—here, the metropolis referred to the Western imperialist (and, after decolonization, neoimperialist) powers and the periphery to the colonies (and the Third World). In their view, local elites acted as agents of neocolonialism and underdevelopment. Third World economic and cultural dependency and social crisis were due to collusion between external actors, the colonial powers, and internal ones, the local elites. By extension, scientific, and technological development in peripheral economies was severely limited by external interests defended by the local elite.[38]

While these studies have been praised for stimulating a critical discussion of capitalist development at the periphery, they have equally been criticized for their "ideological" bias and lack of empirical evidence.[39] Dependency theorists have also been criticized for the lack of empirical studies that show in detail the dependency character of global knowledge. Sociologists of science and technology who became important science administrators, such as Francisco Sagasti, were enthusiastic supporters of the North American "systems approach" but stayed close to dependency theory. Even his works "remained a largely formal, abstract, reductionism analysis of science and technology development which was difficult to translate into action."[40] Reacting to these criticisms, the heirs of the dependency theories who became engaged in science studies have been investigating case studies, for instance in relation to the Green Revolution.

Dependency theory was perhaps the most original contribution to science studies by and about Third World scientists. Several intellectuals in this region rebelled against the assertiveness of the North and demonstrated the possibility of offering alternative solutions to the problem of the role of science in the Third World.[41] These studies marked a turning point in the history of ideas in Latin America in particular and the Third World in general.

Dependency theory also produced a lively debate within the scientific communities, particularly on the question of the social use of "pure" science in *developing* contexts. During the 1970s, politicians and administrators, especially in countries under military regimes such as Argentina, eager to cut funds for research in universities and to close or reduce other research institutions, strategically invoked the irrelevance of pure science for Third World development. In several Third World countries, resources for research in pure science, already scarce, were radically reduced. The research that did continue was redirected toward projects with "a social utility." The industrialized countries shared this view, discouraging scientific research in the South as well as international cooperation in subjects "not directly related to development." Since the 1950s, science has been considered a luxury that Third World nations cannot afford. For instance, negotiations to create an international center for the promotion of theoretical physics in the Third World met open hostility from delegations of virtually all industrialized countries.[42] Nonetheless, it is worth emphasizing that, in spite of such attitudes—and the concomitant difficulties due to the lack of resources—non-applied research is carried out in Third World countries. Yet, the bulk of the literature is concerned with technology transfer, while science is seldom mentioned.[43]

In our discussion thus far, we referred to scientific institutions. National research centers in the Third World developed interesting and complex intellectual, political, technical, and economic links with institutions in the North. Although the number of works is growing, we are far from having a good map of these institutions and their mutual relations.[44] We need to learn more about the role of academies and scientific societies in the South in the consolidation of local elites who used the science for development discourse and becoming local agents of aid programs offered by the industrialized countries. Political and scientific elites in the Third World often received training in Europe and the United States. However, detailed investigations of the globalization of knowledge after the Second World War are scarce. Although it is obvious for some, we must recall that institutional histories cannot be studied independently from cognitive aspects, for they provide the resources necessary for understanding research and pedagogical practices. In particular, by scrutinizing centers that promote South–North cooperation, we are able to increase the understanding of the role of scientific institutions in the construction of development programs and, concomitantly, learn about the global distribution of knowledge. Science, technology, and training programs were enthusiastically supported by philanthropic foundations. Nevertheless, the existing studies show also the enormous diversity of motivations, mechanisms, and strategies deployed by applicants and foundations alike. Indeed, if we really want to learn about the patterns of funding by American philanthropic foundations, and the kind of knowledge they eagerly promoted, we must focus on their activities in the Third World, where these bodies invested more than twice their total budget for institutions in Europe.[45]

At a different level, we have scientific disciplines. We need to remind ourselves that development discourses and practices produced varying images of science. For many years, the image of modern science and progress was represented by the theoretical physicist rather than by the agronomist. The diffusion of such representations and certain practices associated to them were closely related to the models of development and the role ascribed to technoscience. Of course, local cultures and traditions influenced such images.[46] Thus we can ask: what is the relation between such ideas about science and technology and modernity projects in different cultural settings in the South? Financial support was invariably conditioned to demonstrate that the projects contributed to development. Hence, some areas of research became more "pertinent" than others. Why did governments support certain scientific projects, such as theoretical physics or corrosion, and what was expected from them? What did scientists do to fulfill those expectations or at least give that impression? Analyses of the different "discursive strategies" employed by scientists in their countries and abroad would shed light on the problem of establishing scientific disciplines in specific cultural environments, showing that development was—like technoscience—essentially a cultural phenomenon. Finally, one must ask whether and how development requirements shaped research. Some areas of research, in both scientific and industrial settings, that came from the North had to be adapted to the South, as cultural traditions, infrastructural facilities, human and natural resources, and so forth were different there. These processes of adaptation are in effect "new uses" of material and conceptual artifacts. The study of such uses may open a very different picture of innovation in the South.[47]

The old and new Green "Revolutions"

Food, poverty, and—since the 1970s—the environment have been the central issues of most development programs. The Green Revolution is perhaps one of the most discussed cases in the literature on science and development. As one environmental historian has explained, it was "a technical and managerial package exported from the First World to the Third beginning in the 1940s but making its major impact in the 1960s and 1970s."[48] In 1970, the American botanist Norman Borlaug, Director of the Division for Wheat Cultivation at the Centro Internacional del Mejoramiento de Maíz y Trigo in Mexico, was awarded the Nobel Peace Prize. He was the main promoter of a worldwide agriculture development program based on the genetic manipulation of seeds to improve production—the Green Revolution. The program was introduced in several Asian countries in 1965. Five years later, the program covered 10 million hectares of cultivated area. The program was promoted and supported by several institutions from the United States, France, Canada, Germany, Brazil, India, Nigeria, and others that constituted the Consultative Group of International Agricultural Research. Philanthropic foundations, such as Rockefeller and Ford, participated decisively in the program.

The actual impact of the Green Revolution has stimulated major debates. On the one hand, its effects on national production of wheat and rice became apparent. A number of countries in South America and Asia achieved record harvests. By the end of the 1970s, India became self-sufficient in wheat and rice, tripling its wheat production between 1961 and 1980. This is the positive side of the Green Revolution, according to its apologists.[49] On the other hand, since the 1970s, the Green Revolution has been subjected to severe criticisms. The main one was that for the program to be profitable, it was necessary to employ rich soils, optimal irrigation, intensive use of fertilizers, and chemical pesticides. In addition, by the early 1980s, environmentalists found that intensive fertilization stimulated by the Green Revolution led to eutrophication of rivers and lakes. Although some countries increased their agricultural productions, other regions with little water and lack of credit markets—such as sub-Saharan Africa—suffered. Even in those countries where it was successful, some authors found flaws. J. K. Bajaj argues that rather than improving the agricultural system, it devastated its productivity and increased hunger. Economic dependence increased, for the reduction in imported cereals was offset by imports of fertilizers and knowledge dependence on "experts." Thus Bajaj questions the assertion that the Green Revolution made India self-reliant in agricultural production.[50] From the environmental point of view, the speed and scale of dissemination of new breeds made the Green Revolution the largest crop transfer in world history, reducing dramatically biodiversity.[51] Socially, the Green Revolution widened the gap, favoring large landholders who had access to Western education. The drop in the prices of wheat displaced small farmers, which led to the development of urban slums.

These studies of the political dimension of the Green Revolution reveal important aspects of twentieth-century tensions. It created a promised land of efficient export-oriented agriculture, which in turn would lead to rapid industrialization—the key to development, according to the economic theories of the day. The Green Revolution was the epitome of a technoscientific solution, an alternative to social revolution. In regions close to the communist border, such as Turkey and Korea, its introduction was a result of the American fear of the spread of Chinese communism.[52] However, the genius of this revolution was presenting itself as apolitical. Analysts such as

Edmund Oasa, who was commissioned by the Consultative Group to make an evaluation of the program, concluded that class lines and conflicts worsened as a result of the "inherent contradictions in CG [Consultative Group] policies and the *politically neutral stance that the Group has adopted, at least superficially.*"[53] Promoters of the Green Revolution assumed that a technical solution could solve deep social problems such as land distribution and the exploitation of the work force.[54] The Group isolated itself from political debates, instead of incorporating them as a crucial element of the problem. Vandana Shiva's argument is even more radical. Punjab, on the border between Pakistan and India, was supposedly the Green Revolution's major success. However, the socioeconomic conditions of this region are deplorable, and violence continues to be endemic. This tragedy is presented as an endogenous situation, caused by ethnic conflict between religious groups, and therefore independent of the Green Revolution. Shiva offers an alternative interpretation: "it traces aspects of the conflicts and violence in contemporary Punjab to the ecological and political demands of the Green Revolution as a scientific experiment in development and agricultural transformation." Moreover, Shiva brilliantly demonstrates how science "was offered as a 'miracle' recipe for prosperity. But when discontent and new scarcities emerged, science was delinked from economic processes."[55] Such power of science to vanish from the political scene when things go wrong cements the faith in technoscience as the engine of progress. It erases the contradictions between theory and practice of development. More case studies on South–North exchange programs will be helpful to understand issues on science and democracy today.

Genetic engineering (GE) and its products, Genetic Modified Objects (GMOs), is widely regarded as yet another the new technological promise to alleviate hunger in the Third World. Furthermore, this technology involves not only transfer of plants, knowledge, techniques, and processes from North to South, but it also looks for genes to manipulate and "improve." Thus, the GE firms require germplasms from regions with vast genetic resources, such as the Amazon forest. In other words, the relation is bidirectional: GMOs are moved from the North to the South, while genes are drained in the opposite direction. As far as technology transfer to the South is concerned, some have argued that the Green Revolution has served as a point of reference to identify the issues at stake.[56] Indeed, GE cannot be understood without a deep analysis of the Green Revolution. Hitherto, the Green Revolution has been studied mainly in the Indian case. But GE companies have interests in other Third World countries. Hence, we need to learn more about the process of inception and the impact of the Green Revolution in other parts of the world. In the Amazon region, for instance, the cases of Colombia, Peru, and Ecuador are virtually unexplored. What lesson did GE firms extract from the Green Revolution? This is an important question. The conclusions come from the critics. We need to know more about those who consider it a success and, therefore, justify GE as an improved version of that first experiment. In terms of the exploitation of Third World genetic resources, there are urgent questions to tackle. As we mentioned earlier, the appropriation of natural resources was a central element of imperialist policies in the late eighteenth century. What kind of technoscientific practices may or may not lead to domination relations? For instance, much analysis, discussion, and debate are needed regarding access to intellectual property and patent regulations. Compared to the development years, the center of power has shifted to the private sector. What are the implications of the leading role of corporate powers, especially in those regions where the state has been endemically weak? International institutes of research that participated in the Green Revolution are engaged in genetic research in associations with partners in the North. For instance, Lawrence Surendra argues that the International Rice Research Institute in the Philippines has been instrumental in the "gene drain" from the South to the North. This trend requires serious attention, analysis, and action.

Big science and the Cold/Hot War: a South–North perspective

The Cold War and development ideologies, programs, and discourses overlapped. Furthermore, "development" was an instrument of domination and a constitutive factor of the Cold War. In the Third World this phenomenon was particularly evident, for internationalism became a powerful ideological tool to win the people's hearts and minds. However, the final objective was not winning their hearts and minds but the political control of geopolitically strategic territories and resources. Yet the twentieth-century historiography of international relations sees the Cold War as an East–West confrontation, while examines the South–North relations in terms of their economic exchanges, despite the fact that the War in the South was not cold. The Cold War burned entire alternative political projects (such as Allende's in Chile) and produced millions of deaths and refugees (Guatemala, Vietnam, or Congo to mention just a few examples).[57]

It is perhaps not surprising that the literature on scientific internationalism has focused on the first liberal globalization period (1870–1914) and the crisis generated by the Great War and the interwar years.[58] It is more difficult to explain why the historiography of science has shown such little interest in these issues in relation to political attitudes after 1940. The rhetoric of scientific internationalism took a new and perhaps more dramatic turn after the war because of the increasing importance attributed to science and technology, catalyzed by the threat of a nuclear conflict. Hence the postcolonial period offers an excellent and under-used context in which the phenomenon of international science and the ideology of scientific internationalism can be studied.

Indeed, as some studies suggest, in areas such as nuclear armament and space research after 1957, the role of scientists in the formulation of foreign policy was crucial.[59] One of the most interesting aspects of the Cold War was the establishment of international scientific forums. Creating environments for scientific exchange between the superpowers was never a trivial problem either for foreign policy makers or for scientific advisors concerned with issues of national security.[60] But what about organizations for South–North and South–South exchanges? Among the numerous questions one could ask: what were the positions of different governments toward initiatives sponsored by a neutral organization such as the United Nations? What kind of political dividends or costs did they see in this kind of initiatives? To what extent did those scientists who acted as advisers to international forums reflect the interests of their own delegations? As the Third World is not a monolithic unit, we will find very different stances before the creation and establishment of those organizations. The case of the International Centre for Theoretical Physics (ICTP) is a case in point. The negotiations to create the ICTP generated tensions between Third World countries on the one hand and Western and Communist countries on the other. After a few years, the Centre had been visited by scientists from virtually all Third World countries who came for training for research, but political and economic troubles persisted throughout its history.[61] Analysis of similar cases would allow us to investigate in detail different conceptions of the role ascribed to science and technology as an instrument of ideological penetration.

For scholars concerned with questions on technoscience in the South, it is always disappointing how little research has been done on Cold War technoscience outside the United States and Western Europe. The literature on science, technology, and the Cold War concentrates on the production of knowledge and technological goods and, in particular, how the Cold War "distorted" science and technology.[62] Big science occupies a privileged place in the social studies of science, focusing on the industrial–academic–military complex in the United States. In short, nuclear weapons, the space race, and high-technology military gadgets dominate the literature. Several scholars have argued that Big Science is a phenomenon that transcends the obvious question of scale. It affected the way in which scientists interacted with power, the public image of power, interactions between scientists, engineers, technicians, and administrators, and pedagogical techniques.[63] Physics institutions in the United States and Europe that possessed no big instruments became laboratories of theoretical physics, developing new techniques, concepts, and theoretical technologies.[64]

Big Science had, in fact, a significant impact in the image and practice of science and technology in several Third World countries. The most immediate example is the nuclear programs in some Third World countries. The public's astonishment before the nuclear capability of nations such as Pakistan, Iraq, or Iran is a consequence of how little attention was paid to nuclear research programs that started some thirty years ago with the active assistance of countries such as the United States, Canada, France, and the United Kingdom.[65] The establishment of a regional nuclear hegemony was certainly a motivation, though not the only one. In his short but sharp book, Itty Abraham suggests that the Indian program was a modern fetish that served to consolidate the state. The Indian project was, he argues, a strategy that must be seen in the frame of postcolonial culture. He also shows how the Indian Atomic Energy Commission was able enough to negotiate simultaneously with nuclear providers in the North in order to produce an atomic explosion *on Indian soil.*[66] Similar negotiations took place in other countries interested in building an atomic arsenal, such as Iraq and Israel. Of course, nuclear diplomacies varied enormously from country to country. However, those states engaged in building nuclear capacities, either for peaceful or other uses, such as India, Pakistan, Argentina, Brazil, Spain, Iraq, and Israel, adopted an active search for new providers in the North.

Those cases have gained deserved attention in the past few years, but there is a universe to explore in this direction. In the mid-1950s, framed in the "Atoms for

Peace" initiative, the United States started an atomic policy toward some Third World countries. The logic of this assistance followed the Cold War geopolitical interests. Atomic energy commissions were set up in virtually all nations, waiting for the arrival of the promised technology. The United States donated several small research reactors. Somewhat reluctantly, political elites in those countries who had showed no interest in developing nuclear capability—Colombia and Paraguay, for instance—accepted the American "gift" because of the symbolic meaning of the nuclear dream. It did not represent the development of local knowledge, but an imported modernity. Thus, these nations became passive recipients of an unexpected, and perhaps useless, artifact. In contrast to India or Pakistan, they did not look for other exchanges. From the American perspective, the reactor was a political instrument to compel governments to sign bilateral treaties with that government. Meanwhile, scientists, who saw it as an opportunity for the institutionalization of physics, developed the necessary skills, bringing into the debate human and nonhuman allies—such as the reactor itself—in order to break the skepticism.[67] Through an analysis of this complex web of interests and negotiations, one could learn about the role of Third World scientists in diplomatic exchanges, the relation of science and technology and the militaries in the Third World,[68] the intertwining of Cold War science and the development ideology, and industrialized countries' criteria for collaborating with certain regimes rather than others. In relation to the latter question, a comparative study would answer a crucial question: did countries that exported nuclear technology to the Third World have a coherent policy or simply respond to the local nuclear companies' interests? Who actually benefited from programs such as "Atoms for Peace"? One scholar has recently suggested that the current view that the United States was successful in using its technology as a political instrument to prevent nuclear proliferation in the Third World is superficial. He demonstrates that the United States pumped resources to countries that never planned to develop nuclear weapons, while those engaged in such projects moved in various international circuits simultaneously. In contrast, the scientific communities without any nuclear infrastructure learned how to use the American program for their own interests.[69]

The participation of the South in Big Science projects is another unexplored issue. Reconfigured research programs in the North shaped the interests and practices of technoscience in the South, not least due to the numerous physicists trained in the North. At their return, these scientists struggled to set up research groups in order to continue their participation in big experiments. International collaborations are imbued in cultural exchanges and multiple tensions, where national stereotypes, class, ethnicity, and gender play a role.[70] In other words, the dynamics of such exchanges are also driven by power struggles. Authors like Andrew Pickering argue that the ability of groups to adapt to different contexts is constrained by theoretical, experimental, and instrumental expertise as well as career strategies. He emphasizes the microsociology of scientific research carried out in elite institutions, forgetting "marginal" players.[71] We stress that access to symbolic capital is determined by social structures. Therefore, the geo-institutional location of the different groups seems crucial to participate effectively in what he has called "opportunism in context";

epistemological contexts depend on social contexts. Peter Galison teaches us that in order to separate noise from signal—the goal of an experiment—experimenters must develop great familiarity with the instrument through long manipulations.[72] What is the role assigned to Third World scientists who visit the lab once a year for a few months? How is the division of labor decided, and what are the effects? The concept of "trading zone" is useful if we realize that these are spaces of interactions between different scientific subcultures to coordinate different global meanings, but also spaces of cultural and political negotiation in a broader sense.[73] And, finally, what about pedagogical regimes? It would be interesting to investigate how, when, and why theoretical techniques developed in the Third World.[74]

The active role of local actors forces us to ask about other international initiatives. If we know little of the American and Soviet foreign scientific policy toward their allies, we know even less about their activities in "unfriendly" nations. For instance, the Soviet Union provided technical assistance to several countries in Latin America, apart from Cuba. It also hosted several students from nations under the American orbit who returned to their home countries after finishing their graduate studies. What was the logic behind such initiatives? What was the impact on research and pedagogical practices? How was the interaction of these scholars with those coming from the Western block? Such issues deserve special attention if we wish to learn about technoscientific international relations in a broader perspective.

Cold War science has been associated with Big Science, limiting our appreciation of technoscience in the Third World. Although nuclear weapons were central in East–West negotiations, it is important to realize that the main battlefield of the Cold War was the Third World. The military actions of the second half of the twentieth century took place in Asia, Africa, and Latin America. While nuclear weapons had a deterrent effect, small weapons were widely used in this period. Vietnam was a painful episode in American history. However, the number of Vietnamese civilian casualties was about sixty times the number of American deaths. Emphasis continues to be on the East–West confrontation and the implications the war had on the social, political, economic, and, to a lesser degree, scientific and technological consequences in the United States, Europe and, more recently, the Soviet Union. We have to make a step forward to look at the victims of the Cold War. Development and Cold War discourses conflated the Northern neocolonial ambitions in the South. Thus the Cold/Hot War can be seen as a phase of the South–North exchange, modulated by the East–West tension. Furthermore, if we adopt this perspective, the new century's terrorism could be seen as another phase of this conflicting relationship.

This perspective would allow us to shift our attention to other problems related to South–North exchange.[75] Twentieth-century warfare was transformed by the innovation and new uses of conventional weaponry and aerial technology. We should investigate the participation of scientists and engineers in such endeavors and their negotiations with the militaries. Following a liberal ideology, most historians believe that the collaboration between scientists and militaries is contingent and unfortunate.[76] This idea also has permeated Third World historiography. The rise of military dictatorships has been associated with massive scientific emigration and the destruction of scientific communities in countries such as Argentina. However, it would be naïve to think that the militaries lacked any interest in technoscience. Some countries in the South, such as Brazil, produce and export military technology—thanks, among other factors, to the participation of qualified personnel whose skills, education, research areas, relative power in the military, and political status are ignored. Knowledge transfer and adaptation to local conditions must have occurred. Such transfer, both South–North and South–South, certainly entailed weapons but also included repression instruments that became routine anti-communist techniques like torture. The other side of the conflict needs to be investigated too. We know almost nothing about innovation and new uses of technology in the insurgent forces. We have to warn those interested in this line of research: if we focus on innovation, as historiography of technology has done, the result is likely to be deluding.[77] On the other hand, if we concentrate on the new uses of technology, there is a fertile field: the adaptation and use of gas pipes as bombs by insurgent groups in Colombia or the bicycle in Vietnam are two examples. The question can be extended to study terrorism, as the reinvention of kamikazes in New York demonstrates.[78]

Final thoughts

We began this chapter referring to the importance that South–North exchanges had for the generation who rethought science and society relations in the North. The "Science for the People" movement took the lead in demonstrations against the participation of scientists in the Vietnam War. However, as time went on, this interest decreased. We know of only one study of the Jason Group (or similar bodies) in the anti-communist wars fought in the South.[79] In order to understand the dynamics of these conflicts, we must learn the conception, representation, and actions of those scientists who participated in governmental decisions or acted as consultants.[80] However, our research cannot be confined to the role of scientists involved in politics. We must also investigate scientific practices themselves in order to make visible their political consequences. This would be an opportunity for social studies of science to rediscover and vindicate its political vocation in an age of growing gaps and dangerous confrontations between North and South.

We have presented what we consider some of the most promising and relevant lines of research in history and sociology of science. The topics and problems reviewed cover a wide variety of fields and some of them have been approached from different literatures and perspectives, but they all share common grounds for analysis.

The study of North–South scientific exchange can profit from the recent debates raised by both postcolonial historiography and by sociologists of science. Traditional dichotomies such as "scientific–social," "technical–social," "science–technology," "external–internal," "political–epistemological," "pure–applied," "scientific production–scientific diffusion," and "power–knowledge" prevent a thorough understanding of science, technology, or society. The idea of science and technology as autonomous enterprises, independent from politics, has been one of the major obstacles for a critical explanation of the role of science in the shaping of the modern world. In particular, the study of South–North scientific exchange needs to take seriously the political character of science and technology. Our task, borrowing from

Roy Macleod's provocative phrase, is to study science not within political history but to explain science as political history.

Acknowledgment

For their encouragement, careful reading, and suggestions, we wish to thank Ron Doel, Thomas Söderqvist, Stefania Gallini, and David Edgerton. We are also grateful to the following institutions for their support: Universidad Nacional de Colombia, Universidad de Los Andes, and COLCIENCIAS. Each author of this joint publication contributed in equal measure to the research and writing.

Notes

1 Roshdi Rashed, "History of Science and Diversity at the Beginning of the 21st Century," in *Science and Cultural Diversity: Proceedings of the XXIst International Congress of Science, Vol. 1*, ed. Juan José Saldaña (Mexico City: Sociedad Mexicana de Historia de la Ciencia y la Tecnología-UNAM, 2003), 15, 15–29.

2 Ibid., 27.

3 Ibid., 22–23.

4 Ibid., 27.

5 Ludwik Fleck, *Genesis and Development of a Scientific Fact* (Chicago, IL: The University of Chicago Press, 1979); Thomas S. Kuhn, *The Structure of Scientific Revolutions* (Chicago, IL: University of Chicago Press, 1970); David Bloor, *Knowledge and Social Imagery* (London: Routledge & Kegan Paul, 1976); Steven Shapin, "Pump and Circumstance: Robert Boyle's Literary Technology," *Social Studies of Science*, 14 (1984): 481–520; Bruno Latour, *Science in Action: How to Follow Scientists and Engineers through Society* (Cambridge: Harvard University Press, 1987); Bruno Latour, *Pandora's Hope: Essays on the Reality of Science Studies* (Cambridge: Harvard University Press, 1999).

6 For instance, see Hillary Rose and Steven Rose, *Science and Society* (London: Penguin Press, 1969); Paul Forman, "Weimar Culture, Causality, and Quantum Theory, 1918–1927: Adaptation by German Physicists and Mathematicians to a Hostile Intellectual Environment," *Historical Studies in the Physical Sciences*, 3 (1971): 1–115; Marcello Cini, *L'ape e l'architetto. Paradigmi scientifici e materialismo storico*, (Milan: Felitrinelli, 1976); and Marcello Cini, "The History and Ideology of Dispersion Relations: The Pattern of Internal and External Factors in a Paradigmatic Shift," *Fundamenta Scientiae*, 1 (1980): 157–172.

7 The term "Third World," which for its political origins we find more accurate, was coined by the French demographer and economic historian Alfred Sauvy in 1952 (Alfred Sauvy, "Trois Mondes, Une Planète," *L'Observateur*, August 14, 1952) and caught on after the Afro-Asian Conference in Badung in 1955. We use the terms South–North and East–West in the way they are currently used in political science. North refers to the industrialized countries, sometimes also called "Atlantic countries"; East to the Soviet bloc; West to the United States and Europe; and South to the "Third World."

8 Every "review paper" is a deliberate attempt of the authors to create, consolidate, or close a field of research, including and excluding subjects, authors, and questions in order to set an agenda for future researches; see Olga Restrepo Forero, "On Writing Review Articles and Constructing Fields of Study" (PhD dissertation University of York, 2003).

9 This literature is too vast to adequately cover here, but see especially Michel Foucault, *The Order of Things* (New York: Random House, 1970); Patrick Petitjean, Catherine Jami, and Anne Marie Moulin, eds, *Science and Empires* (Boston, MA: Kluwer Academic Publishers, 1992); Stephen Jay Gould, *The Mismeasure of Man* (London: Penguin Books, 1981); David Philip Miller and Peter Hanns Reill, eds, *Visions of Empire: Voyages, Botany, and Representations of Nature* (Cambridge: Cambridge University Press, 1996).

10 George Basalla, "The Spread of Western Science," *Science* 156 (1967): 611–622; A. Lafuente, A. Elena, and M. Ortega, eds, *Mundialización de la Ciencia y la Cultura Nacional* (Madrid: Doce Calles, 1993).

11 On colonial science and imperialism, see the recent debate between Paolo Palladino and Michael Warboys, "Science and Imperialism," *Isis* 84 (1993): 91–102; and Lewis Pyenson, "Cultural Imperialism and Exact Sciences Revisited," *Isis* 84 (1993): 103–108.

12 David Edgerton, "From Innovation to Use: Ten Eclectic Theses on the Historiography of Technology," *History and Technology* 16, no. 2 (1999): 111–136.

13 For instance, Mauricio Nieto, *Remedios para el Imperio: Historia Natural y la Apropiación del Nuevo Mundo* (Bogotá: ICANH, 2001).

14 Sandra Harding, *Is Science Multicultural? Postcolonialisms, Feminisms and Epistemologies* (Indianapolis, IN: Indiana University Press, 1998), 25.

15 Sandra Harding, *Whose Science? Whose Knowledge?: Thinking from Women's Lives* (Ithaca, NY; Cornell University Press, 1991).

16 Roy Macleod, "On Visiting the Moving Metropolis: Reflections on the Architecture of Imperial Science," *Historical Records of Australian Science*, 5, 3 (1982): 1–16.

17 See for instance Ziauddin Sardar, ed., *The Revenge of Athena: Science, Exploitation, and the Third World* (London and New York: Mansell Publishing Ltd, 1988), Part Three.

18 See for instance Michael Bravo and Sverker Sörlin, eds, *Narrating the Arctic: A Cultural History of Nordic Scientific Practices* (Canton, MA: Science History Publications, 2002); also, Fikret Berkes, *Sacred Ecology: Traditional Ecological Knowledge and Resource Management* (Philadelphia, PA: Taylor and Francis, 1999).

19 Arif Dirlik, "History without a Center? Reflections on Eurocentrism," in *Across Cultural Borders: Historiography in Global Perspective*, ed. Eckhardt Fuchs and Benedikt Stuchtey (Lanham, MD: Rowman and Littlefield, 2002).

20 Gayatri Chakravorty Spivak, "Can the Subaltern Speak?" in *Colonial Discourse and Post-colonial Theory: A Reader*, ed. Patrick Williams and Laura Chrisman (New York: Columbia University Press, 1994).

21 Arturo Escobar, *Encountering Development: The Making and Unmaking of the Third World* (Princeton, NJ: Princeton University Press, 1995).

22 V. Y. Mudimbe, *The Invention of Africa: Gnosis, Philosophy, and the Order of Knowledge* (Bloomington, IN: Indiana University Press, 1988); Chandra Mohanty, Ann Russo, and Lourdes Torres, eds, *Third World Women and the Politics of Feminism* (Bloomington: Indiana University Press, 1991); Homi K. Bhabha, *The Location of Culture* (London and New York: Routledge, 1994).

23 Escobar,*Encountering Development*, 107.

24 Gilbert Rist, *The History of Development: From Western Origins to Global Faith*, trans. Patrick Camiller (London and New York: ZED Books, 1999), 22.

25 Ibid., 24.

26 John A. Agnew, "Technology Transfer and Theories of Development," *Journal of Asian and African Studies* 17 (1982): 16–31.

27 Escobar, *Encountering Development*, 151.

28 Rist, *History of Development*, 23.

29 Michael Adas, *Machines as the Measure of Men: Science, Technology, and Ideologies of Western Dominance* (Ithaca, NY and London: Cornell University Press, 1989).

30 Marcos Cueto, *Excelencia Científica en la Periferia: Actividades Científicas e Investigación Biomédica en el Perú, 1890–1950* (Lima, OH: Grade-Concytec, 1989).

31 J. D. Bernal, *The Social Function of Science* (Cambridge: The MIT Press, 1964 [1939]).

32 C. P. Snow, *The Two Cultures* (Cambridge: Cambridge University Press, 1959), 48.

33 Michael Moravcsik and J. M. Ziman, "Paradisia and Dominatia: Science and the Developing World," *Minerva* 53 no. 4 (July 1975): 699–724.

34 Thomas F. Glick, "Science in Twentieth-Century Latin America," in *Ideas and Ideologies in Twentieth Century Latin America since 1870*, ed. Leslie Bethell (Cambridge: Cambridge University Press, 1996), 287–359, on 348–349.

35 Celso Furtado, *La Economía Latinoamericana* (Mexico City: Siglo XXI, 1993).

36 Amílcar Oscar Herrera, *Ciencia y Política en América Latina* (Mexico City: Siglo XXI, 1971).

37 Fernando H. Cardoso and Enzo Faletto, *Dependency and Development in Latin America*, trans. M. Urquidi (Berkeley, CA: University of California Press, 1979). The school has roots in the United States (Paul Baran, Paul Sweezy), in Chile (Oswaldo Sunkel), in Brazil (Cardoso, Faletto, and Celso Furtado), in Colombia (Orlando Fals Borda), and in Mexico (Rodolfo Stavenhagen).

38 Hebe Vessuri, "The Social Study of Science in Latin America," *Social Studies of Science*, 17 (1987): 519–554; Glick, "Science," 347–355.

39 See Dudley Seers, ed., *Dependency Theory: A Critical Reassessment* (London: Pinter, 1981).

40 Vessuri, "Social."

41 Terry Shinn, Jack Spaapen, and Venni Krishna, "Science, Technology and Society Studies and Development Perspectives in South-North Transactions," in *Science and Technology in a Developing World*, ed. Terry Shinn, Jack Spaapen, and Venni Krishna (Dordrecht: Kluwer, 1997), 1–34, on 11.

42 Alexis De Greiff, "The Tale of Two Peripheries: The Creation of the International Centre for Theoretical Physics in Trieste," *Historical Studies in the Physical and Biological Sciences* 33, Part 1 (2002): 33–60.

43 See Wesley Shrum, Carl L. Bankston III, and D. Stephen Voss, *Science, Technology, and Society in the Third World: An Annotated Bibliography* (Metuchen, NJ and London: The Scarecrow Press Inc., 1995).

44 Dong-Won Kim, "The Conflict between Image and Role of Physics in South Korea," *Historical Studies in the Physical and Biological Sciences* 33, Part 1 (2002): 107–130; Ana Maria Ribeiro de Andrade, *Físicos, Mésons e Política: A Pinamica da Ciencia na Sociedade* (Sao Paulo and Rio de Janeiro: Editora HUCITEC; Museu de Astronomia e Ciencias Afins, 1999).

45 Alexis De Greiff, "Supporting Theoretical Physics for the Third World Development: The Ford Foundation and the International Centre for Theoretical Physics in Trieste (1966–1973)," in *American Foundations and Large-Scale Research: Construction and Transfer of Knowledge*, ed. Giuliana Gemelli (Bologna: CLUEB, 2001), 25–50.

46 Kim, "Conflict."

47 We are following Jorge Katz's *De la Importación de Tecnología al Desarrollo Local* (Mexico: Fondo de Cultura Económica, 1976); and David Edgerton, "Innovation."

48 John McNeill, *Something New under the Sun: An Environmental History of the Twentieth Century* (London: The Penguin Press, 2000), 219. On the Green Revolution see also: John H. Perkins, *Geopolitics and the Green Revolution: Wheat, Genes, and the Cold War* (New York: Oxford University Press, 1997); N. Cullather, "Miracles of Modernization: The Green Revolution and the Apotheosis of Technology," *Diplomatic History* 28, no. 2 (March 2004): 227–254.

49 McNeill, *Something New*, 219–227; Bernhard Glaeser, ed., *The Green Revolution Revisited: Critique and Alternatives* (London: Allen & Unwin, 1987), 1–9.

50 J. K. Bajaj, "Science and Hunger: A Historical Perspective on the Green Revolution," in *The Revenge of Athenia: Science, Exploitation, and the Third World*, Ziauddin Sardar ed., (London: Mansell, 1988): 131–156.

51 McNeill, *Something New*, 224. See also Vandana Shiva, *The Violence of the Green Revolution: Third World Agriculture, Ecology. and Politics* (London and New York: Zed Books Ltd., 1991), Chapter 2.

52 McNeill, *Something New*, 222.

53 Glaeser, *Green Revolution Revisited*, 3.

54 James C. Scott, *Seeing Like a State: How Certain Schemes to Improve the Human Condition Have Failed* (New Haven, CT: Yale University Press, 1999).

55 Shiva, *Violence*, 20.

56 B. Sorj and J. Wilkinson, "Biotechnologies, Multinationals and the Agrofood Systems of Developing Countries," in *From Columbus to ConAgra: The Globalization of Agriculture and Food*, ed. Alessandro Bonanno, Lawrence Busch, William H. Friedland, Lourdes Gouveia, and Enzo Mingione (Lawrence, KS: University Press of Kansas, 1994), 85–104.

57 See for instance Noam Chomsky, *Deterring Democracy* (New York: Hill and Wang, 1992).

58 Brigette Schroeder-Gudedus, "Nationalism and Internationalism," in *Companion to the History of Modern Science*, ed. R. C. Olby, J. R. R. Christie, and M. J. S. Hodge (London: Routledge, 1990), 909–919.

59 Lawrence S. Wittner, *One World or None: A History of the World Nuclear Disarmament Movement through 1953* (Stanford, CA: Stanford University Press, 1993); Lawrence S. Wittner, *Resisting the Bomb: A History of the World Nuclear Disarmament Movement, 1954–1970* (Stanford, CA: Stanford University Press, 1997).

60 Yakov M. Rabkin, *Science between the Superpowers* (New York: Priority Press, 1988).

61 De Greiff, "Two Peripheries"; Alexis De Greiff, "The International Centre for Theoretical Physics, 1960–1980: Ideology and Practice in a United Nations Institution for Scientific Co-operation and Third World Development" (PhD dissertation, University of London, 2003).

62 David A. Hounshell, "Epilogue: Rethinking the Cold War; Rethinking Science and Technology in the Cold War; Rethinking the Social Study of Science and Technology," *Social Studies of Science* (Special Issue: *Science in the Cold War*) 31, no. 2 (April 2001): 289–297.

63 Peter Galison and Bruce Hevly, eds, *Big Science: The Growth of Large-Scale Research* (Stanford, CA: Stanford University Press, 1992).

64 David Kaiser, "Making Theory: Producing Physics and Physicists in Postwar America" (PhD dissertation, Harvard University, 2000).

65 On the nuclear disarmament movement see Wittner, *One World*; Wittner, *Resisting*. On "Atoms for Peace" see Richard G. Hewlett and Jack M. Holl, *Atoms for Peace and War, 1953–1961: Eisenhower and the Atomic Energy Commission* (Berkeley: University of California Press, 1989).

66 Itty Abraham, *The Making of the Indian Atomic Bomb: Science, Secrecy and the Postcolonial State* (London and New York: Zed Books, 1998).

67 Juan Andrés León, "Los Inicios del Programa Nuclear Colombiano 1955–1965: Diplomacia y ayuda internacional en la formación de una comunidad científica del Tercer Mundo durante la era del desarrollo" (undergraduate thesis, Universidad de Los Andes (Bogotá), 2004).

68 See Diego Hurtado de Mendoza, "Autonomy, even Regional Hegemony: Argentina and the 'Hard Way' toward the first Research Reactor (1945–1958)," *Science in Context* 18, no. 2, (2005): 285–308.

69 León, "Los Inicios," 112–113.

70 Sharon Traweek, *Beamtimes and Lifetimes: The World of High Energy Physicists* (Cambridge: Harvard University Press, 1988).

71 Andrew Pickering, *Constructing Quarks: A Sociological History of Particle Physics* (Chicago, IL: University of Chicago Press, 1984).

72 Peter Galison, *How Experiments End* (Chicago, IL: University of Chicago Press, 1987).

73 On the trading zone concept see Peter Galison, *Image and Logic: A Material Culture of Microphysics* (Chicago, IL: University of Chicago Press, 1997).

74 Kaiser, "Making Theory."

75 See "Special Issue on 9/11," *History and Technology* 19, no. 1 (2003).

76 David Edgerton, "Science and War," in *Companion to the History of Modern Science*, ed. R. C. Olby, J. R. R. Christie, and M. J. S. Hodge (London: Routledge, 1990), 934–945.

77 Edgerton, "Innovation."

78 See Miriam R. Levin and Rosalind Williams, eds., "Forum on Rethinking Technology in the Aftermath of September 11," *History and Technology* 19, no. 1 (2003): 29–83.

79 The existence of the Jason Group, where young theoretical physicists participated, was exposed in the *New York Times*; see *The Pentagon Papers* (New York: Bantam Books, 1971). On the origins of the Jasons see Finn Aaserud, "Sputnik and the 'Princeton Three': The National Security Laboratory that was not to be," *Historical Studies in the Physical and Biological Studies* 25 (1995): 185–239.

80 Mark Solovey, "Project Camelot and the 1960s Epistemological Revolution: Rethinking the Politics-Patronage-Social Science Nexus," *Social Studies of Science* (Special Issue: *Science in the Cold* War) 31, no. 2 (2001): 171–206.

14 Witnessing the witnesses

Potentials and pitfalls of the witness seminar in the history of twentieth-century medicine

E. M. Tansey

"Witness: One who is or was present and is able to testify from personal observation; one present as a spectator or auditor."[1]

The History of Twentieth Century Medicine Group, home to the long-running "Witness Seminars," was established in 1990 by the British medical research charity, the Wellcome Trust. Its goals include promoting the historical study of recent medicine and medical science, creating and strengthening synergistic links between professional medical historians and members of the biomedical research community, and emphasizing the potentials of working jointly.[2] A Programme Committee of historians, biomedical scientists, and practitioners oversees the Group's activities.[3] An important part of the Group's mission has been to encourage the deposit of conventional archives related to the history of recent medicine, such as personal and professional papers as well as research artifacts such as computer tapes, visual material, and equipment in appropriate repositories.

But in recent years we have concentrated our efforts on a novel approach to oral history: the Witness Seminar. Involving numerous participants—all recorded simultaneously and able to interact with each other and with the seminar convenors—the Witness Seminars produce group discussions on topics of special interest. As with other products of the History of Twentieth-Century Medicine Group, we make these collective oral history materials available for widespread use.

Oral testimonies, both individual and collective, are important—if contested—tools for historians of recent science, technology, and medicine, although this is not the place to discuss such limitations.[4] We have developed a series of group oral history meetings that arose from one of our routine seminars held in 1992 on the subject of "Interferon." At that meeting, we were intrigued and impressed by the extensive, animated, and illuminating discussions between the chairman and members of the invited audience, many of whom had participated in the discoveries under investigation. Their discussions alerted us to the importance of recording "communal" eyewitness testimonies. We decided to organize a formal group oral history meeting, to which we would invite people associated with a particular set of circumstances or events in recent medical history. Once gathered, we would urge them to discuss, debate, agree, or disagree about their reminiscences and their significance. Whilst doing so we learned that the Institute for Contemporary British

History held similar meetings, called Witness Seminars, to examine modern political, diplomatic, and economic history.[5] This seemed a suitable title for us to use also. In line with our objective to assemble archival material for the use of present and future historians, the entire proceedings of each Witness Seminar are recorded and transcribed. The transcripts are edited for publication, and all primary material is deposited in relevant libraries, usually that of the Wellcome Trust, London.

How these meetings are planned and organized is the subject of this chapter. It also describes how the resultant publications are produced and explores some of the potentials and pitfalls of this approach for the history of recent medicine and medical science.

The first experiment: monoclonal antibodies

The first Witness Seminar was held in September 1993 in the Wellcome Institute, London, on the subject of monoclonal antibodies. Antibodies are part of the defense mechanism of the body against foreign material such as infectious agents or transplanted material, and although all antibodies are structurally similar, the body produces millions of different types. The monoclonal technique provided a mechanism to produce highly specific, pure antibodies.

"Such [monoclonal] cultures could be valuable for medical and industrial use" was the final, perhaps provocative—but certainly prescient—sentence of a paper published in *Nature* in 1975 by the Argentinian-born scientist César Milstein and his German post-doctoral fellow, Georges Köhler. In it they announced the discovery of their new monoclonal technique.[6] The subject seemed to be an appropriate topic for the first Witness Seminar for two principal reasons: it was a major scientific breakthrough for which the authors shared the Nobel Prize in Physiology or Medicine for 1984 (together with Niels Jerne) and it also seemed to be a stimulating exemplar around to discuss the nature of scientific discovery.

However, we had additional reasons for pursuing the subject. They are illustrated by the following quote: "It seems incredible that in the mid-1970s the two British [sic] scientists who discovered how to make monoclonal antibodies decided not to patent their invention. In accordance with a long scientific tradition they felt they had no right to benefit commercially and personally by obtaining a patent."[7] This was a well-established and oft-repeated story about monoclonal antibodies: Milstein and Köhler had signally failed to protect their own interests and those of their employer, the Medical Research Council (MRC), in the commercial exploitation of their discovery. Debates, accusations, and continuing recriminations about this failure of British science and technology gave monoclonal antibodies a public prominence above and beyond their considerable scientific importance. Indeed, that scandal has, in many ways, overshadowed the significance of the original research and obscured the context in which it was undertaken.[8] Therefore, the scientific discovery itself— and the subsequent controversies about the exploitation and ownership of that discovery—made it an ideal subject to explore in more depth. Key participants at the seminar were Köhler and Milstein themselves; also included were several of their laboratory colleagues, many of whom were key eye-witnesses to the events under discussion. Representatives of the MRC and the National Research Development Corporation, the body responsible from 1949 for developing the commercial potential of scientific discoveries made in Government laboratories, also took part, as did medical and scientific administrators and journalists from the period.

What did this first Witness Seminar achieve? An area of contemporary biomedical history, shrouded in controversy and touched by conflict, was dissected and scrutinized in a semi-public forum. As the tape rolled, scientists, administrators, historians, and other interested parties debated, disagreed, analyzed, and argued. A clearer understanding emerged from the scientific and national history of the discovery and exploitation of monoclonal antibodies. In a revealing discussion, Milstein and Köhler recalled their scientific relationship and how their ideas developed in the lab. Several witness accounts from this seminar were later used by the Canadian historians Alberto Cambrosio and Peter Keating in their extensive studies of "the monocloncal antibody revolution" to explore the initial reception by the scientific community of Milstein and Köhler's work.[9] Our Seminar was held just in time: a little over a year after the meeting, Köhler died at age 48, and Milstein died in 2002 at 75 years of age. This Witness Seminar thus provided a unique record of the two laureates discussing their work and its importance and subsequently provided the basis for at least one of Köhler's obituaries and also for a reflective article on his work by Milstein himself.[10]

We do not, of course, believe that we are recording some consensual "truth." That first meeting clearly demonstrated—as did every subsequent one—that no seminar arrives at such a consensus nor produces a set of complete contradictions. Like many of our contributors, we acknowledge that such raw historical material is often an untidy confusion. Nevertheless, we were impressed by the candor, interaction, and genuine enthusiasm of most of the witnesses, in both their arguments and agreements. We were also aware that we were observing open peer-review, as opinions and reminiscences were immediately open to challenge from others who were "there at the time." Positive responses from those who participated in the meeting and those who attended as observers, plus subsequent requests for the transcript from both historians and practitioners, convinced the Programme Committee that this forum should be further developed. That we have done.

Procedures and planning for the seminars

From 1995, we have included such seminars routinely in the Group's program (see Table 14.1 for a list of all Witness Seminars held to the end of 2003) and have developed a set of routine organizational procedures. We have learned that a number of seemingly trivial issues, such as the seating arrangement (we use a semicircular format whenever possible) and having only two roving microphones at a time (more than that has encouraged people to contribute simultaneously), are of enormous importance in ensuring the success of the meeting. Some meetings have been specifically planned to tie in with on-going historical research programs, including Oral Contraceptives, Obstetric Ultrasound, British Contributions to Medicine in Africa, and Foot and Mouth Disease.[11] An important consideration from the very beginning was that the

Table 14.1 Witness seminars, 1993–2003

Date	Subject and publication details[12]
1993	Monoclonal Antibodies[13]
1994	The Early History of Renal Transplantation
	Pneumoconiosis of Coal Workers[14]
1995	Self and Non-self: A History of Autoimmunity[15]
	Ashes to Ashes: The History of Smoking and Health[16]
	Oral Contraceptives
	Endogenous Opiates[17]
1996	Committee on Safety of Drugs[18]
	Nuclear Magnetic Resonance and Magnetic Resonance Imaging[19]
1997	Research in General Practice[20]
	Drugs in Psychiatric Practice[21]
	The MRC Common Cold Unit[22]
	Early Heart Transplant Surgery in the UK[23]
1998	Haemophilia: Recent History of Clinical Management[24]
	Obstetric Ultrasound: Historical Perspectives[25]
	Post-penicillin Antibiotics[26]
	Clinical Research in Britain, 1950–1980[27]
1999	Intestinal Absorption[28]
	The MRC Epidemiology Unit (South Wales)[29]
	Neonatal Intensive Care[30]
	British Contributions to Medicine in Africa After the Second World War[31]
2000	Childhood Asthma and Beyond[32]
	Maternal Care[33]
	Peptic Ulcer: Rise and Fall[34]
2001	The MRC Applied Psychology Unit[35]
	Foot and Mouth Disease: The 1967 Outbreak and its Aftermath[36]
	Leukaemia[37]
	Genetic Testing[38]
2002	Environmental Toxicology[39]
	Innovation in Pain Management[40]
	Cystic Fibrosis[41]
2003	Thrombolysis
	The Rhesus Factor and Disease Prevention[42]
	Beyond the Asylum: Anti-psychiatry and Care in the Community
	Platelets in Thrombosis and Other Disorders

material gathered at these meetings should be disseminated to wider audiences. Thus, the transcripts of most meetings, edited for style and annotated for comprehension, are published in hard copy and also placed on the Centre's web-site. To date all or part of 26 such meetings have been published, and four are currently being edited.

Suitable topics for Witness Seminars were (and are) proposed by, and to, members of the Programme Committee of the Group. In the past five years, the Committee has had to consider more than twenty suggestions each year. The viability of each is considered. Is the subject of sufficient interest, both historically and scientifically, to justify the expenditure? Would a Witness Seminar add to the extant secondary literature, and would it provide a helpful guide to primary material? Is a variety of key people available, and are they willing and able to take part? Another critical concern: will it be within our budget?[43] Financial constraints mean that, by and large, our meetings focus on British work and contributions. We acknowledge the limitations, but also the benefits, of that restriction and hope that similar initiatives elsewhere will contribute to a broader, comparative picture.[44]

Once a subject is under serious consideration, we invite specialist advisers to act as co-organizers with members of the History of Twentieth Century Medicine Group, usually professional historians and/or medical or scientific practitioners.[45] Simultaneously, we identify a suitable chairman, usually a senior person from a closely related field, and we start to discuss the meeting with possible contributors. These approaches usually lead to other suggested participants. Although we rely heavily on such proposals, we try to avoid introducing a large sampling error into the proceedings, by seeking advice as widely as possible. We also undertake additional literature searches and gradually compile a detailed database of contacts.[46]

Some time before a seminar—each of which normally lasts for about four hours—the organizers plan a flexible outline program for the meeting with the designated chairman.[47] We discuss the areas that are important to address and invite some participants to act as lead witnesses. We are acutely aware that meetings cannot be "scripted." Indeed, if they could, there would be little point in holding them. However, we have learned that some semi-formal structure is necessary—at the very least to focus participants' minds on the topic, and to give some coherence to the meeting. Clearly, however, Witness Seminars have to be flexible enough to take account of unexpected hiatuses or effusions. We also invite a professional historian in an appropriate field, if available, to make some introductory remarks about the background to the meeting and the subject matter.

At this point, we sometimes discover that proposals cannot be developed because of the lack of suitable witnesses. But other topics remain viable. Those invited to introduce particular subjects are asked to speak for about 5 minutes on a particular theme to initiate and stimulate further discussion. Their job is to "set the ball rolling" for others to follow. We emphasize to everyone attending, however, that this is an invitation to *introduce*, not dominate a topic. Every participant, whether invited to contribute in this way or not, is encouraged to take part throughout the meeting. Before the meeting, the outline program and a list of all those attending—which offers the courtesy of telling everyone who else will be at the meeting—is sent to every participant. Whilst we are anxious to obtain "raw" rather than "rehearsed" memoirs, we recognize that forethought and effort may be necessary for our witnesses to recall some of the events under discussion.[48] Throughout, we are careful to emphasize the essentially informal nature of the meeting. We discourage participants from bringing prepared scripts, and we do not (except in special circumstances, such as the meeting on Obstetric Ultrasound) allow slides or other visual material to be shown, simply because of the disruption this can introduce to the meeting.[49] These arrangements may differ in detail from meeting to meeting, as each develops its own particular shape and dynamics, largely determined by the subject matter, the personalities of the participants and the chairman, and the relationships amongst those taking part.

Figure 14.1 Photograph of a Witness Seminar in progress, June 2002. Dr Archie Norman (facing camera, on right) spoke of his personal experience of gaining knowledge about and treatment of cystic fibrosis from 1945 to 1955, at the meeting on Cystic Fibrosis (published 2004). Seated on the far left are the Chairman, Professor John Walker-Smith, and Dr James Littlewood, one of the meeting's organizers.

Source: © The Trustee of the Wellcome Trust, reproduced courtesy of the Wellcome Library.

Publishing procedures

The complete proceedings of each meeting are recorded and transcribed and photographs taken for use in future publications and to augment the archival record (see Figure 14.1). After the meeting, members of the Programme Committee consider whether the transcript should be edited for publication. This decision is guided by two factors: the overall coherence of the meeting and the significance of new material to the existing published historical record.

Ultimately, we have decided against publishing some meetings. The proceedings of Renal Transplantation and Oral Contraceptives exemplify those we have refrained from publishing thus far. In both meetings, some of the better-known speakers repeated familiar anecdotes and accounts, adding little in the way of fresh information or interpretation. The tapes, transcripts, and associated correspondence are nevertheless deposited in the Wellcome Library in London (Archives and Manuscripts) and are available for bona fide scholars to consult.[50] Our experiences with these two meetings emphasizes another difficulty: in group meetings, as in one-to-one interviews, one can be regaled with familiar stories (as we frequently encounter in printed materials). It can be awkward, sometimes impossible, for historians to shift interviewees from well-trodden paths. This can be a particular problem with well-known witnesses who are interviewed frequently, such as Nobel Laureates. In most cases, however, the presence of contemporaries, old friends, colleagues, and rivals at a Witness Seminar is a positive benefit in such circumstances, as they are more likely to express irritation at, and challenge, well-worn anecdotes and reminiscences.

Once a decision has been made to publish the proceedings, we embark on a lengthy editorial process. This is influenced by several factors. First for legal and copyright reasons, all participants take responsibility for their own remarks, the copyright of which is assigned to the publishers. We send a copy of the unedited transcript to all participants for them to amend their own, and only their own, contributions. Changes usually involve spelling, grammar, or style, or the correction of simple errors of memory, dates, or names; all such minor comments are incorporated into the master text. At this stage we do not allow extensive alterations (which some participants have sought), the addition of new material, or for participants to "correct" others' contributions (which has been tried by some). We make much use of footnotes to accommodate further comments, information and remarks, on the proviso that all participants have the opportunity to view the final version and, if necessary, respond to any insertions made by others.

The editors continue to annotate the manuscript, often with considerable help from the participants, removing untidy conversational repetitions, asides, and incomplete sentences, elucidating jargon and professional shorthand, and gradually transforming the spoken word into readable text. We equip the transcripts with the scholarly apparatus of bibliographical and biographical footnotes and provide explanatory glossaries or appendices of specialized terminology or molecular structures where deemed appropriate, usually for the more technical subjects, including Nuclear Magnetic Resonance, Obstetric Ultrasound, Intestinal Absorption, and Peptic Ulcer. These additional footnotes, appendices, and glossaries are intended to facilitate the entry of others—including present and future historians—into the events and developments that the participants themselves consider to be the most influential, or significant, in their field. They also explore and explain the context and value of such work. Thomas Kuhn, for example, has suggested that although scientists often cite crucial key experiments, in isolation these are of little help to historians concerned with describing and explaining the development of scientific belief.[51] Witness Seminars explicitly encourage participants to go beyond such bald listings and to consider the broader context and significance of their work.

Throughout this painstaking editorial process, participants are regularly asked for amplification of details or for assistance with technical terminology. This is necessary because an early decision we made was not to interrupt contributors during the meeting to seek clarification of technical details or jargon but to permit interchanges and discussions to continue unimpeded. The consistent aim of the editors is for the sense and significance of the transcripts to be clear, even if some of the precise technical detail remains obscure to the general reader. After incorporation of all these changes and footnotes, edited transcripts are again sent to every contributor, for comments on the complete text, the additional footnotes, and editorial material.[52] Each participant is asked to assign copyright to the publishers and to agree to deposit and access agreements with the Wellcome Library.

Absences and silences

There are obvious disadvantages to this sort of meeting. Some witnesses may be unable or unwilling to participate. Those who do attend may not contribute; others may deafen with their axe-grinding. It can be particularly difficult if one member of a group is absent, as happened for example in Post-Penicillin Antibiotics when Dr George Rolinson was unable to join Dr Peter Doyle and Dr Ralph Batchelor, surviving colleagues of a famous scientific quartet from the pharmaceutical firm Beecham's, which developed several major antibiotic drugs. He was however able and willing to read the transcript of the meeting and to add pertinent observations and comments in appropriate footnotes.

Despite the fact that we go to considerable lengths beforehand to let people know the format, some participants misunderstand the purpose of the seminar and arrive determined to have their say with little regard to the rest of the meeting. Similarly, if some are unwilling to attend, even after further encouragement about the purpose and conduct of the meeting, then there is little we can do—equally, if individuals choose not to contribute during the meeting, there is little we can realistically do.[53] Verbosity, misunderstandings, prejudices, and axe-grinding are all retained in the published text, for these are revealing in their own right.

An obvious problem is to ensure that we invite all the appropriate witnesses. Suggestions may, deliberately or inadvertently, reflect an individual's own laboratory, university, or specialty. We try to minimize such bias by casting our enquiries widely. Having identified important contributors, however, there is no guarantee that they will be able or willing to attend. Because many of those we invite are elderly; this is a constant problem. It is, conversely, also a problem if a subject is still scientifically "hot," and workers in the field are so actively involved that it is not possible to collect them all together in one room for one afternoon. Occasionally meetings have been abandoned or postponed at this stage because key people are unable to attend. Apoptosis, In-Vitro Fertilization, Epstein-Barr Virus, and Fifty Years of the National Health Service all had to be withdrawn from our program relatively late in the planning process because of the unavailability of several significant witnesses. Sometimes we continue with a meeting despite such absences: Endogenous Opiates was held although one of the key scientists in the discovery, Professor Hans Kosterlitz of the University of Aberdeen, was hospitalized and could not participate. In the event, although the meeting would have been different if Kosterlitz had been present, we cannot say whether it necessarily would have been better or worse. Indeed, we seriously have to consider that the absence of a key figure can be beneficial in some circumstances, although we have had occasion to cancel a meeting at very short notice when it became apparent that more than one key individual was unable to attend. Sometimes too, the most appropriate person from an organization is not available or willing to attend a meeting. Monoclonal Antibodies would have been markedly different if the National Research Development Corporation had sent a representative who had been personally involved with the patenting issues that were raised. Instead it provided a spokesman who was unable to respond to detailed questions. This contrasted markedly with the meeting on Post-Penicillin Antibiotics, when the National Research Development Corporation was represented by exactly the same person as previously, but this time entirely appropriately. If individuals or organizations, whom we are obviously unable to subpoena, will not accept our invitations, we confine ourselves to commenting upon their absence in our publications.[54]

Meetings can also be disrupted at the very last minute by accidents or ill-health, and we have adopted different strategies to cope. Two hours before Drugs in Psychiatric Practice was to start, a key participant pulled out. In that situation we unhesitatingly continued. A more significant challenge was presented by the Pneumoconiosis meeting, when Professor Charles Fletcher suffered an incapacitating fall just days before he was due to join Dr Philip D'Arcy Hart as one of the two principal witnesses. In this case, we faced an immediate and serious dilemma. Hart had pioneered medical surveys in the South Wales coalfields in the mid-1930s and was the only survivor we had been able to locate from that period. Fletcher had become the head of the Medical Research Council's Pneumoconiosis Research Unit in 1945, which was created as a result of Hart's reports. Other participants at this Witness Seminar were either from the latter part of his directorship or even later. Thus, without Fletcher, there was no "bridge" between the early surveys and the later research work. At the meeting itself we attempted to make up this deficit by showing two brief video excerpts of prerecorded interviews: one of Fletcher talking about the establishment of the Pneumoconiosis Research Unit; the second of his successor, Dr Archie Cochrane, discussing the subsequent work of the unit.[55] These helped cover Fletcher's absence and provided chronological continuity, although the contributions could neither be challenged nor integrated into the meeting in the same way as could personally delivered reminiscences. When the transcript of the meeting was prepared, however, the deficiency created by Fletcher's absence was very apparent. Several key issues had not been addressed, including the ways in which such a specialized Research Unit—representative of a new way of funding research in the UK at that time—had developed its research agenda and negotiated relationships with both its governing body, the Medical Research Council, and with local doctors and academics at the University of Wales. A particularly intriguing issue that we had hoped to address was the way in which the urbane Eton-educated Englishman (Fletcher) had established effective and long-lasting relationships with the local population of miners, one that led to 97–99 percent compliance with the epidemiological surveys conducted by the Unit in the Welsh valleys. To fill the gap, we planned a separate interview with Fletcher, although his continuing ill-health prevented that, and he died soon thereafter. Even had we been able to undertake such an interview, the insertion of additional material in this way would undoubtedly have provided fresh editorial problems. The dilemma remained that Hart's portion of the meeting provided a valuable record of a unique episode in industrial medicine, and eventually an edited transcript was published separately.[56] Since then a successful meeting has been held on the successor body to the Pneumoconiosis Unit, The MRC Epidemiology (South Wales) Unit, which incorporated unpublished material from that earlier meeting.

Individuals unable to travel to London to take part in meetings have also presented us with fresh problems. Should we allow the inclusion of material during the meeting that cannot be challenged by those present? A number of devices have been tried. An important witness for The Committee on Safety of Drugs was Professor Bill Inman who had been Senior Medical Officer to that Committee and later Medical Assessor to the Witts Subcommittee on Adverse Reactions and was a key figure in the early days of regulating new drugs in the United Kingdom. Wheelchair-bound, Inman was unwilling to travel to London from his home in Southampton to take part in the seminar. The chairman of the meeting, Dr Stephen Lock, a former editor of the *British Medical Journal*, therefore visited him to record a special videointerview that was shown during the meeting. As with the use of videotapes in the Pneumoconiosis meeting, the inclusion of such material in an interactive meeting presented particular problems. Inman could neither respond to others' comments nor could he take questions. However, the fact that the interview was specifically made as a contribution to the meeting, with the overall agenda of that meeting in mind and with the chairman of the meeting conducting the interview, did make it of more direct relevance than the use of non-tailored material. It was included in the published transcript and Inman contributed fully in the editorial process, as did the other participants.

Haemophilia: Recent History of Clinical Management provided a similar problem. Dr Rosemary Biggs, a pioneer haematologist and former director of the Medical Research Council's Haemophilia Research Unit, was unable to attend the London meeting from her home in Oxford. She was interviewed before the meeting by the Oxford haematologist Dr Charles Rizza and the seminar's chairman Professor Christine Lee, their discussion being specifically directed toward some of the themes expected to emerge during the seminar. Biggs' interview was not included in toto during the meeting; rather, the chairman quoted from its transcript as seemed appropriate. Neither was the complete interview added to the published transcript: those sections read out during the meeting were included and subjected to the rigorous and transparent editorial process adopted for all our publications, whilst the full interview was separately published elsewhere.[57] Although none of these mechanisms is completely satisfactory, we believe that a focused interview conducted by one of the organizers of a meeting, used judiciously throughout the meeting, is a useful technique to cover an otherwise unavoidable lacuna. Internet or videoconferencing facilities have not yet been used.[58]

Several witnesses have also brought relevant documents with them, immediately enriching both the seminars and the resultant publications. This was particularly so in Clinical Research in Britain, 1950–1980 and Heart Transplant, and participants were and are encouraged to deposit all papers in appropriate repositories either at the time of the meeting or subsequently. Many do so. Other meetings have also included contemporary film sequences (MRC Common Cold Unit), and equipment and photographs (Haemophilia: Recent History of Clinical Management), all brought to meetings by witnesses. Such papers and artifacts can stimulate many unexpected reminiscences and debates at the meeting, as well as being of archival value.[59]

We have taken other approaches, as for example, with the meeting on the MRC Epidemiology (South Wales) Research Unit. This was a far-ranging meeting, covering the forty years of the Unit's existence and including an immense review of the epidemiological surveys conducted by the Unit. The organizers of the meeting decided that the seminar itself had raised so many unanswered questions that supplementary work should be undertaken. Funded by a separate research grant, we made a number of additional interviews with participants from the meeting and also with those unable to attend. Extracts from these interviews were incorporated in the same volume as the transcript of the meeting (although clearly distinguished from the latter text) and the whole volume was illustrated with more than seventy historical photographs. All the original tapes and transcripts from the supplementary study were included with the archives of the Witness Seminar.

Conduct, collusion, and content

Is oral history irredeemably flawed? The core materials collected in a Witness Seminar—recorded conversations among practitioners—clearly raise questions about its reliability and usefulness. An inherent problem in oral history is bias and distortion, occasionally deliberate and more frequently unintentional. For this reason, many historians distrust such "unchallenged" memory and dismiss it as unreliable— although other, more conventional sources can be equally unreliable and open to manipulation. At the beginning of the twenty-first century we are more aware of how readily written and visual material can be distorted, and the increasing importance of electronic sources will only exacerbate this problem.[60] Historians will increasingly have to consult, interrogate, and ultimately accept or reject a range of different sources.[61] What, therefore, can Witness Seminars contribute to the history of recent biomedicine?

One important check on the conduct and content of the Witness Seminar meetings is the presence of other participants. We consider the Witness Seminar format to be a form of open peer-review, with all remarks and opinions immediately subject to rejoinder, agreement, or dispute from others who were there. Contemporaries often have intimate, insider knowledge to challenge and question accounts and to prompt memories, that even well-researched historians are unable to elicit in one-to-one interviews. They also act as a valuable check on unreliable personal testimony. Will there be collusion? An historian colleague insisted, as we started this program, that we should interview every participant separately before a Witness Seminar because "they will all agree on a story when they are together." Our experience from the Witness Seminars strongly contradicts that suggestion. Disagreements and discrepancies are neither minimized nor eradicated but are fully maintained. In our seminars, the historian is often presented not with a conspiracy of agreement but an abundance of conflicting accounts. Some interesting differences of memory and interpretation are found in our published Witness Seminars. One example happened during the meeting on Endogenous Opiates, which examined the discoveries of the naturally occurring opiate-like peptides, including endorphin and the enkephalins, and their medical and social impact. One contributor, Derek Smythe, insisted that during one of his research seminars, the biochemist Howard Morris (another witness) had jumped up in a typical "Eureka" moment, on suddenly seeing a different significance to his own research data. Morris vigorously denied the story and offered an alternative scenario of the two men talking in his laboratory after the lecture. No agreement was reached either during the meeting or afterwards about what had actually happened. Voluminous correspondence to the editors from several of the seminar participants ensued on this point and its importance to the subsequent experimental work of both scientists. Again, there was no agreement on events, others "recalled" both scenarios. This fascinating additional material could not be incorporated into the text or footnotes of the published transcript, but everything is deposited with the rest of the meeting's papers. Somewhat differently, in The Committee on Safety of Drugs, several simultaneous explanations were given of how the "report cards"—used by medical practitioners to inform the central regulatory authority of adverse reactions to drugs—came to be yellow. When confronted with their disparate accounts, the participants all expressed amazement at others' versions, but each steadfastly continued to believe their own story—a telling demonstration of the fallibility and variability of human memory. These are not, of course, problems peculiar to oral history.[62]

An important distinction must be made here. These "group" interviews do not explore "collective" memory. Rather, they expose many *overlapping* memories, collective and individual, which frequently span a wide spectrum of recollections and opinions.[63] Although to an outside eye it may appear that these meetings are predominantly comprised of homogeneous, elite biomedical practitioners, the truth is far from that. For instance, Nuclear Magnetic Resonance recorded debates between physicists, radiologists, engineers, administrators, and civil servants. Cardiac surgeons, cardiologists, anaesthetists, basic scientists, and ethicists all participated in Early Heart Transplant, while Intestinal Absorption witnesses included biochemists, physiologists, and gastroenterologists. Obstetric Ultrasound participants included physicists and obstetricians, as well as the engineers who designed and built the early scanners. Each of these different professional groups, and the individuals within them, have very different memories and accounts of the events to which they were witnesses. Thus the Witness Seminar format allows contributors to voice a wide range of professional and related non-professional recollections. As the sociologists of science Nigel Gilbert and Michael Mulkay have commented in a study of several individual interviews, it is "not surprising that speakers in different contexts, justifying different beliefs, offer different histories."[64]

Our ultimate goal is to provide the most comprehensive version of the Witness Seminars possible to the historical community. All opinions expressed during the meeting are published, unless a participant specifically asks us to exclude a remark, for example, on the grounds of confidentiality.[65] Sometimes participants have asked us to remove or to edit part or all of another contributor's comments. Such requests are firmly rejected, although they are included with the meeting's archives.[66]

Promoting interactions between historians and scientists

We have occasionally expanded the constituency of participants beyond distinguished practitioners and scientists, long the sole focus of elite interview programs, to include others who contributed to the history of particular subjects. One example is particularly illuminating: although we considered Oral Contraceptives, at which Dr Carl Djerassi was the principal witness, unsuitable for publication, local family planning doctors— the very physicians who had witnessed the enormous impact of the Pill on the daily lives of women—had also been present. Some of their accounts were so compelling that we considered developing a second Witness Seminar to include these doctors and their patients, to allow us to examine more of the social aspects that proved impossible to explore during a meeting dominated by medical and scientific issues. Unfortunately because of extreme pressure on time, it has not yet been possible to include this in our program. At Ashes to Ashes, we included representatives of anti-smoking and industry pressure groups; volunteers and administrative staff added new perspectives and opinions to The Common Cold Unit meeting, which in turn encouraged the production of a radio documentary.[67] Patients and support groups contributed significantly to both Haemophilia and Cystic Fibrosis, and the contributions of specialist nurses and midwives were important to the discussions in the seminars on Neonatal Intensive Care and Maternal Care. But throughout these seminars, we have remembered our original focus on the history of recent medical science, and our remit of exploring and developing interactions between the biomedical community and historians of medicine, to be our principal responsibility.

Witness Seminars also have a direct effect on the community from which the participants are drawn. Our contributors are frequently amazed to discover that "history" embraces their own working careers. This realization can have several consequences. One is vigorous participation not only in the meeting but also subsequently in the editorial process and indeed beyond.[68] Awareness of historians' interest in their lives and work can also increase the deposit of conventional papers, both individual and group archives such as those of organizations and institutions. After our seminars, we have encouraged the deposition of written, photographic, and film archives, mainly in the relevant collections in the Wellcome Library. We have also been able to direct instruments to suitable Museum curators. This has resulted in several connected collections now being available for future analysis. For example, several of the cardiac surgeons at Early Heart Transplant have deposited their personal papers in the Wellcome Library; equipment illustrated at Obstetric Ultrasound has been added to a historical collection at the British Medical Ultrasound Society. Additionally, as reviews in the medical press have emphasized, these volumes encourage current and recent practitioners to become engaged with their own history.[69]

We now have come to appreciate two unanticipated uses of our Witness Seminars. One was mention of the Foot and Mouth Disease meeting in a House of Lords debate on animal health. One of the participants in the Seminar, the Duke of Montrose, recalled to the House what he had learned there of the effective local responses to the 1967 epidemic, in comparison to those in 2001 when an outbreak yet again hit the United Kingdom.[70] A second, more substantial use, was of the Neonatal Intensive Care seminar in a reevaluation of the influential work of American scientists Julius Comroe and Robert Dripps. Comroe and Dripps' classic paper in *Science* assessed the impact of, and funding for, key work in cardiovascular medicine, as assessed by panels of experts. They judged that over 60 percent of key articles were the result of basic, not clinical, research, a statistic frequently used in subsequent discussions about the relative support for basic and clinical research.[71] Using the Neonatal Intensive Care meeting to provide a core group of experts and a focussed body of evidence, policy analysts from the Health Economics Research Group sought to examine the key advances in the field. Their conclusions, that Comroe and Dripps' methodology was flawed and unrepeatable, and that the mechanisms whereby basic research supports clinical advances need more detailed examination and understanding, point also to the important role that Witness Seminars can play in investigating and elucidating these links.[72]

Sources for present and future historians

An immediate question for anyone involved in oral history is whether—and how much—to edit the transcripts. Purist oral historians once objected even to transcription, as "interfering" with the prime historical source, the tape recording, let alone editing such a transcript; others continue to voice concerns about the "perils" of transcribing.[73] An important part of our program however is to reach out, beyond the participants of the meeting, and to create material resources of use to scientists and historians. Therefore we feel responsible for the promulgation of the proceedings of these meetings, as sources in their own right and as stimulators of future historical research. Transcription, editing and annotation are the tools we have selected to assist us.

We do however deposit the original tapes of all the meetings, including those that are not published, in the Wellcome Library, and permission is obtained from all participants for them to be consulted. So far, no one has asked to listen to the original recordings. Additionally, the first "raw" transcripts from each meeting, representative versions of manuscripts from each editorial stage, plus all the relevant correspondence that accrues during the editorial process are deposited with the records of the meeting. Throughout, we are explicit and transparent about our interventions in the published document and are meticulous that all original material is retained, and available, for future scholars. Certainly we intend that the publications resulting from these meetings will serve as archival resources in their own right.[74] No meeting however is a "stand-alone" project. They are supplemented by primary and, sometimes, secondary literature and can also stimulate the creation of further records.

We also hope that the published transcripts may help to guide historians through the mass of published and archival sources and alert them to subjects and sources of which they were unaware.[75] First, there is the problem, clearly articulated by Sir Peter Medawar in his famous paper "Is the Scientific Paper a Fraud?" There he points to the deficiencies of the modern scientific paper as an accurate record of the process by which science is actually carried out. The dissections and expositions that emerge at the seminars frequently elucidate such processes. Witness Seminars therefore are yet another mechanism whereby historians can get behind that published record. Second is the sheer profusion of scientific papers.[76] Calculations indicate that by the end of the twentieth century the biomedical literature was doubling every 12–15 years, which is difficult enough for practitioners focused on particular research fields to keep up with—and an almost impossible task for the historian who is usually working within a broader framework. The discussions at these meetings can quickly suggest influential publications to the historian and pathways into and through the primary literature.[77]

The edited transcript therefore represents a record produced by the participants— one that is simultaneously the creation of the historians who have organized the meeting, invited the participants, helped formulate the interrogative framework of the meeting, and edited the proceedings. They provide the traditional benefits of oral history interviews, revealing material not found in conventional documentary sources; they also facilitate the interpretation of events, personalities, and documents and offer additional sources as participants provide further material. However, the presence of a number of witnesses not only offers unique challenges to an individual's testimony, it can also be synergistic in promoting recall. As Saul Benison, one of the doyens of oral history, has remarked, such records "mark a beginning of interpretation, not an end."[78]

Acknowledgments

Mrs Wendy Kutner assists me in organizing and running the Witness Seminars, as do Mrs Lois Reynolds and Dr Daphne Christie, who also help edit the transcripts. I am particularly grateful to all those who have participated in our Witness Seminars and gratefully acknowledge the Wellcome Trust for its financial support.

Notes

1 Oxford English Dictionary online, http://dictionary.oed.com/entrance.dtl
2 Since 1990 we have, *inter alia*, organized seminars and symposia on a wide range of topics in twentieth-century medical history, held a Summer School and published a regular *Newsletter*. For further details, http://www.ucl.ac.uk/histmed/modern/20thcentury.html, accessed on March 21, 2006 – *updated url added in proof*.
3 Members for 2001–2003 were Dr Tilli Tansey (Wellcome Centre, Chair), Sir Christopher Booth (Wellcome Centre), Dr Robert Bud (Science Museum), Professor Hal Cook (Wellcome Centre), Professor Mark Jackson (University of Exeter), Professor Ian McDonald (Harveian Librarian, Royal College of Physicians), Dr Jon Turney (University College London), and Dr Daphne Christie (Wellcome Centre, Secretary). Previous members have included Professor Bill Bynum (Wellcome Institute), Dr Gordon Cook (formerly St Pancras Hospital for Tropical Diseases), Dr David Gordon (Wellcome Trust), Dr Stephen Lock (former Editor, *British Medical Journal*), Dr Lara Marks (Imperial College), Professor Vivian Nutton (Wellcome Institute), Professor Chris O'Callaghan (University of Leicester), Professor Roy Porter (Wellcome Institute), and Professor Tom Treasure (St George's Hospital, London).
4 In addition to Hoddeson (this volume), a useful summary is Soraya de Chadarevian, "Using Interviews to Write the History of Science," in *The Historiography of Contemporary Science and Technology*, ed. Thomas Söderqvist (Amsterdam: Harwood Academic Publishers, 1997), 51–70; see also Paul Thompson, "Introduction," in *Oral History, Health and Welfare*, ed. Joanna Bornat, Robert Perks, Paul Thompson and Jan Walmsley (London: Routledge, 2000), 1–20; and Ulla-Maija Peltonen "The Return of the Narrator" in *Historical Perspectives on Memory*, ed. Anne Ollila (Helsinki: SHS, 1999), 1–20, for views on the growing acceptance of the important role of oral history in contemporary historiography.
5 The Institute for Contemporary British History became the Centre for Contemporary British History in October 2002. See http://icbh.ac.uk for more details of its activities, accessed on July 21, 2004.
6 G. Köhler and C. Milstein, "Continuous Cultures of Fused Cells Secreting Antibody of Predefined Specificity," *Nature* 256 (1975): 495–497.
7 Harry Schwartz, "Talking Points: American Protectionism—The Schwartz View," *Scrip Magazine* (June 1993): 6–9.

8 See also S. de Chadarevian, *Designs for Life: Molecular Biology after World War II* (Cambridge: Cambridge University Press, 2002), esp. "The Monoclonal Antibody Scandal," 353–362.

9 See Alberto Cambrosio and Peter Keating, *Exquisite Specificity: The Monoclonal Antibody Revolution* (Oxford: Oxford University Press, 1995) for the inclusion of witness seminar contributions by Ita Askonas (41); David Secher (78); and Sir James Gowans (125).

10 César Milstein, "Georges Jean Franz Köhler," *The Biochemist* 17 (1966): 13–14; César Milstein, "With the Benefit of Hindsight," *Immunology Today* 21 (2000): 359–363.

11 These subjects respectively were part of projects being undertaken by Dr Lara Marks (Imperial College, London), Dr Malcolm Nicolson (Glasgow), Dr Mary Dobson and Dr Maureen Malowany (Oxford), and Dr Abigail Woods (Manchester).

12 Early transcripts were published by the Wellcome Trust and later by the Wellcome Trust Centre for the History of Medicine, both in London. All are published in the series *Wellcome Witnesses to Twentieth Century Medicine* (hereafter WW20), except for S. Lock, L. A. Reynolds, and E. M. Tansey, eds, *Ashes to Ashes: The History of Smoking and Health* (Amsterdam: Rodopi, 1998).

13 E. M. Tansey and P. P. Catterall, eds, *Technology Transfer in Britain: The Case of Monoclonal Antibodies*, in WW20, vol. 1 (London: Wellcome Trust, 1997), 1–34. See also E. M. Tansey and P. Catterall, "Monoclonal Antibodies: a Witness Seminar in Contemporary Medical History," *Medical History* 38 (1994): 322–327.

14 A. Ness, L. A. Reynolds, and E. M. Tansey, eds, *Population-based Research in South Wales: The MRC Pneumoconiosis Research Unit and the MRC Epidemiology Unit*, WW20, vol. 13 (London: Wellcome Trust, 2002); P. D'A. Hart, "Chronic Pulmonary Disease in South Wales Coalmines: An Eye-witness Account of the MRC Surveys (1937–1942)," ed. E. M. Tansey, *Social History of Medicine* 11 (1998): 450–468.

15 E. M. Tansey, S. V. Willhoft, and D. A. Christie, eds, *Self and Non-Self: A History of Autoimmunity*, in WW20, vol. 1 (London: Wellcome Trust, 1997), 35–66.

16 Lock, Reynolds, and Tansey, *Ashes*, 198–220.

17 E. M. Tansey and D. A. Christie, eds, *Endogenous Opiates*, in WW20, vol. 1 (London: Wellcome Trust, 1997), 67–101.

18 E. M. Tansey and L. A. Reynolds, eds, *The Committee on Safety of Drugs*, in WW20, vol. 1 (London: Wellcome Trust, 997), 103–135.

19 D. A. Christie and E. M. Tansey, eds, *Making the Body more Transparent: The Impact of Nuclear Magnetic Resonance and Magnetic Resonance Imaging*, in WW20, vol. 2 (London: Wellcome Trust, 1998), 1–74.

20 L. A. Reynolds and E. M. Tansey, eds, *Research in General Practice*, in WW20, vol. 2 (London: Wellcome Trust, 1998), 75–132.

21 E. M. Tansey and D. A. Christie, eds, *Drugs in Psychiatric Practice*, in WW20, vol. 2 (1998), 133–207.

22 E. M. Tansey and L. A. Reynolds, eds, *The MRC Common Cold Unit*, in WW20, vol. 2 (London: Wellcome Trust, 1998), 209–268.

23 E. M. Tansey and L. A. Reynolds, eds, *Early Heart Transplant Surgery in the UK*, WW20, vol. 3 (London: Wellcome Trust, 1999).

24 E. M. Tansey and D. A. Christie, eds, *Haemophilia: Recent History of Clinical Management*, WW20, vol. 4 (London: Wellcome Trust, 1999).

25 E. M. Tansey and D. A. Christie, eds, *Looking at the Unborn: Historical Aspects of Obstetric Ultrasound*, WW20, vol. 5 (London: Wellcome Trust, 2000).

26 L. A. Reynolds and E. M. Tansey, eds, *Post-Penicillin Antibiotics: From Acceptance to Resistance?* WW20, vol. 6 (London: Wellcome Trust, 2000).

27 L. A. Reynolds and E. M. Tansey, eds, *Clinical Research in Britain, 1950–1980*, WW20, vol. 7 (London: Wellcome Trust, 2000).

28 D. A. Christie and E. M. Tansey, eds, *Intestinal Absorption*, WW20, vol. 8 (London: Wellcome Trust, 2000).

29 This meeting was incorporated into a larger project that included both additional and supplementary material. Ness, Reynolds, and Tansey, *South Wales*.

30 D. A. Christie and E. M. Tansey, eds, *Origins of Neonatal Intensive Care in the UK*, WW20, vol. 9 (London: Wellcome Trust Centre for the History of Medicine, 2001).

31 L. A. Reynolds and E. M. Tansey, eds, *British Contributions to Medical Research and Education in Africa after the Second World War*, WW20, vol. 10 (London:, Wellcome Trust Centre for the History of Medicine, 2001).

32 L. A. Reynolds and E. M. Tansey, eds, *Childhood Asthma and Beyond*, WW20, vol. 11 (London: Wellcome Trust Centre for the History of Medicine, 2001).

33 D. A. Christie and E. M. Tansey, eds, *Maternal Care*, WW20, vol. 12 (London: Wellcome Trust Centre for the History of Medicine, 2001).

34 D. A. Christie and E. M. Tansey, eds, *Peptic Ulcer: Rise and Fall*, WW20, vol. 14 (London: Wellcome Trust Centre for the History of Medicine, 2002).

35 L. A. Reynolds and E. M. Tansey, eds, *The MRC Applied Psychology Unit*, WW20, vol. 16 (London: Wellcome Trust Centre for the History of Medicine, 2002).

36 L. A. Reynolds and E. M. Tansey, eds, *Foot and Mouth Disease: The 1967 Outbreak and its Aftermath*, WW20, vol. 18 (London: Wellcome Trust Centre for the History of Medicine, 2003).

37 D. A. Christie and E. M. Tansey, eds, *Leukaemia*, WW20, vol. 15 (London: Wellcome Trust Centre for the History of Medicine, 2003).

38 D. A. Christie and E. M. Tansey, eds, *Genetic Testing*, WW20, vol. 17 (London: Wellcome Trust Centre for the History of Medicine, 2003).

39 D. A. Christie and E. M. Tansey, eds, *Environmental Toxicology: The Legacy of "Silent* Spring," WW20, vol.19 (London: Wellcome Trust Centre for the History of Medicine, 2004).

40 L. A. Reynolds and E. M. Tansey, eds, *Innovation in Pain Management*, WW20, vol. 21 (London: Wellcome Trust Centre for the History of Medicine, 2004).

41 D. A. Christie and E. M. Tansey, eds, *Cystic Fibrosis*, WW20, vol. 20 (London: Wellcome Trust Centre for the History of Medicine, 2004).

42 D. Zallen, D. A. Christie and E. M. Tansey, eds, *The Rhesus Factor and Disease Prevention*, WW20, vol. 22 (London: Wellcome Trust Centre for the History of Medicine, 2004).

43 Costs include travel, accommodation and hospitality, transcription of the tapes, and publishing. We cannot invite overseas witnesses to the meetings, although there have been the occasional, often self-financed, exceptions. The focus of each meeting, and the subsequent publication, is predominantly on British experiences. However a current project at the University of Amsterdam, under the chairmanship of Professor Eddy Houwaart and modeled on our work, is for Witness Seminars on some of the same subjects as those held in London. Ultimately there will be comparative volumes on aspects of recent medicine and medical history in both Britain and the Netherlands. See http://www.metamedicavumc.nl/mge/index.htm, accessed on March 21, 2006 – *updated url added in proof.*

44 In addition to the Amsterdam initiative (Note 43), Dr Janet McCalman supervises a Witness Seminar series at the University of Melbourne. See "Witness to Australian Science," http://www.chs.unimelb.edu.au/programe/jnmhu/witness.html, accessed on March 21, 2006 – *updated url added in proof.*

45 The edited transcripts detail all the advisers and those who otherwise assist with each project.

46 Whilst trying to avoid what A. J. P. Taylor considered the problem of "old men drooling about their youth" (quoted in Brian Harrison, "Oral history and Recent Political History," *Oral History* 1 (1972): 30–46, on 46), we are aware that younger people may well be too close to the events under discussion to contribute meaningfully. See Anthony Seldon "Interviews" in *Contemporary History: Practice and Method*, ed. Anthony Seldon (Oxford: Blackwell, 1988), 3–6. Younger observers who were witnesses to their elders' discoveries, achievements, or failures have, however, proved to be very informative, see for example, *Monoclonal Antibodies*, *Early Heart Transplant* and *Clinical Research*.

47 The meetings on "Pneumoconiosis," "Autoimmunity," "Endogenous Opiates," and "Committee on Safety of Drugs" were each two hours long. The meeting on "Peptic Ulcer: Rise and Fall" was spread over an entire day. When several participants may be quite elderly, a full day is extremely demanding and emotionally and intellectually draining.

48 See Lillian Hoddeson, this volume.

49 Any necessary visual material such as graphs or tables are photocopied, subject to copyright law, and distributed to every attendee, so they can be introduced into the discussion as and when appropriate. In "Obstetric Ultrasound," we arranged for a slide projector to be available in the meeting room, so that participants could insert a slide of a particular piece of equipment or clinical scan, as and when appropriate.

50 Some are already catalogued and available for study; see Wellcome Library catalogue, http://catalogue.wellcome.ac.uk, accessed on July 21, 2004. Witness Seminars are in Manuscripts and Archives at class mark GC/253. For mention of the renal transplant meeting, see Thomas Schlich, "How Gods and Saints Became Transplant Surgeons: The Scientific Article as a Model for Writing History," *History of Science* XXXIII (1995): 311–331.

51 Thomas S. Kuhn, *The Essential Tension* (Chicago, IL: Chicago University Press, 1977): 327.

52 As far as we know, we are unusual in providing this level of editorial care/intervention. The Witness Seminars published by the Institute of Contemporary British History are much shorter and more sparsely footnoted, although clearly they are edited for style. See, for instance, the three sequential Witness Seminars edited by Brian Brivati and David Wincott, "The Campaign for Democratic Socialism, 1960–64," *Contemporary Record* 7 (1993): 363–385; Brian Brivati and David Wincott, "The Labour Committee for Europe," *Contemporary Record* 7 (1993): 386–416; Brian Brivati and David Wincott, "The launch of the SDP 1979–81," *Contemporary Record* 7 (1993): 417–464. Similarly, published round-table discussions rarely describe how the meetings were set up, the participants selected and briefed, or the discussions edited and offer little in the way of explanatory footnotes. See, for example, Laurie M. Brown and Lillian Hoddeson, eds, *The Birth of Particle Physics* (Cambridge: Cambridge University Press, 1983), 261–293; D. J. Bradley, K. Kirkwood, and E. E. Sabben-Clare, eds, *Health in Tropical Africa during the Colonial Period* (Oxford: Clarendon, 1980).

53 Such lacunae are also found in text-based archives, and all we can do is explicitly acknowledge the gaps of which we are aware.

54 This situation is completely analogous to the difficulties that historians can have with access to archives, with the partiality of archival material, or with permission to quote or cite material being withheld. It is not unique to oral history. These all provide challenges to the historian's traditional investigative and analytical skills.

55 Interviews by Dr Max Blythe, of the Oxford Brookes University Medical Biography Video History Project, http://www.brookes.ac.uk/schools/bms/medical/index.html, accessed on July 21, 2004.

56 Hart, "Chronic Pulmonary."

57 This interview has also been published separately: C. Lee and C. Rizza, "Witnessing Medical History: An Interview with Dr. Rosemary Biggs," ed., E. M. Tansey, *Haemophilia* 4 (1998): 769–777.

58 This is a large and distinct field. On this issue see Terri A. Schorzman, ed., *A Practical Introduction to Videohistory: The Smithsonian Institution and the Alfred P. Sloan Foundation Experiment* (Malabar, FL: Krieger, 1993).

59 See Doel and Henson, this volume.

60 For a particularly brazen example of visual distortion, see David King, *The Commisar Vanishes: the Falsification of Photographs and Art in Stalin's Russia* (New York: Metropolitan Books, 1997).

61 A point provocatively addressed by Arne Hessenbruch, this volume.

62 See Lillian Hoddeson, this volume.

63 Witness Seminars differ markedly from both the "focus group discussions" and "community interviews" described in Hugo Slim and Paul Thompson, with Olivia Bennet and Nigel Cross, "Ways of Listening" in *The Oral History Reader*, ed. Robert Perks and Alistair Thomson (London & New York, Routledge, 1998), 114–125, esp. 118–119.

64 G. Nigel Gilbert and Michael Mulkay, "Experiments are the Key: Participants' Histories and Historians' Histories of Science," *Isis* 75 (1984): 105–125, on 107.

65 The original tapes and transcripts are available for consultation, but restrictions may be applied to the quotation or citation of certain sections of the material, depending on the

deposit agreements signed by each participant. So far, no such restrictions have been applied.

66 See, for example, Christie and Tansey, *Neonatal*, 44.

67 "To Catch a Cold," BBC Radio 4, first broadcast February 6, 2004.

68 For example, establishing historical and archival programs and/or Witness Seminars within their professional societies, beginning formal historical studies themselves, or starting history of medicine courses for their students.

69 "Few books are so intellectually stimulating or uplifting," in J. N. Blau, "Book of the Month: Review of *Wellcome Witnesses to Twentieth Century Medicine*, Vols 1 and 2," *Journal of the Royal Society of Medicine* 29 (1999): 206–208, on 208.

70 Speech by the Duke of Montrose, January 14, 2002, House of Lords, *Hansard* (London: Her Majesty's Stationery Office, 2002), Column 928.

71 J. H. Comroe and R. D. Dripps, "Scientific Basis for the Support of Biomedical Science," *Science* 192 (1976): 102–111.

72 Jonathan Grant, Liz Green, and Barbara Mason, "From Bedside to Bench: Comroe and Dripps Revisited," (Brunel University: Health Economics Research Group, 2003), Report 30. (*Note added in proof:* even more helpful is Steve Hanney, Miranda Mugford, Jonathan Grant and Martin Buxton, "Assessing the Benefits of Health Research: Lessons from Research into the Use of Antenatal Corticosteroids for the Prevention of Neonatal Respiratory Distress Syndrome," *Social Science & Medicine* 60 (2005): 937–947.)

73 Whilst this may not be such a major issue nowadays, it was raised by oral historians when the Twentieth Century Group was formed. A hint of the debates within the UK medical history community about the necessity, or not, of transcription is in Zineta Sabovic and David Pearson, *A Healthy Heritage: Collecting for the Future of Medical History* (London: The Wellcome Trust, 1999), 57. See also Raphael Samuel, "Perils of the Transcript," *Oral History* 1 (1971): 19–22.

74 Access to transcript material is also available via the World Wide Web, currently http://www.ucl.ac.uk/histmed/publication/wellcome_witness/index.html, accessed on March 21, 2006 – *updated url added in proof.*

75 "This is oral history at its best...all the volumes make compulsive reading...they are, primarily, important historical records," in Irvine Loudon, "*Wellcome Witnesses to Twentieth Century Medicine*: A Review," *British Medical Journal* 325 (2002): 1119.

76 For a brief review of the problems posed by the immense proliferation of modern scientific literature see E. M. Tansey, "'The Dustbin of History,' and Why so much of Modern Medicine should End Up There," *Lancet* 354 (1999): 811–812. For a specific example of these problems, see D. J. Cook, M. O. Meade, and M. P. Fink, "How to Keep Up with the Critical Care Literature and Avoid Being Buried Alive," *Critical Care Medicine* 24 (1996): 1757–1768.

77 Peter Medawar, "Is the Scientific Paper a Fraud?" in *The Threat and the Glory: Reflections on Science and Scientists* (Oxford: Oxford University Press, 1991), 228–233.

78 Saul Benison, "Oral History: A Personal View," in *Modern Methods in the History of Medicine*, ed. Edwin Clarke (London: Athlone Press, 1971), 286–305, on 291.

15 "The mutt historian"

The perils and opportunities of doing history of science on-line

Arne Hessenbruch

Historians of contemporary science face a profound and unprecedented problem: the sheer volume of information and the difficulty of coping with the esoteric technicalities within each scientific specialization.[1] Recent science is a huge and very diverse terrain. The number of practicing scientists is vast; the number of research fields is greater than at any previous time in history, and the number of potential sources for recent developments is enormous.[2] For historians, the task is daunting.

A three-year project running from 2000 to 2003—based at the Dibner Institute for the History of Science and Technology and funded jointly by the Dibner Fund and Sloan Foundation—explored this problem.[3] It was entitled the History of Recent Science and Technology on the Web (HRST). One important prong of the project was to investigate the opportunities provided by electronic data storage and communication, because this medium has obvious advantages over printed text with regard to accessibility and searchability. Another important prong was to foster a collaborative network of geographically dispersed scientists and historians, using a website as the meeting ground and archive.

There were five sub-projects within HRST: Apollo Guidance Computer, Bioinformatics, Materials Research, Perspectives on Molecular Evolution, and Physics of Scale. I participated in the Materials Research sub-project with the historically minded philosopher Bernadette Bensaude-Vincent and the scientist Hervé Arribart. Over the course of three years we interviewed several dozen scientists and engineers, designed a website, wrote histories on institutions, instrumentation, specific materials, disciplinary boundaries, and funding. We also scanned in texts and materials to create archives and encouraged the direct participation on the website in the writing of the history, including criticism of our designs and texts.

In this chapter I describe the project in greater detail and draw some lessons learned from the experience. I begin by addressing a recent contribution in the *American Historical Review* about the potential for the digital medium to upset the apple cart of historical practice in general. One of the main issues is the tremendous promise of space saving, accessibility, and searchability—set against the ominous threat of technical obsolescence and with it the loss of historical data. I argue that while the permanence and reliability of the digital medium is being problematized, that of paper is taken for granted. We need to remind ourselves of the infrastructure that currently props up paper records in order to understand the promise of electronic data for historical archives in general. Discussions about any technology have to be seen in connection with social structures actually or potentially built up around it.

Our project contains lessons that are specific to history of recent science. We do well to remember that not all historical practice requires the same tools, just as not all electronic data are equally likely to become obsolete. Much of the argument in the *American Historical Review* and elsewhere turns on the problem of archiving ephemeral webpages, where we face much more severe obsolescence issues than the storage of plain text. But in the history of recent science, from my perspective, the main problem is very different; it is more akin to drawing a general map of a complex and highly interconnected activity, already characterized by a superabundance of historical data. This is the problem we contended within our project.

Indeed, our experiment highlighted the usual problems of networking but also—and this is important—the need for a change in the sociability of the historian. In the second part of this chapter, I argue that the historian of recent science necessarily comes up against values held dear by many historians of science: s/he must collaborate with Whiggish scientists to a compromising extent. This goes against the grain of the historian's integrity and is revealed in our discipline's tacit acceptance of an ethos of objectivity at the same time as our discipline seeks to deconstruct the objectivity in the natural sciences.[4]

The permanence and reliability of the digital medium

In a 2003 article in the *American Historical Review*, Roy Rosenzweig called historians to arms.[5] Appropriate to such a call, he painted an alarmist picture: historians of the future will be either swamped by an abundance of historical sources (because we can store so much digitally) or else starved of the same (because the digitally stored will be wiped out when technologies change rendering our current standards obsolete). Rosenzweig argues that only government action will prevent us from falling into the darkness of a lost history. He points to the ephemeral nature of many current webpages, requiring plug-ins and other software that may indeed be obsolete by the time this book is published. And indeed, if we aspire to conserve all current digital sources forever, then the task is daunting. But this amounts to the aspiration of saving a large part of oral culture in a previous age. We must parse out two very different obsolescence problems: first, that of the digital medium performing the task of paper, and second, the digital medium's expansion into areas with which paper could never and will never compete.

Rosenzweig's conflation of the two issues feeds a skeptical attitude to the digital medium. Much of the data collated by scientists as late as the 1980s is held in a format that is already unreadable by more recent technologies. Indeed, our skepticism arises primarily because of our day-to-day experience with the web. The webpages we know best, such as web-newspapers, are being constantly changed. In several places within this chapter I refer to Uniform Resource Locators (URLs); are you confident that in fifty years time they will still be valid? By contrast, our current experience with the print medium shows it to be reliable and safe from obsolescence. But it behooves us to remember just why the printed publication is considered trustworthy and permanent. It is not just a question of the technology but the combined effect of technology and a social infrastructure. To reveal the role of the social in propping up print as a medium for reliably fixing historical data, it is useful to examine the history of the book in an earlier age when fixity and trustworthiness had not yet been established. Adrian Johns, a historian of science at the University of Chicago, has examined the development of conventions of handling and investing credit in textual materials in early modern England. The very first paragraph of his book makes us view many aspects of books that we normally take for granted as actually quite surprising:

> Pick up a modern book. This one will do: the one you are looking at right now. What sort of object is this? There are certain features about it of which you can be reasonably confident. Its professed author does indeed exist and did indeed write it. It contains information believed to be accurate, and it professes to impart knowledge to readers like you. It is produced with its author's consent, and it is indeed the edition it claims to be. If the dust jacket announces that it is the product of a given organization—in this case the University of Chicago Press—then this too may be believed. Perhaps you may even say to yourself that that fact vouches for the quality of its content. You may safely assume that the book you now hold will have been printed in many copies, and a copy of the same book bought in Australia, say, will be identical in all relevant respects to one bought in the United States or in Great Britain.[6]

Johns describes how we came to place trust in all the aforementioned aspects of books over a long period of time, and he argues that trust in books developed as a new kind of civility and sociability emerged. The convention of authorial copyright is the best-known convention to emerge in Johns's account, but it is in fact just one small part of a quite sweeping change. For that reason Johns's book is a voluminous 754 pages, and one lesson we might take from that fact is that the establishment of textual fixity is generally a complex and long drawn-out process.

The reason that we have faith in a textual reference is that books are printed in at least dozens of copies and distributed amongst just as many libraries across the world. Even if a dozen libraries were to burn down, we would still be able to locate a copy in one of the surviving ones. Furthermore, we allocate resources to keep books in controlled environments (temperature, humidity) and have an infrastructure of searching, the most prominent of which is the library catalogue. The bottom line here is that the print medium also would not exhibit the fixity that we are accustomed to had it not been for the constant maintenance work by librarians and building services.

Digital media require the same kind of maintenance, predicated on the continuous allocation of resources. Of course digital data have to be backed up (in multiple copies) just as multiple copies of books are deposited in a range of libraries, so that several physical copies of the digital data exist. And of course digital storage media need care and maintenance and need to be kept away from high humidity and extreme temperatures, just as books do. No difference there. But digital media have the additional problem of obsolescence. On our site, for example, we have many pages kept in html (hyper-text mark-up language) format; they contain text, sound (in Real streaming audio format, the back-ups kept as MP3), image (usually as jpg, sometimes with back-ups kept in higher resolutions; only during the project did pdf become a more common standard), and video (in Real streaming video format, with back-ups kept as avi). We also have databases in various formats. Html is the language in which the code of most plain webpages is written. While html will very likely cease to be used, this will not entail obsolescence. To be sure, the many different plug-ins currently used to make webpages more visually interesting (often those with moving elements, to which Rosenzweig referred) might be more of a problem. Text on today's websites is mostly in html format or in pdf (a proprietary format of Adobe Acrobat) and as such are standardized. The same is true of the jpg format for images. The video formats are likely to pose greater obsolescence risks. The point is that while we cannot rely on these formats being usable in the technology of the future, in fact, we can predict with some confidence that the formats currently in use will not be the formats of the future, and consequently the material stored in these formats will become unreadable with future technology.

Still, it is technologically comparatively easy to migrate data from one format into another. The problem lies in the labor-intensive nature: if we amass a great deal of data, it might well take a lot of work to migrate it into new formats. Put another way: the infrastructure needed to maintain digital archives must include provision for routine migration of data into new formats, just as the maintenance of paper requires physical buildings, humidity and temperature control, and librarians. And furthermore, the infrastructure required to keep current html, pdf, and jpg files usable in the future is not large, whereas the infrastructure required to keep all current formats usable, including the webpage plug-ins that Rosenzweig worries about, *would* be very large.

Since it took a very long time to establish the social structures propping up the fixity of print, one would expect the same to be the case for digital data. It requires the development of new conventions—new forms of civility and sociability, as Johns puts it. The address of our website (http://hrst.mit.edu) contains the reference to the Massachusetts Institute of Technology, which, you may say to yourself, vouches for the quality of its content. There is of course a difference between a University of Chicago Press book and an MIT website. The University of Chicago Press will have asked referees to assess prospective authors whose submissions will have been further evaluated by readers deemed independent; drafts will have been criticized and amended; and several proof-readings likely will have taken place. By contrast, an MIT website has much less of an infrastructure in checking and rechecking uploaded material. You may of course trust that a website on the history of recent science under the aegis of MIT will not be downright flippant. I have been careful to state the credentials of staff and advisors on the site, ranging from PhDs to a Nobel Prize winner. But the infrastructure (sociability) remains less elaborate than that of the University of Chicago Press. So readers may be forgiven for placing less trust in our webpages than they do in a University of Chicago Press book. The same difference also explains why much less credit accrues to the author of a website than to the author of a book published by a reputable university press. But with the development of the same kind of infrastructure, I do not see why a website should not become just as credible as a printed text. Clearly some webpages are meant to be ephemeral (on our website, for instance, the "news" page announcing new contributions). But the pages of scanned documents, for example, are intended as archival documents, and their fixity is in principle as feasible as the fixity of a piece of paper in a conventional archive. In fact, one of the stated aims of the Sloan Foundation, one of our patrons, is that one day PhDs may just as well be given for a website as for a printed linear narrative.

We also know how to deal with copyright in printed material, whereas in the case of the Internet it is not yet so clear. Because the multiplication of digital material can be done with such ease, it presents different legal problems from those posed by the photocopier and the book. Many publishers are currently investigating how profit could be made with publications on the World Wide Web. MIT Press has experimented with the use of the Internet as advertising by publishing their books simultaneously in a print format (books for sale) and on the web (accessible for free). Directors at the Press have evidence to show that the web-publication actually enhanced sales of the printed version—an outcome that is perhaps surprising. Amazon.com has begun to publish parts of books online. Others are experimenting with encoded data files sold to users with a built-in blocking mechanism disabling the copying of such files.[7] On our site, we host contributions from many individuals, and we have also had to establish procedures to deal with copyright issues. All contributors are required to click through (the Internet equivalent of the legal signature) a statement containing the following:

> You agree to conduct yourself on the Program Site in a manner that will not interfere with any other Contributors' use and enjoyment of the Program Site or their ability to participate in a respectful and polite discourse about ideas. You will not post or transmit material that is libelous, defamatory, obscene, racist, fraudulent, harmful, abusive, threatening or hateful, that contains nudity or pornography, that violates the copyright and other intellectual property rights of others, or that is in violation of applicable law.[8]

Such procedures constitute an element of civility or sociability strengthening general trust that the web-authors do indeed exist, that they did indeed author the material posted, and that that material is of a certain quality. In practice, it is less of a hassle to click through these legal statements (every time we download software we do so—mostly without reading the legal statement) than to sign on the dotted line.

To repeat, fixity in web-publications cannot come overnight but requires an infrastructure and a set of behaviors, along with a set of punitive practices to deter those who do not play along—just as was the case for the book. Hence the management of digital data is not fundamentally different from that of print media: both require an infrastructure and resources for routine maintenance. The management of digital media is basically unproblematic.

Some strengths of the digital medium

Our project, HRST, was conceived as an experiment in writing history of recent science using web tools. Software for this project was developed in-house, with two main concerns.[9] The first involved the gathering of information, the second its management and display. The software for gathering information consisted of a set of modules that function as interviews, discussion groups, collaborative timelines, and the like. The aim was to have individuals contributing to the writing of history regardless of their location. For example, an interview was posted with the following questions:

- Is materials research a discipline?
- How should materials research be taught?
- What should a history of materials research focus on?

Any visitor—having logged in, registered an email address, and clicked through the legal statements—was enabled to reply to the questions. These replies are posted on the same page—almost automatically. Replies can include text and images.

One can make endless variations upon this theme: for example, restricting access so that only a select group is empowered to reply. One might want to target only the individuals at a particular department, and set the interview such that only individuals logging in with the appropriate email address would be allowed to actually answer the interview questions. If so desired, one could enable all registered users to add comments to the replies, displayed in a smaller font. The software is very flexible so that many permutations are conceivable. Historians themselves must decide which module is best suited to specific historical tasks. The invitation extended to draw in visitors was not restricted to specific pages but to the entire website. For example, in our introduction to Materials Research, we were aware that even our description of the field might be contentious, and we wanted to use that as a way of welcoming other perspectives:

Materials science is an hybrid entity coupling fundamental research with engineering application of the end-product. Knowing and producing are never separated. The interplay of cognitive purposes and technological interests is seen in the very definition of materials. Unlike matter, the notion of materials refers to a substance which is useful or of value for human purpose. Materials are usually defined as substances having properties which make them useful in machines, structures, devices, and products. Because of this dual aspect, materials science has contributed to a deep transformation of the overall organization of research and teaching. Whereas in 1900, it was taken for granted that there were two kinds of sciences, the pure or basic on the one hand and the applied on the other, this is no longer the case. By pursuing simultaneously cognitive and practical interests emerging disciplines such as materials science have subverted this clearcut distinction. One major objective of materials science and engineering is to create materials by design, i.e. structures tailored for specific purposes, whose properties are adapted to a set of specific tasks. This goal was achieved only once instruments provided access to the microstructure of materials. X-ray diffraction was the first of the new techniques for imaging the microworld that fostered the development of materials science and technology. Each new technique opened up new windows on the microstructures. The generic concept of materials presupposes that there is something common to such diverse things as metals, polymers, glass, or semi-conductors. All these different materials are submitted to a common approach typical of materials science. Instead of adopting a linear sequence, materials scientists and engineers have to embrace conceptually in a single approach structures, properties, functions, processes and end-uses. Do you agree with this intro? If not, go to discussions.[10]

The last word was hyperlinked, leading to a discussion page.

One of the great strengths of the web format is the searchability of words and the ability to connect relevant information with the help of the hyperlink. All contributions to the website are deposited in a database and can be searched efficiently. One outcome we hope to achieve is that with the growth of the database, a search will reveal unexpected cross-links between fields as diverse as bioinformatics and materials research. Our software team put some effort into the issue of automatic hyperlinking. They built on software by the classics-based Perseus Project[11] that can automatically tag words (turning them into hyperlinks) in a new document posted on the website (and thus added to the database). If one were to click on that word, one might, for instance, get a menu containing a list of the webpages on which that same word is used. With a further click, one can be taken to any of those pages. Perseus initially developed their software for philological purposes, especially ancient Greek texts, where it makes sense to tag every single word. But in history, some kind of human intervention is required. For example, a list of all pages using the word "and" is not very interesting, whereas searches for "simulation," "microscope," or "Nobel" might be. The pragmatic solution could be that an historian would provide a master list of words: if the system finds any of those words in a newly submitted text, it is tagged. Of course, in a sense the visitor's control is diminished just as the reader of a book has no control over the entries in an index. It is a give and take, and there's no obvious best way. But to display information on a collaborative history site, one has to adapt to the nature of the medium—and hence the long, one-dimensional narrative will lose out. In return, searchability and the easy linkup with relevant material will improve.

While the digital medium has its strengths, it is not a panacea. In our HRST project, we probably began with an exaggerated expectation of the use of the Internet as a collaborative tool. But our failures have forced us to explore novel sociabilities. The problems themselves are interesting and suggest some avenues that the history of recent science can pursue to increase its relevance beyond our narrow disciplinary confines.

Materials research

As I mentioned, our HRST project consisted of five sub-projects, devoted to large and small topics: the Apollo Guidance Computer,[12] Bioinformatics,[13] Molecular Evolution,[14] Materials Research, and Physics of Scales.[15] I can speak most authoritatively for the sub-project on materials research, and the nature of the topic of investigation differentiates the historiographical issues from, say, physics of scales, which is comparatively small and well-defined.

Materials research is a very large field and highly interdisciplinary. In that respect it is an excellent topic to study with the aim of learning about problems in historiography of contemporary science: recent science in general is increasingly characterized by interdisciplinarity and vast scales of operation. As I write, at the beginning of the twenty-first century, the Materials Research Society (MRS) sees the field as encompassing the following: photonics, metastable materials, superconductivity, surface science, porous materials, nanotechnology, nuclear waste, intercalation science, biotechnology, and compound semiconductors. Practitioners see themselves as simultaneously engaged in science and engineering. Because of this, it is a field that is finely attuned to the market. Its history over the last half-century is thus not merely an ivory tower academic history. This means that the historian's attention ought ideally be directed just as much to global markets and research funding structures as to the conventional topics of history of science, such as institutions, theory and experiment, publication infrastructures, and individuals. This section will introduce the sheer size and complexity of the field—the fact that entails the necessity of a collaborative historical effort.

Since the early 1960s, there have been institutions that include the word "materials" in their titles. We have thus confined ourselves to the period after 1960.[16] Of course, materials were also investigated before: the fatigue of metals was certainly an important area of investigation during the railway boom of the nineteenth century, and one can successfully find research on materials in just about any age.[17] What happened around 1960? There are at least two complementary explanations. Bernadette Bensaude-Vincent has shown that in the wake of the Sputnik crisis, military funding created new interdisciplinary laboratories at US universities, many of which achieved permanence. In other words, for her, the discipline of materials science and engineering (MSE) was a top-down creation by funding agencies, generating the infrastructure within which a new disciplinary identity was (and continues to be) forged.[18] However, Bernhardt Wuensch, Professor of Ceramics at the Department of MSE of MIT, puts the emphasis elsewhere.[19] Around 1960, researchers in such diverse fields as metallurgy, organic chemistry, semiconductor physics, and ceramics were developing conceptual tools highlighting relationship between structure and properties, tools that could be applied in all fields and thus led naturally to a materials-generic perspective.[20] A number of instruments, such as x-ray diffraction and the electron microscope, also gained widespread usage in all fields. For Wuensch, these diverse disciplines began to merge because the internal development was conducive to increased collaboration and sharing. When the US Department of Defense's Advanced Research Project Agency and other funding agencies supported this development, they were merely enhancing a natural development.

Indeed, what seems clear is that the first textbooks of MSE began by treating families of materials (metals, ceramics, polymers) separately, before addressing properties.[21] More recently, some textbooks have taken the materials-generic perspective more seriously by starting with general structure–property relations: only later, if at all, do they move on to the classes of materials that correspond to older disciplines.

A recent textbook by Samuel Allen and Edwin Thomas proceeds from very general statements about symmetry and chemical bonding to cover in turn three states (non-crystalline, crystalline, and liquid-crystalline), each of which apply to all families of materials.[22] Later in their book these authors cover processing and assembly and finally microstructure.

The materials-generic approach thus spans several sub-disciplines that one would ordinarily have located either within physics, chemistry, or engineering. In the 1990s, the manipulation of biological materials on the nanoscale has been incorporated as well, so that materials science truly ranges across the whole spectrum of the natural sciences.[23] The protean nature of materials research is only constrained within the academic setting. Indeed, the MRS does not adhere to any definition of materials research; it has no compunction about putting on sessions at its meeting that may go beyond materials research as understood up to that point. In fact, it consciously arranges its meetings to focus on the hottest fields. For example, in the 1980s this included a flurry of interest in superconducting materials. By the late 1990s, their sessions included biomaterials. Because the MRS meetings are very successful, they help mutate the field wherever the market is going. By comparison, academic departments are stolid because they have to identify core curricula and adapt to grant-giving bodies using disciplinary categories.

Hence materials research is not adequately described as a discipline gobbling up its neighbors or indeed by disciplinary boundaries. Rather, a crucial aspect of the field is denoted by the title of many departments: Materials Science *and* Engineering. Its self-conscious identity lies in the pursuit simultaneously of pure and applied science—with the implication that the notions of pure and applied have no sense. The field has reacted to the changes in funding over the past forty-odd years. One might even argue that its success has encouraged a development of science attentive to the market. In the next few paragraphs I will sketch this development.[24] In one sense, the market pull is quite straightforward. For example, the demand for electronics and plastics in the last half-century has rendered investment in semiconductor and polymer research very promising. But most research funding is mediated by national governments. The way in which tax money has been channeled toward research is a result of the negotiation between different interests, and the strength of various players has differed from decade to decade (and of course from country to country). Generally speaking, the development has been toward a blurring of the boundary between science and engineering—which is precisely what MSE is about. After Sputnik, US R&D budgets expanded on average by 15% per year, the bulk of which was for military and space research. But the fundamental structure did not yet change the set up established at the end of the Second World War basic research was determined by peer review. A 1963 Organization for Economic Co-operation and Development (OECD) policy doctrine confirmed the academic control of the funding structures, stressing that "expenditures on both education and research represent long-term investment in economic growth."[25] In this period, the first Interdepartmental Laboratories were set up with Department of Defense money. Several of these developed into the largest departments of MSE in the United States.

One might conveniently summarize the history of funding as a negotiation between separate cultures whose muscle has changed over time. During the first few decades after the Second World War a governmental policy culture and a distinct academic culture were the main actors. The former consisted of the state administration with its agencies, committees, councils, and advisory bodies. In many countries, this culture was dominated by the military. The general attitude within these cultures was that science existed for policy. By contrast, the general attitude within the latter was that policy existed for science. Academics have been concerned to preserve traditional academic values: autonomy, integrity, objectivity, and control over funding and organization. In the 1960s, the Department of Defense (governmental culture) and the National Science Foundation (academic culture) each vied for greater control, launching projects that sought, respectively, to undermine or support Vannevar Bush's understanding of the proper role of science policy.[26] In the wake of Sputnik, the military was particularly interested in materials research related to nuclear technologies such as radiation damage, a topic that has remained important over the decades. Semiconductor research also promised to be of military usefulness and was funded handsomely. NASA was funded in the wake of Sputnik, leading to funding for research on ceramics with appropriate refractory characteristics.

Toward the end of the 1960s, at the time of the Vietnam War, a civic culture emerged in opposition to military culture. In its most dynamic form, the civic policy culture was based in popular social movements, and its main concern was that science aid society.[27] The strength of the civil policy culture has depended crucially on the structure of the overall political culture of each country.

One impact of the new civic culture's hostility to the military and its research funding was that from 1967 onward and well into the 1970s, US federal R&D support declined. New social critical movements emerged, including environmentalism and feminism. As this new influential culture for science policy became rooted, it ushered in a new era of social accountability in science. Mission-oriented programs were initiated on unprecedented scales. The War on Cancer was initiated at this time,[28] so was the massive federal program, Research Applied to National Needs.[29] In addition, the Office of Technology Assessment was created to aid Congress.[30] A new OECD doctrine fit this new consensus: it emphasized a need for greater societal control over applied science, but science was still to be planned and directed from above.[31] Whereas science had hitherto been considered unified, it was now divided into sectorial programs, and new concepts were introduced, including mission orientation, technology policy, and social relevance. Another, more hostile, way of expressing the same development is to say that politics was brought back into science and technology (S&T) policy.[32] With the 1974 OPEC oil crisis, the environmentalism debate homed in on energy policy, especially nuclear. This period witnessed a boom of materials related to energy technologies, such as fuel cells and batteries, and it produced an influential report defining the role of materials science in an environmentally conscious world, focusing for instance on the materials cycle through waste and reuse.[33] By about 1980, almost all industrial nations introduced environmental protection agencies.

In Western industrialized nations, the 1980s were dominated by the growing economic and technological challenge from Japan, and then Korea, Taiwan,

Singapore, and Hong Kong. A fourth science policy culture gained considerable muscle as a result: *economic culture*. Its representatives were mostly found in business and management, and they concerned themselves mostly with technological uses of science in an entrepreneurial spirit. The Reagan and Thatcher governments sided with this new culture against civic policy culture, with its emphasis on the social relevance of science. Vannevar Bush's sense that basic science would automatically lead to desired technological advances now clearly took a back seat. Again, a 1981 OECD report gives the flavor of the times: it aimed to stimulate the growth of new technologies by means of active industrial policy by encouraging industrial–academic partnerships. Science policy was given a commercial orientation in all industrialized countries with an emphasis on industrial innovation and technological forecasting. Material Science and Engineering (MSE), its practitioners consciously straddling science and engineering, fit the new paradigm very well. Indeed, one might ponder just how much the success of materials research itself contributed to the new way of thinking.[34]

In the last quarter of the twentieth century, the successful Japanese R&D policy— which integrated science policy and industrial policy, with government-funded research playing a more active supporting role in export industries—increasingly became a model for others. Many national programs supporting advanced, generic, or base technologies of microelectronics, and biotechnology (as well as industrial materials encouraging academic–industrial linkages) came into being. Thus the values of the economic culture came to dominate. Third World countries, such as China, followed suit. Here, S&T policy reforms were strongly tied to the reforms starting in the late 1970s—the reaction to the horrors of the Cultural Revolution. In industrialized countries, the civic policy culture gradually diffused and lost its collective punch with the introduction of alternative energy sources (sometimes the rejection of nuclear), the institutionalization of Green parties, and the marginalization of radical aspects of the 1970s protest movement. The Japanese consensual model was introduced using foresight as a policy methodology, with forums intended to shape a consensus among representatives of the academic, economic, and bureaucratic policy cultures. (A similar process was introduced within industry under the rubric of issue management.) The older Vannevar Bush doctrine of relative autonomy for science was replaced by an orchestration policy: stronger integration of academic science with both public and private sectors while emphasizing importance of basic research.[35]

One important milestone in this development was the Bayh-Dole Act, passed by the US Congress in 1980. Prior to this Act, universities and individual researchers were not allowed to patent any result arising out of federally funded research. The Act successfully redirected academic work toward the generation of intellectual property and R&D for consumer markets. At the same time, processing rose to greater prominence within MSE. At MIT, for example, a Materials Processing Center opened its doors in 1980. Its first director, Tom Eagar, was outspokenly opposed to the perspectives embodied in such older textbooks as Allen and Thomas, which I mentioned earlier. The chief bone of contention for him was not the materials-generic perspective per se but rather the conception of a general theory (e.g. starting with symmetry, as in the above case) that encompassed all of materials research. Instead, Eagar emphasized

Table 15.1 Juxtaposing science before and after 1980

Mode 1 (Old):	Mode 2 (New):
Primacy of university-based curiosity-led research	Discovery in application—blurring of boundaries between pure and applied
Knowledge generated within disciplinary context	Transdisciplinarity
Hierarchical and conservative	Organizational diversity—multisite; networking
Strict rules of quality control	Sensitivity to broader implications of what one is doing

the more craft-like skills involved in processing—skills that cannot be captured by an overarching theoretical perspective.[36] Universities were increasingly shifting away from being producers of scientific theory toward becoming providers of knowledge relevant to private enterprises. Academics increasingly applied for patents and developed contacts with local industries. As the university thus developed into an engine for the local economy, this in turn led to heightened competition for federal research funding. But this occurs now with a twist: because the US representative system has increasingly given equal weight to, say, Montana and Massachusetts, and funding is being spread more evenly among all states.[37] Funding has also diversified geographically in Europe, partly because of the EU's support of cross-national connections of all kinds, partly because of regional structural funds, and partly because of the increasing financial muscle of countries such as Greece, Spain, or Portugal.[38]

During the 1990s, research in general has become more like materials research. Michael Gibbons *et al.* have contrasted the old and new way of doing science in a rhetorically very useful schematic (Table 15.1).[39]

Without having to accept all aspects of this stark and all-encompassing picture, one can at least notice that materials research emerges as quintessentially Mode 2.[40] Indeed, one might speculate that the success of materials research itself has contributed to the more general shift in the direction of Mode 2.

Exploring the collaboration between historian and scientist

The historian of materials research cannot do justice to the field by restricting the attention to academia alone. Markets, funding, and politics are integral parts of its history. To add to the difficulties of the historian, it is also a field with a large and diverse technical literature. Even professional materials researchers with PhDs in the field do not understand neighboring sub-fields very well, despite the fact that they do not consider themselves specialists in the sense of having particular expertise of one small area only. Indeed, in a world of shifting markets, one has to diversify one's knowledge and be ready to expand in the direction of market growth.[41] Ideally then, historians seeking to explore this field should get help and guidance from several materials researchers, each of whom understands his or her disciplinary area.

Because it is a discipline of such great diversity, materials researchers are particularly attentive to resources becoming available to them in neighboring fields. The MRS currently organizes two annual meetings that reflect this aspect of the field: they are giant affairs, with some 4,000 attendees organized into some twenty-five symposia with topics considered hot. A new committee is set up for each meeting to counteract a prevalence of any perspective. These meetings are certainly markets and trading zones, where researchers strut their stuff and keep their eyes peeled for new resources in terms of ideas, methods, tools, and prospective collaborators.[42]

I attended the 2001 Fall Meeting of the MRS in Boston with a poster advertising our history of materials research. We had thought that putting up provocative historical arguments would suffice to bring materials scientists and engineers to our site. That was not the case: despite the support of the MRS, individual attendees at the conference showed no interest whatsoever. I could not entice a single one of the approximately 4,000 attendees to log on to the website. (See Figure 15.1.)
We learned two important lessons from this experience:

1   In order to involve living materials researchers, we have to present them with something that is geared more to their interests. Our own historical questions will not do.
2   The MRS staff is very interested in what we do and actively support us, as long as we do not ask them for money.[43]

I decided to approach the MRS more formally in order to discuss whether we might be able to do something historical that would be of mutual benefit. The MRS staff, fronted by the editor of the *MRS Bulletin*, Betsy Fleischer, was very open. It seemed appropriate to first of all define our interests and to pinpoint any potential

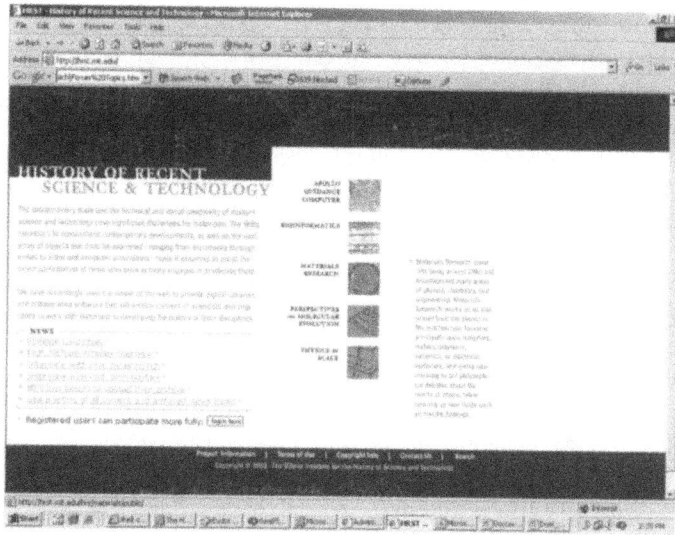

Figure 15.1 Entry web page for the History of Recent Science and Technology on the Web. Here the "Materials Research" sub-project is highlighted.

Source: MIT/Dibner. http://hrst.mit.edu

tensions. The explicit interests of the MRS lay in creating texts of interest to their members. They had a historical section in the *Bulletin* that had been scrapped as an expenditure consolidation measure. They were very interested in having historical material, especially if we could pay for it.[44]

The 2001 president of the MRS had explicitly stated identity as a primary concern of the society in the December 2001 issue of the *Bulletin*:

> I am conflicted about the role and identity of the MRS. [...] What is the nature of MRS? I suspect being a member of the Materials Research Society is not the same as being a member of the American Physical Society (APS) or the American Chemical Society, for example. About half of our members view MRS as their second home; they vacation with us for the meetings, but don't join if they attend neither the Spring nor Fall meeting. [...] Sometimes I think of materials scientists as the mutts of the physical science world. After all, our field is a mélange of science (e.g. physics, chemistry, crystallography) and engineering (chemical, electrical, mechanical). Its existence at the intersection of many pure disciplines (see what I meant about mutts?) is what attracted us to the field in the first place. However, I'd like to point out that the very nature of our field has repercussions for its survival. [...] A strong MRS will provide a strong identity for our field. If we lose our identity, our students and our funding will be backscattered to the departments that spawned our field. In fact many smaller materials departments have already been absorbed by their neighbors.[45]

Our interests could be stated in a similarly raw fashion: to promote our careers by helping the project to succeed, for example by devising a sensible way for materials researchers to utilize the interview webpages. The MRS agreed to brainstorm to find ways of using the modules to create material that also could be used in the *MRS Bulletin* and on the MRS website.[46]

One possible plan emerged: to collaborate in a longer-term project—beyond the short three years available to us within the HRST project—to examine what might be termed the accretion of the field over the past twenty years. Just as biomaterials have become a part of materials research over the past decade, lesser sub-disciplines have also emerged and been brought into the fold. The *MRS Bulletin* is a useful first indicator for this, since each monthly issue focuses on hot topics.[47] We planned to take advantage of the MRS network of people to reach out to the broader materials science community, picking out one prominent researcher to interview him or her in the time-honored face-to-face manner. These interviews would be conducted jointly with the MRS staff, and the questions prepared together. This process would in itself be illuminating. We would then target half a dozen other people in that field and set up an interview module on our website to which only they would be empowered to post responses. Finally, we would advertise to all MRS members and elicit their comments.[48] We, the historians, needed the MRS at several stages of this process. We obviously needed their network and we needed to operate under a joint aegis so as to gain clout. But we also needed to adapt to the kinds of questions that will get materials researchers interested, and they may well turn out to be very different from the ones which historians of science have traditionally employed. The materials researcher also needed historians for a sense of the many possible historical questions, the experience with source criticism, and the sheer experience in the historical craft: how to get from a set of historical sources to a narrative.

The linear model and the identity of the historian

Is there a problem with this approach? Would the integrity of the historian be compromised by sharing the decision on what historical data ought to be gathered? Does this amount to losing the historian's grip on interpreting the past by opening the door to professional scientists who, in the self-understanding of professional historians of science, are prone to Whiggism and self-interested accounts?

There are certainly some historians who followed this new path. As a graduate student I wrote a history of Siemens in the United Kingdom.[49] This history was commissioned by the company, in my judgment, with the aim of generating a corporate identity at a time of expansion and wholesale incorporation of thousands of employees from companies that Siemens had just bought up. Each employee received a copy of the final product. My supervisor at the time, Simon Schaffer at the University of Cambridge, called my remuneration "blood money."

I am content with the conception of science as nonlinear. I am happy to discard the notions of basic and applied science along with that of objectivity. Yet it makes me uneasy when the same boundaries are blurred for historical work. I think my unease is felt by most historians of science. Collaborating with the MRS in the way described earlier goes beyond being paid a salary for writing an in-house history; it goes beyond the compromise of the journalist getting on with informants at, say, the mayor's office.[50] In both those cases, the writer is constrained by having to maintain social relations. For example, the journalist can write something very critical of the mayor's office only once before the source of information runs dry. The constraint lies only in the way in which the mayor's office is portrayed. In our project the compromise is at a much more fundamental level; it is in asking the scientific actors *which questions to ask.*[51]

As a PhD student I was taught to read Whiggish accounts of the history of science with the utmost suspicion.[52] Retired scientists, in particular, were prone to fall into the fallacy of Whiggism—that is, paying no attention to historical context but merely identifying precursors to the present state of affairs. Going back in time, the resemblance with the present grew ever more tenuous, and when that narrative was turned around, one necessarily ended up with a story of progress. Furthermore, scientists were prone to look for progress stories with specific morals supporting a claim for funding or institutional influence. The professional expert, the historian of science, was to rise above Whiggism and interestedness.

In contemporary collaborations between historians and scientists, such adherence to status, expertise, and disciplinary boundaries can erect obstacles. It resembles the situation in Mode 1 research, where attention to the kind of work that properly belongs to engineers, physicists, or chemists may send researchers back to the ivory tower to conduct research that no one but their closest colleagues has any interest in. By contrast, materials researchers on the prowl at an MRS meeting do not ask themselves whether a potential resource belongs to chemistry or engineering and whether their disciplinary identity allows them to take it on board. A physicist also has no problem teaching a chemist a skill that ordinarily would have been a part of the physicist's toolbox. No guild secret is being sacrificed; instead there are a multitude of skills that one can learn and the main way is to learn by teaching something else in return. As such, the disciplinary tags themselves restrict, and hence they have become less important.

In our web-collaboration, the historian had to attend to the demands of our audience of scientists, the very people who are tainted as Whiggish.[53] The success of this narrative was not just whether it accorded with the sources on which it is based but also whether it actually had anything of value to a community of non-historians. Value was determined outside the guild. And this was not only true of the final product but of the many smaller narratives that had to be constructed along the way in order to entice scientists into collaboration.

At various stages of planning, and during the interactions with scientists, the historian had to prepare for the exchange of information. By explaining lucidly the steps of the historical craft and the nature of source criticism, it was possible to prepare scientists to provide information that would be of greater historical use. A Faustian bargain between historians on the prowl and scientists on the prowl thus enables a better exchange—but one where, in a sense, the historian has sold out and lost the virtues of disinterestedness and objectivity. The historian of science would have to abandon any pretense to a pedigree and instead must relish being a *mutt*.[54]

Conclusion

Adrian Johns has rightly emphasized the fact that new sociabilities are developed over long periods of time and that these sociabilities have propped up the fixity of print and are the cause of our trust in them. Similarly, the digital archive cannot emerge as fixed and trustworthy overnight. Even less so can the kind of collaboration described here, dependent on these new technologies, come into being suddenly, because sociabilities change but slowly. We historians of contemporary science would have to jettison attitudes that currently seem fundamental to the identity of the historian: we would have to accept Whiggism back in through the back door, accept that we are in no way superior to the Whiggish scientist. We would have to learn new technologies and new ways of interacting with the historical sources. Historians would no longer simply face the question of identifying sources but rather of collaborating with historical actors to determine what will remain as a historical source in the future. We would become curs. In other words, a web-based project on the recent history of S&T would only be able to succeed after an extended period of time, during which historians and scientists renegotiated their roles and learned new mongrel ways. The advantage would lie in the ability to communicate and collaborate with the subjects of history of science—as scientists, we would learn to better treat as human beings and not as lepers.

Considering the peripheral nature of the discipline of history of science, this advantage is very great. There is a constant scarcity of funding for history of science, which is not a sign of an unfair and uncomprehending world but an indication of a cloistered, inward-looking field.[55] Our project may well give an insight into the avenues that history of science has to pursue in order to expand, or perhaps even survive, as an academic field. This possible future is one in which history of science resembles more closely the complex cross-disciplinary collaborations within materials research and Mode 2 research. But perhaps the most interesting and yet anxiety-provoking aspect is the deliberate search for outside interest (or perhaps marketability) that would take history of science out of its comfortable but hemmed-in present situation where historians address only the select audience of their nearest peers.

Notes

1 For instance, Thomas Söderqvist, "Who Will Sort Out the Hundred or More Paul Ehrlichs? Remarks on the Historiography of Recent Contemporary Technoscience," in *The Historiography of Contemporary Science and Technology*, ed. Thomas Söderqvist (Amsterdam: Harwood Academic Publishers, 1997), 1–17.

2 This is particularly so since we have not yet reached the leveling-off point in the exponential growth of twentieth-century science predicted by the late Derek J. de Solla Price; see his *Big Science, Little Science* (New York: Columbia University Press, 1963). An insightful review of his predictions is Susan E. Cozzens, "Derek Price and the Paradigm of Science Policy," *Science, Technology, and Human Values* 13 (Summer–Autumn 1988): 361–372.

3 There are now many web-based histories, all of them experimenting in different ways. All of them have difficulties getting people to participate, but each and every one is trying something different. The Sloan Foundation seems to deliberately fund different approaches in order to learn from each.

4 See Stephen Shapin and Simon Schaffer, *Leviathan and the Air Pump: Hobbes, Boyle, and the Experimental Life* (Princeton, NJ: Princeton University Press, 1989).

5 Roy Rosenzweig, "Scarcity or Abundance? Preserving the Past in the Digital Era," *American Historical Review* 108 (2003): 735–762.

6 Adrian Johns, *The Nature of the Book—Print and Knowledge in the Making* (Chicago, IL and London: Chicago University Press, 1998), 1.

7 For example, http://www.docurights.com/, accessed on May 2, 2004.

8 http://hrst.mit.edu/hrs/public/terms.htm, accessed on May 2, 2004. The statement was formulated by our lawyer, Bonnie Edwards.

9 The software development team consisted of Babak Ashrafi (at the time a PhD student in MIT's Science, Technology, and Society Program), Daniel Tsai (a professional software developer), James Voelkel (PhD in history of science), and Paul Warner (also a professional software developer).

10 http://hrst.mit.edu/hrs/materials/public/general.htm, accessed on May 2, 2004.

11 http://www.perseus.tufts.edu/, accessed on May 2, 2004.

12 The Apollo Guidance Computer (AGC) provided reliable real-time control for the Apollo spacecraft that carried US astronauts to the moon from 1969 to 1972. It was designed by the MIT Instrumentation Laboratory (now the Charles Stark Draper Laboratory, Inc.) and manufactured by Raytheon Corporation. The AGC was significant for its tight coupling of human and machine, its early use of integrated circuits, and its reliable, mission critical software. The history of the AGC project provides a window into the history of technology in America during the space race and the Cold War.

13 Two decades before the Human Genome Project, a new breed of hybrid biologist began to emerge: part molecular biologist, part computer scientist. The field these scientists created was called computational molecular biology or bioinformatics.

14 Beginning in the 1960s, evolutionary biology was significantly transformed by the incorporation of ideas and techniques from molecular biology. This led to many novel views (and as many controversies) about phylogenetic relationships, rates and mechanisms of evolutionary change, and standards of inference and hypothesis testing.

15 In the early 1970s two fields of physics intersected in a set of techniques that have since been used to explore both the connections between the "fundamental" forces and to gain insight into character of "critical phenomena." These techniques entailed working out the relationship between phenomena at different scales.

16 The team working on the history of materials research consists of Bernadette Bensaude-Vincent, Hervé Arribart, and Arne Hessenbruch.

17 We have "Ages of Man based upon materials: The Golden and Silver Ages of the Greeks; the Stone, Bronze and Iron Ages of the archeologists. (Oddly, these classifications do not include, as they should, a ceramic age . . .)"; see Cyril Stanley Smith, *Metallurgy as a Human Experience—An Essay on Man's Relationship to his Materials in Science and Practice throughout History* (Metals Park, OH: American Society for Metals, 1977), 5. For a history of this classification, see Bruce Trigger, *A History of Archaeological Thought* (Cambridge and New York: Cambridge University Press, 1989).

18 Bernadette Bensaude-Vincent, "The Construction of a Discipline: Materials Science in the United States," *Historical Studies in the Physical Sciences*, 31 (2001): 223–248.

19 Wuensch presents this internalist (theory-focused) perspective with great clarity in http://hrst.mit.edu/hrs/materials/public/Wuensch_interview.htm, accessed on May 2, 2004. Buried within a great many other themes, it may also be found in Robert Cahn, *The Coming of Materials Science* (Amsterdam: Pergamon, 2001).

20 Structure most often refers to structure at the atomic scale of a solid (the nanoscale), but also includes micro-, meso-, and macro-. The properties referred to include the mechanical, electrical, magnetic, optical, and thermal, basically the properties of interest to an engineer deploying materials for a particular purpose.

21 For example, Richard A. Flinn and Paul K. Trojan, *Engineering Materials and their Applications* (Boston, MA: Houghton Mifflin Company, 1975).

22 Samuel M. Allen and Edwin L. Thomas, *The Structure of Materials* (New York: John Wiley & Sons, 1999).

23 This may be seen with great clarity by browsing the *MRS Bulletin* (first published as the *MRS* Newsletter in 1975).

24 The main source is Aant Elzinga and Andrew Jamison, "Changing Policy Agendas in Science and Technology", in *Handbook of Science and Technology Studies*, ed. Sheila Jasanoff, Gerald E. Markle, James C. Peterson, and Trevor Pinch (Thousand Oaks, CA: Sage, 1995), 572–597. Cf. also Daniel Lee Kleinman, *Politics on the Endless Frontier: Postwar Research Policy in the United States* (Durham, NC: Duke University Press, 1995); Bruce L. R. Smith, *American Science Policy since World War II* (Washington, DC: Brookings Institution, 1990); Daniel Greenberg, *Science, Money, and Politics: Political Triumph and Ethics Erosion* (Chicago, IL: University of Chicago Press, 2001).

25 Organization for Economic Co-operation and Development—Advisory Group on Science Policy, *Science and the Policies of Governments: the Implications of Science and Technology for National and International Affairs* (Paris: OECD, 1963), 30.

26 For further background, see David M. Hart, *Forged Consensus: Science, Technology, and Economic Policy in the United States, 1921–1953* (Princeton, NJ: Princeton University Press, 1998).

27 Discussions of this issue appear for instance in Smith, *American Science Policy*, and in Samuel P. Hays, *Beauty, Health and Permanence: Environmental Politics in the United States, 1955–1985* (New York: Cambridge University Press, 1989).

28 The National Cancer Act, 1971; see http://www3.cancer.gov/legis/1971canc.html, accessed on May 2, 2004.

29 President Lyndon B. Johnson amended the National Science Foundation charter in 1968 specifically to expand the agency's mission to include problems directly affecting society. Now "relevance" became the new by-word, embodied in the 1969 launch of a new, engineering-dominant program called Interdisciplinary Research Relevant to Problems of Our Society (IRRPOS), which funded projects mostly in the areas of the environment, urban problems, and energy. IRRPOS gave way in 1971 to a similar but much expanded program called Research Applied to National Needs (RANN). Materials research also redirected itself, the clearest sign of which is the COSMAT Summary Report, *Materials and Man's*

Needs (Washington, DC: National Academy of Sciences, 1974 http://www.nap.edu/html/materials_and_man/NI000883/HTML/51-52.HTML, accessed on May 2, 2004).

30 The Office of Technology Assesment OTA was set up in 1972; see http://www.wws.princeton.edu/~ota/ns20/cong_f.html, accessed on May 2, 2004.

31 On this issue see Harvey Brooks, *Science, Growth and Society: A New Perspective* (Paris: OECD, 1971).

32 Bernhardt Wuensch, for example, expressed it thus, cf. http://hrst.mit.edu/hrs/materials/public/Wuensch_interview.htm, accessed on May 2, 2004.

33 Morris Cohen, ed., *Materials Science and Engineering: Its Evolution, Practice and Prospect* (Lausanne: Elsevier, 1989); and COMSAT, *Materials and Man's Needs.*

34 Aant Elzinga and Andrew Jamison, "Changing Policy Agendas in Science and Technology," in *Handbook of Science and Technology Studies*, ed. Sheila Jasanoff, Gerald Markle, James Petersen and Trevor Pinch (Thousand Oaks, CA: Sage, 1995), 572–597.

35 Elzinga and Jamison, "Changing Policy."

36 Interview with Thomas Eagar, by George Smith and Arne Hessenbruch, Dibner Institute, MIT, October 20, 2002; see http://hrst.mit.edu/hrs/materials/public/Eagar/Eagar.htm, accessed on May 2, 2004.

37 Greenberg, *Science, Money, and Politics.*

38 *Towards a European Research Area—Science, Technology and Innovation Key Figures, 2000* (Luxembourg: Office for Official Publications of the European Communities, 2000).

39 Michael Gibbons, Camille Limoges, Helga Nowotny, Simon Schwartzman, Peter Scott, and Martin Trow, *The New Production of Knowledge—The Dynamics of Science and Research in Contemporary Societies* (London: Sage, 1994), esp. the introduction, 1–16.

40 The Model/Mode 2 argument has not met with enthusiasm all round; see for instance Terry Shinn, "The Triple Helix and New Production of Knowledge: Prepackaged Thinking on Science and Technology," *Social Studies of Science* 32, no. 4 (2002): 599–614; and Jane Calvert, *Goodbye Blue Skies? The Concept of 'Basic Research' and its Role in a Changing Funding Environment* (DPhil thesis, University of Sussex, 2002).

41 For example, the Danish physicist Flemming Besenbacher started his career as a surface scientist physicist (a field consisting of the analysis of metallic and semiconductor surfaces analyzed with such tools as diffraction using x-rays, neutrons, or electrons—surface physicists conducted esoteric research of no immediate industrial relevance, but of course the field grew to a large extent because large private laboratories such as Bell Labs invested heavily in it). In the late 1980s, he was able to build up a research institute doing research with the then novel scanning tunneling microscope. He had no choice but to work on materials of industrial relevance, such as catalysts (which had not been his field before). Because Danish research structures are geared to push such scientists as Besenbacher on, the funding for his institute could not be renewed after some dozen years. In order to get new funding and feed the many mouths in his institute, he developed new contacts in the life and medical sciences and proceeded into the field of biocompatible materials. Cf. http://hrst.mit.edu/hrs/materials/public/SPM_in_Denmark/SPM_in_DK.htm, accessed on May 2, 2004.

42 Peter Galison, *Image and Logic: A Material Culture of Microphysics* (Chicago, IL: University of Chicago Press, 1997). However, it has to be said that the notion of a trading zone presupposes other zones where trade does not occur—zones of disciplinary unity. In materials research there are only trading zones taking much of the usefulness out of that concept.

43 This is markedly different from the experience of historians of high-energy physics. One might speculate about the nature of funding in the two cases: maybe high-energy physicists are sensitive about their public image at a time when public funding and their general status are declining. By contrast, materials researchers are not threatened. On the case of high-energy physics, see Armin Hermann, John Krige, Ulrike Mersits, Dominique Pestre, *History of CERN* (Amsterdam & New York: North Holland Physics Publications, Volume I–III, 1987–1996); and Galison, *Image and Logic.*

44 For a related perspective see Cantor, this volume.

45 Martin L. Green, "Letter from the President: What Kind of Society is MRS, Anyway?" *MRS Bulletin* (December 2001): 963–964.

46 http://www.mrs.org, accessed on May 2, 2004.

47 The titles are collated on http://www.mrs.org/publications/bulletin/archives.html, accessed on May 2, 2004.

48 There is an obvious hierarchy inherent in this set up and it blocks a history paying equal attention to the professor and the humble staff. But how to avoid it?

49 Arne Hessenbruch, *Sir William Siemens—A Man of Vision* (Bracknell, England: Siemens UK plc & Caric Press, 1993).

50 Susan Lindee, "The Conversation: History and History as it Happens," in *The Historiography of Contemporary Science and Technology*, ed. Thomas Söderqvist (Amsterdam: Harwood Academic Publishers, 1997), 39–50.

51 Interestingly, one may point to anthropology, where one (politically correct) attitude is that the anthropologist should ask the subjects of investigation for the appropriate questions to ask: Vine Deloria, *Red Earth, White Lies: Native Americans and the Myth of the Scientific Fact* (New York: Scribner, 1995). I am grateful to Hugh Gusterson and Debra L. Martin for this reference.

52 Sir Herbert Butterfield, *The Whig Interpretation of History* (London: G. Bell & Sons, 1931).

53 In the introduction to Cahn, *Materials Science*, metallurgist Cahn indeed defends himself against the anticipated criticism of Whiggism.

54 In a footnote to his letter as MRS president (cit. n. 44), Martin Green declared,

> A mutt is a mixed as opposed to a pure bred (pedigreed) dog. Don't take my analogy in the wrong way. I own an adorable mutt, and she's everything I've ever wanted in a dog. Mutts are the best dogs because they are generally healthier and in possession of more "dog sense" than pure breeds. Also, they are usually free.

That might just as well apply to the mutt historian.

55 Other historians are raising this point; see Kai-Henrik Barth and John Krige, "Introduction," in John Krige and Kai-Henrik Barth, eds., *Global Power Knowledge: Science and Technology in International Affairs, Osiris* 21 (2006): 1–21 and also Ken Alder's comments on receiving the Watson Davis and Helen Miles Davis Prize (best book in the history of science intended for a broad audience) for his *The Measure of All Things* (New York: Free Press, 2002) at the annual meeting of the History of Science Society, Cambridge, MA, November 2003.

Index

Note: Page numbers in italics indicate illustrations.

For Product Safety Concerns and Information please contact our EU
representative GPSR@taylorandfrancis.com
Taylor & Francis Verlag GmbH, Kaufingerstraße 24, 80331 München, Germany

www.ingramcontent.com/pod-product-compliance
Lightning Source LLC
Chambersburg PA
CBHW071411180526
45170CB00001B/54